9-24-92
Cetacea

Handbook of Marine Mammals

Volume 2 Seals

Handbook of Marine Mammals

Volume 2 Seals

Edited by

SAM H. RIDGWAY
*Naval Ocean Systems Center,
San Diego, USA*

and

RICHARD J. HARRISON, F.R.S.
*Anatomy School,
University of Cambridge, UK*

1981

ACADEMIC PRESS
A Subsidiary of Harcourt Brace Jovanovich, Publishers
London New York Toronto Sydney San Francisco

ACADEMIC PRESS INC. (LONDON) LTD.
24/28 Oval Road,
London NW1

United States Edition published by
ACADEMIC PRESS INC.
111 Fifth Avenue
New York, New York 10003

Copyright © 1981 by
ACADEMIC PRESS INC. (LONDON) LTD:

All Rights Reserved
No part of this book may be reproduced in any form by photostat, microfilm,
or any other means, without written permission from the publishers

British Library Cataloguing in Publication Data

Handbook of marine mammals.
  Vol. 2: Seals
  1. Marine mammals
  I. Ridgway, Sam H.   II. Harrison, Richard J.
  599.092   QL713.2   80-42010
  ISBN 0-12-588502-4
  LCCCN 80-42010

Printed in Great Britain at
The Lavenham Press Ltd., Lavenham, Suffolk, England

Contents

Contributors	vii
Preface to the Series	ix
Preface to Volume 2	xiii
1 Harbour Seal—*Phoca vitulina* and *P. largha* MICHAEL A. BIGG	1

2   Ringed, Baikal and Caspian Seals—
    Phoca hispida, *Phoca sibirica* and *Phoca caspica*
    KATHRYN J. FROST and LLOYD F. LOWRY
	29

3 Harp Seal—*Phoca groenlandica* K. RONALD and P. J. HEALEY	55
4 Ribbon Seal—*Phoca fasciata* JOHN J. BURNS	89
5 Grey Seal—*Halichoerus grypus* W. NIGEL BONNER	111

6	Bearded Seal—*Erignathus barbatus* JOHN J. BURNS	145
7	Hooded Seal—*Cystophora cristata* RANDALL R. REEVES and JOHN K. LING	171
8	Monk Seals—*Monachus* KARL W. KENYON	195
9	Crabeater Seal—*Lobodon carcinophagus* GERALD L. KOOYMAN	221
10	Ross Seal—*Ommatophoca rossi* G. CARLETON RAY	237
11	Leopard Seal—*Hydrurga leptonyx* GERALD L. KOOYMAN	261
12	Weddell Seal—*Leptonychotes weddelli* GERALD L. KOOYMAN	275
13	Southern Elephant Seal—*Mirounga leonina* JOHN K. LING and M. M. BRYDEN	297
14	Northern Elephant Seal—*Mirounga angustirostris* SAMUEL M. McGINNIS and RONALD J. SCHUSTERMAN	329
Index		351

Contributors

MICHAEL A. BIGG, Department of Fisheries and Environment, Pacific Biological Station, Nanaimo, British Columbia V9R 5K6, Canada

W. NIGEL BONNER, British Antarctic Survey, Natural Environment Research Council, Madingley Road, Cambridge CB3 0ET, UK

M. M. BRYDEN, School of Anatomy, University of Queensland, St Lucia, Queensland 4067, Australia

JOHN J. BURNS, Alaska Department of Fish and Game, 1300 College Road, Fairbanks, Alaska 99701, USA

KATHRYN J. FROST, Alaska Department of Fish and Game, 1300 College Road, Fairbanks, Alaska 99701, USA

P. J. HEALEY, College of Biological Science, University of Guelph, Guelph, Ontario, N1G 2W1, Canada

KARL W. KENYON, 11990 Lakeside Place, NE Seattle, Washington 98125, USA

GERALD L. KOOYMAN, Physiological Research Laboratory, Scripps Institution of Oceanography, La Jolla, California 92037, USA

JOHN K. LING, South Australian Museum, Department for the Arts, North Terrace, Adelaide, South Australia 5000

LLOYD F. LOWRY; Alaska Department of Fish and Game, 1300 College Road, Fairbanks, Alaska 99701, USA

SAMUEL M. McGINNIS, Department of Biology, California State University, Hayward, California 94542, USA

G. CARLETON RAY, Department of Environment Sciences, Clark Hall, University of Virginia, Charlottesville, Virginia 22903, USA

RANDALL R. REEVES, CETAP, Box 119, Hudson, Quebec, H0P 1M0, Canada

KEITH RONALD, College of Biological Science, University of Guelph, Guelph, Ontario, N1G 2W1, Canada

RONALD J. SCHUSTERMAN, Departments of Psychology and Biology, California State University, Hayward, California 94542, USA

Preface to the Series

The idea of producing a Handbook of Marine Mammals of this type was the result of many discussions between the authors and the late Mr John Cruise of Academic Press during his visits to Cambridge over ten years ago. It would be, it was hoped, a comprehensive account of all marine mammal species with each chapter written by someone who had actually worked on a particular form. We felt that this would give a more personal flavour to each chapter, provide opportunity to include original observations, and increase accuracy. Some species of marine mammal are worldwide in their distribution and known to many investigators, others are restricted to certain rivers, lakes, seas and even to relatively short ranges along a coast. We know of no one person who claims to have seen every extant species alive in its environment. We hoped that such a series of chapters written by experts on each species would naturally draw attention to subtle differences in form, coloration, behaviour and many other characteristics, many of which might not be known to an ordinary reviewer.

There are obvious difficulties and problems to vex the editors of multi-author works and they all affected this one. We knew it would take time; active marine mammalogists spend long periods away from home on expeditions and at conferences. The majority of contributors has remained loyal to the project but a few have had to withdraw with inevitable difficulties over replacement and with continual postponement of completion dates. New information has become available about many species over the last ten years, due mainly to efforts by countries to find out more about their local species but also to enterprising attempts to display little-known species in captivity. These last ten years have also seen a much increased interest in conservation of marine mammals and an awareness by the public of their importance as interesting animals from many points of view. There has been much debate about whether a moratorium should be introduced by international agreement over the taking of large or even any cetaceans. The arguments have affected many of the classical views held on harvesting, culling, controlling or managing stocks of all marine mammal species. They have exposed our ignorance on many basic facts concerning reproduction, school competition and behaviour, and even what is really meant by many of the terms and concepts used so glibly in marine mammalogy in the past.

The last decade has seen much burgeoning in scientific investigation of marine mammals, though not always in a direction leading to important results. For example, the delightfully satisfying construction of mathematical models has occupied many folk. The results give joy to the constructor, the administrator, the politician and this whole game is not only fun but really quite cheap. What the computer tells us is only as good as the data put into it which are often incomplete or inadequate for detailed analysis. At the very best the results are always based on past events. What is happening now in the seas is still as mysterious as it always was.

Many advances have been made in our knowledge of how to keep marine mammals in captivity, about their diseases and the associated pathological changes, and on treatment. The life styles, life history, expectation of life, growth rates, vulnerability and responses to adverse effects of disease and pollution are becoming better known and understood. Formulae have been constructed to provide the optimum size of pool for holding marine mammals of varying number and size. Codes of practice have been issued to ensure the welfare of marine mammals during transport and in captivity. This concern has increased life expectation of captive animals, and has improved their lot from being mere objects of curiosity and participants in circus acts. The list of different species displayed to the public has steadily lengthened; to an extent that many countries demand permits to be issued for taking of rare and even not so rare animals from local stocks, and records must be kept about captive animals.

Film and television companies have realized the increasing popularity of productions involving marine mammals, and this has spread knowledge more than any other factor, and unfortunately in some cases misinformation, about the various species. Conservationists have also promoted public awareness of which nations still hunt whales and the purposes to which the carcasses are put. These groups have put pressure on national and international bodies to provide more accurate information about endangered species and the state of stocks generally. Again and again it has had to be confessed that many species are still little known, their habits obscure, and assessment of their numbers mere guesswork.

Problems of nomenclature are bound to arise in presenting any group as diverse and widely dispersed as marine mammals. We have avoided strict rules and have left decisions as to the use of particular names to each expert author. Authors were asked to summarize the scientific and common nomenclature of the genera and species about which they have written. As a guide to the various genera and species Dale W. Rice's "A List of Marine Mammals of the World" (NOAA Technical Report NMFS SSRF-711, US Department of Commerce, Washington, D.C., 1977, 15p) is especially useful.

This Handbook is intended as a guide to marine mammal types and is all about them as animal forms. It is meant for use in the field and laboratory as a practical aid to identification and to provide useful basic information. It is not concerned with management, husbandry or treatment of disease, stock levels or whether there should be a moratorium on the taking of certain marine mammals. We have both worked for many years on various aspects of the biology of what we have found an entirely fascinating group of mammals. We know that all contributors have an interest equal to ours and far more expertise, so we have tried to keep editorial intervention to a minimum. We dare to hope that only marine mammals will know about themselves what is not in this Handbook, or that it is the editors' fault that useful information has been overlooked.

We thank Academic Press for their patience and guidance.

Sam H. Ridgway
Richard J. Harrison

Preface to Volume 2

Volume 2 of this Handbook series is devoted to one family—the Phocidae or true seals—the seals without a protruding external ear or pinna. True seals cannot run along the beach or stand raised on their fore flippers; they generally haul or hump themselves caterpillar-like across sand, mud or ice.

There are eight genera in the family Phocidae. *Phoca* is the largest genus in terms of species, with the smallest species in terms of body size, and the most diverse species in terms of size, habitat, colour pattern and many other characteristics. Because of this diversity we have invited several authors to give their accounts of the species included in the genus *Phoca*. Harbour seals *Phoca vitulina* and *Phoca largha* are covered in Chapter 1 by Michael Bigg. In Chapter 2, Kathryn Frost and Lloyd Lowry describe the ringed seal, the Baikal seal and the Caspian seal, three species that were formerly recognized as admitted to the genus *Pusa*. These include the smallest of seals, all of which live in water that freezes at certain times of the year. Ringed seals inhabit the Arctic ocean and adjacent seas dwelling chiefly on fast ice, but some live in freshwater lakes. Baikal seals are found far inland in the USSR's Lake Baikal—fresh water that freezes in winter. The Caspian seal inhabits the inland sea of that name.

The harp seal or Greenland seal, currently a major concern of the authorities in Canada, is the subject of Chapter 3. Thousands of its white-coated baby seals are harvested yearly. Persons active in environmental affairs confront the sealers and the Canadian government each year over the seals' slaughter. Keith Ronald and P. J. Healy discuss this problem in the perspective of what is known about the biology of this species. The ribbon seal is perhaps the most interesting of the northern pinnipeds. John Burns of the Alaska Department of Fish and Game is one of the few scientists to have studied them. His chapter contains much new information on the species.

The second genus *Halichoerus* is described in detail in Chapter 5 by Nigel Bonner. The grey seal competes with fishermen for food all over the North Atlantic and there is controversy over the factor by which its numbers should be reduced. A recluse among seals, and seldom seen, the bearded seal is the subject of John Burns's Chapter 6; it is not at the heart of controversy, but all the more intriguing because there is still so much unknown about this seal.

The large, ice-dwelling, hooded seal is the topic of Chapter 7 by Randall Reeves and John Ling. In Chapter 8 Karl Kenyon gives information about a truly endangered genus, *Monachus,* those monk seals of tropical islands, desert shores and rocky Mediterranean coasts.

In the southern hemisphere the most abundant seal is the crabeater. discussed by Gerald Kooyman, a veteran of many seasons in the Antarctic. In Chapter 10, Carleton Ray, considers perhaps the least known of all seals— the Ross seal which produces interesting sounds with the aid of a special "mouth pouch".

Gerald Kooyman deals with two more Antarctic seals: the leopard seal in Chapter 11 and the Weddell seal in Chapter 12. The most menacing of seals, the leopard seal patrols the edge of the Antarctic ice pack. The Weddell seal, as far as is now known, is the consummate diver among pinnipeds and makes dives to 600 m in depth. Shallow dives lasting over an hour have been recorded while these seals explore for new breathing holes under the ice. Dr Kooyman has achieved an international reputation based largely on his studies of the Weddell seal and of its remarkable capability for diving.

Diving is also a strong suit of the elephant seals—genus *Mirounga*—largest of pinnipeds. The elephant seal of the southern hemisphere is the subject of Chapter 13 by John Ling and M. M. Bryden and the northern elephant seal is discussed in Chapter 14 by Samuel McGinnis and Ronald Schusterman.

It is among the Phocidae that remarkable and fascinating adaptations to diving have evolved. These include anatomical, physiological and biochemical capabilities not yet fully quantified and not by any means understood. The members of the family are so skilled at diving that one wonders why their geographical distribution is so relatively limited. They are undoubted masters of the underwater environment, and yet have not abandoned land. We suspect that they have close social affinities, a characteristic not uncommon in marine mammals. This aspect of the marine mammal life style is now a matter of international investigation.

Thanks are due to Denis A. McBrearty for help with proof reading and indexing and to I. R. Bishop of the British Museum Natural History for invaluable guidance on nomenclature.

Naval Ocean Systems Center,   Sam H. Ridgway
San Diego, USA

Anatomy School,   Richard J. Harrison
Cambridge, UK

June, 1981

1
Harbour Seal
Phoca vitulina Linnaeus, 1758
and *Phoca largha* Pallas, 1811

Michael A. Bigg

Genus and Species

Common names
At least 80 local names exist. The most commonly used in the English literature are harbour, common and spotted seal. Also, insular and Kuril seal are used for *Phoca vitulina stejnegeri* and larga seal for *P. largha*.

Taxonomy
For many years much confusion has existed as to whether one, two or three species occur. Current studies point to an almost arbitrary assignment of two closely related species or one species, *P. vitulina*, with a distinctive subspecies, *largha*. Evidence seems strongest for two species. *Phoca vitulina* differs from *P. largha* in morphology, physiology and

behaviour due to adaptations for breeding on land while *P. largha* has adapted to breeding on ice. Both species belong to the Family Phocidae.

Scheffer (1958) reviewed the taxonomic history and considered them as one species with five subspecies designated on the basis of geographical distribution: *vitulina*—eastern Atlantic; *concolor*—western Atlantic; *mellonae*—Seal Lakes, Ungava Peninsula; *richardsi*—eastern Pacific; and *largha*—western Pacific. McLaren (1966) and Chapskii (1967, 1969) argued for two species by elevating *largha* to a separate species. Burns and Fay (1974), however, did not recognize this new status and Shaughnessy (1975) was unable to separate *largha* from *richardsi* on the basis of blood proteins. From Scheffer's *largha* Belkin (1964) proposed a third species *Phoca insularis* which McLaren (1966) later renamed *P. kurilensis*. However, Chapskii (1969) and Burns and Fay (1974) reported the latter as only a large form of *P. v. richardsi*. Now, in the latest and most complete review of North Pacific taxa, Shaughnessy and Fay (1977) tentatively recognize *largha* as a species, rename and give subspecific status to the kuril seal, *P. v. stejnegeri*, and correct the long-standing mis-spelling of *P. v. richardi* to *richardsi*.

The validity of the four *P. vitulina* subspecies is uncertain. *P.v. vitulina* and *P.v. concolor* may be identical while skulls of *P.v. mellonae* can occasionally be distinguished from *P.v. concolor* (Doutt, 1942; Mansfield, 1967) as can those of *P.v. richardsi* from *P.v. concolor* (Doutt, 1942). *Phoca vitulina stejnegeri* is distinguished from *P.v. richardsi* by its generally larger size and darker pelage although the two forms may intergrade in the Aleutian Islands (Shaughnessy and Fay, 1977).

Evolution

Harbour seals probably evolved in the North Pacific from an ice-breeding ringed seal-like ancestor at least 2-3 million years ago (McLaren, 1966; Chapskii, 1969; Barnes and Mitchell, 1975). *Phoca largha* is thought to be most like the ancestor and to have given rise to *P. vitulina* which later spread across the Arctic to the Atlantic. The landlocked *P.v. mellonae* was separated either from *P.v. richardsi* (Doutt, 1942) or *P.v. concolor* (Davies, 1958) perhaps 4000 years ago.

External Characteristics and Morphology

Size

The harbour seal is a small phocid which Allen (1880) aptly described as large-headed, short-bodied and short-limbed. Scheffer and Slipp

TABLE 1 Average standard length (cm) and body weight (kg) of harbour seals from different populations. Sample size is in parentheses.

Population		New-born Length	New-born Weight	Length M	Length F	F/M	Adult Weight M	Adult Weight F	M wt / F length	Source
P. largha	Bering Sea			160(6)	143(9)	0.89				Chapskii (1967)
	Okhotsk Sea (?)			150(14)	146(7)	0.97				Belkin *et al.* (1969)
	Tatar Strait			158(16)	158(35)	1.00	100(17)	89(34)	0.63	
	Peter the Great Bay			159(3)	157(10)	0.99	113(3)	114(10)	0.71	Kosygin and Goltsev (1971)
	Bering Sea			149(12)	141(13)	0.95	85(22)	66(16)	0.57	
	Kuril Islands			150(14)	146(6)	0.97	90(13)	75(7)	0.60	
	Okhotsk Sea			153(19)	142(16)	0.93				
	Hokkaido	85		170(8)	159(28)	0.94				Naito and Nishiwaki (1972)
	Bering Sea (?)	84	9-12							Burns (1970)
P. v. stejnegeri	Hokkaido	98(20)		186(3)	169(8)	0.91				Naito and Nishiwaki (1972)
	Kuril Islands		19	174(8)	160(17)	0.92	87-170	60-142		Belkin (1964)
	Kuril Islands	82(9)	10(9)	161(11)	148(50)	0.92				Belkin *et al.* (1969)
P. v. richardsi	British Columbia-Tugidak Island						87(10)	65(42)	0.54	Bigg (1969b)
P. v. concolor	Sable Island (M)	81(13)	11(8)							Boulva (1971)
	(F)	79(18)	11(6)							

(1944) and Fisher (1952) provide details of body proportion. Both species attain the same size although show different growth characteristics (Table 1). *P. v. stejnegeri* is largest. Males tend to be slightly longer than females but less so in *P. largha* and *P. vitulina* appears to weigh relatively less than *P. largha*. Also, pups of *P. largha* more than triple their birth weight by weaning while those of *P. vitulina* only double their weight (Tikhomirov, 1968; Bigg, 1969b; Boulva, 1971). The few data available on growth in *P.v. mellonae*, *P.v. concolor* and *P.v. vitulina* suggest similar sizes to *P.v. richardsi* (see Havinga, 1933; Doutt, 1942; Mohr, 1966; Boulva, 1971).

Pelage

The two species can be distinguished on the basis of pelage morphology and pattern. *P. largha* is apparently always born with a whitish-coloured lanugo hair shed two to four weeks after birth (Tikhomirov, 1964; Naito and Nishiwaki, 1972). *Phoca vitulina* usually moults this hair before birth and is born with the spotted adult-type hair. Shaughnessy and Fay (1977) feel that the incidence of lanugo hair in the forms *stejnegeri* and *richardsi* is less than 5%, although Boulva (1975) has recorded 12-15% in *concolor* at Sable Island. The lanugo hair of *largha* is more woolly and longer than *richardsi* and *concolor* (Stutz, 1966; Boulva, 1975).

The adult-type hair of *P. vitulina* consists of coarse over-hairs up to 9 cm long and finer, shorter under-hairs (Montagna and Harrison, 1957; Scheffer, 1964). Each hair follicle contains one bristle-like flattened over-hair and 3-6 slightly flattened under-hairs. The density and thickness of over-hairs apparently increases with age, resulting in a progressively coarser pelage.

There is considerable individual variability in pattern and colour of the adult type of pelage (Fig. 1). The small markings, shaped like spots, rings and blotches, range in colour from white-grey to dark brown and black against a background of similar colour variability. Shaughnessy and Fay (1977) recognize a light- and dark-colour phase in the forms *stejnegeri* and *richardsi*, dismissing the "muddy" type described by Stutz (1967a) as an artefact of moult. They note that the dark phase in *richardsi* predominates in southern California—Baja California and western Aleutian Islands and the light phase in central California to eastern Aleutian Islands. The dark phase predominates in *stejnegeri*. The pelage of *P. largha* is similar to the light phase but is less variable and lacks whitish rings on the back and has other subtle differences described in detail by Shaughnessy and Fay (1977). While descriptions of pelage in the other forms *vitulina, concolor* and *mellonae* (see Allen, 1880; Doutt, 1942) are similar to *stejnegeri* and *richardsi*, adequate detailed comparisons have not yet been made.

FIG. 1 Adult harbour seal *(Phoca vitulina)* (photograph by Steve Jeffries, University of Puget Sound, Washington, USA.)

Distribution

Harbour seals inhabit temperate, subarctic and some arctic waters of the North Atlantic and North Pacific, giving them one of the largest distributions of any pinniped (Fig. 2). The distributions are as follows: *Phoca vitulina vitulina*—northern Portugal (rare) to the Barents Sea (rare) and Iceland but not in the White Sea, Gulf of Bothnia, north Baltic Sea nor Gulf of Finland (Scheffer, 1958; Hook, 1961; Curry Lindahl, 1975); thought by Ivashin *et al.* (1972) to be abundant on the Faeroe Islands but

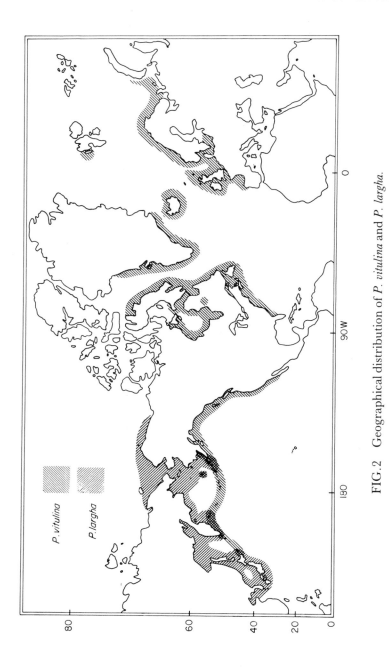

FIG. 2 Geographical distribution of *P. vitulina* and *P. largha*.

found to be absent by Bigg (1969a) and Smith (1965); present in Spitsbergen (Krog and Bjarghov, 1973); *P.v. concolor*—northern Florida (rare south of Massachusetts) to Hudson Bay, to Admiralty Inlet on Baffin Island and to south-eastern and western Greenland (Scheffer, 1958; Mansfield, 1967; Caldwell and Golley, 1965; Caldwell and Caldwell, 1969); absent west of Melville Peninsula and northern Baffin Island to the MacKenzie River delta; *P.v. mellonae*—land-locked in the Upper and Lower Seal Lakes 90 miles east of Richmond Gulf, Hudson Bay (Doutt, 1942); *P.v. richardsi*—central-west coast of Baja California to Aleutian Islands, Bristol Bay, and Pribilof Islands (Scheffer, 1958; Mohr, 1965; Burns and Fay, 1974; Shaugnessy and Fay, 1977); *P.v. stejnegeri*—north-eastern Hokkaido, the Kuril Islands to southern and eastern Kamchatka and Commander Islands, unclear range on the Aleutian Islands (Belkin, 1964; Marakov, 1967; Chugunkov, 1970; Naito and Nishiwaki, 1973; Shaughnessy and Fay, 1977); *P. largha*—Po Hai Sea, Korea, Sea of Japan, Hokkaido, Honshu, Sea of Okhotsk, Kamchatka, Commander Islands, the Bering Sea and Chukchi Sea, eastern Aleutian Islands, although the extent here is not known (Tikhomirov, 1961; Huang, 1962; Fay, 1974; Naito, 1976; Shaughnessy and Fay, 1977); but not off central Kuril Islands (Belkin, 1964).

Throughout the year *P. vitulina* is littoral in distribution and non-migratory. Generally it hauls out on protected tidal rocks, sandbars and reefs, frequently ascending rivers and lakes for many miles, and even remaining there all year (see Beck *et al.*, 1970). Local movements are known to be associated with food and breeding (Fisher, 1952; van Bemmel, 1956; Spalding, 1964; Vaughan, 1971; Paulbitski and Maguire, 1972). Tagging studies in The Wash, England, show that young seals disperse up to 300 km and cross the English Channel to France and Holland (Bonner and Witthames, 1974).

Phoca largha is also littoral in distribution during summer when the fast ice has melted along the shore. However, in fall and winter, as fast ice forms along the shore, it migrates off-shore to the edge of the ice pack, hauling out on scattered floes (Fay, 1974). Shaughnessy and Fay (1977) describe eight off-shore breeding areas, three in the Bering Sea, two in the Okhotsk Sea, and one in Tatar Strait, Peter the Great Bay and Po Hai Sea. The seals remain here until the end of the breeding season and ice break-up in spring, then follow the retreating ice and disperse along the shore. The migration direction and timing varies regionally. For example, in the Po Hai Sea, migration is to the south-east for shore haul-outs by June and July (Huang, 1962), while after breeding in the Sea of Okhotsk, seals migrate in a variety of directions appearing along the shores of western Kamchatka in April, Yamsh and Tayuish in May and early June and Shanta in late July-August (Tikhomirov, 1961; Fedoseev, 1971). In the Bering Sea after pupping they disperse north along the shores of Alaska and Siberia with some reaching Wainwright, Alaska by mid-August (Burns, 1970; Fay, 1974).

Abundance and Life History

Only crude estimates are available on total abundance. Large-scale censusing is difficult as herds are scattered, small, and easily frightened into the water. Scheffer (1958) estimated the world abundance at 150 000-450 000 and Lockley (1966) at a minimum of 140 000 excluding pups. Recent surveys indicate that the two species probably number closer to 760 000-950 000 with *P. vitulina* comprising slightly more than half. Estimates for each form are: *P.v. vitulina,* 29 000-100 000 (Scheffer, 1958; Lockley, 1966; Bonner, 1972; Ivashin *et al.*, 1972); *P.v. concolor,* 40 000-100 000 (Scheffer, 1958; Lockley, 1966; Ivashin *et al.*, 1972); *P.v. mellonae,* 500-1000 (Scheffer, 1958); *P.v. richardsi,* 50 000-200 000 (Scheffer, 1958; Lockley, 1966) although probably closer to 320 000 with 275 000 in Alaska (J. Burns, personal communication, 1978), 40 000 from British Columbia to Baja California (Bigg, 1969b; Frey and Aplin, 1970; Pearson and Verts, 1970; Newby, 1973a; Brownell *et al.*, 1974); *P.v. stejnegeri,* perhaps 5000-8000 (Belkin, 1964; Marakov, 1967; Chugunkov, 1970; Naito and Nishiwaki, 1973); *P. largha,* 20 000-50 000 (Scheffer, 1958; Lockley, 1966), although probably closer to 370 000-420 000 with 200 000-250 000 in Alaska (J. Burns, personal communication, 1978) and 168 000 in the Sea of Okhotsk (Fedoseev, 1971); apparently few in Peter the Great Bay (Tikhomirov, 1961) and an unknown number in the Po Hai Sea.

Populations of *P. vitulina* in Alaska, and *P. largha* are probably near historical levels. The reason for low numbers of *P.v. stejnegeri* is unknown. However, some populations of *P. vitulina* in Shetland, Norway, Waddensea, eastern Canada, Oregon and Washington have decreased markedly in recent years through hunting, pollution or disturbance (Øynes, 1964; Mansfield, 1965; Pearson and Verts, 1970; Bonner *et al.*, 1973; Newby, 1973a; Reijnders, 1976).

Population composition

Some aspects of population dynamics appear to differ between the species. In *P. vitulina,* pups comprise about 18-20% of the population

(Venables and Venables, 1955; Bigg, 1969b; Boulva, 1975) whereas in *P. largha* they are reported to comprise 24-25% (Goltsev and Fedoseev, 1970). Only a single pup is usually born in both species. Other reproductive parameters appear similar, males maturing at 3-6 years and females at 2-5 year with 85-92% of mature females producing a young each year (Bigg, 1969b; Goltsev and Fedoseev, 1970; Burns and Fay, 1973). In *P. vitulina* Bigg (1969b) estimated mortality during the first year to be 21% while Goltsev and Fedoseev (1970) calculated it in *P. largha* to be 42-43%. Longevity in both species is at least 30 years. In *P. vitulina* males die at a greater rate than females after reaching sexual maturity.

Food

Newly weaned pups of *P. vitulina* feed primarily on bottom-dwelling crustacea for one and a half to three months with the shrimp *Crangon* being particularly important (Bigg, 1973). Pups of *P. largha* feed on small amphipods found around the ice floes (Goltsev, 1971). Older seals of both species feed opportunistically on a wide variety of fish, some cephalopods and crustaceans. Commonly eaten fish include herring, anchovy, trout, smelt, codfish, scorpionfish, rockfish, prickleback, greenling, sculpin, sandlance and flounder. In The Wash, England, mainly whelks are eaten, as few fish are present, while off nearby Scroby Sand and Blackeney, fish are eaten (Sergeant, 1951). In Holland the principal food is common flounder, and off Scotland gadoids and clupeoids (Havinga, 1933; Rae, 1973). In British Columbia chub and pink salmon are eaten during the fall, octopus, rockfish and salmon in summer and perhaps eulachon in winter (Spalding, 1964).

Feeding occurs during the day but its occurrence at night is uncertain (Havinga, 1933; Spalding, 1964; Goltsev, 1971). Seals tend to eat one prey species per feeding period with small fish eaten whole below the surface and large fish bitten into small pieces and consumed at the surface (Spalding, 1964; Goltsev, 1971). About 5-6% of the body weight is eaten per day (Spalding, 1964).

The feeding habits can conflict with local fisheries. In British Columbia commercial fish comprise 54% of the diet and in Holland 75% (Havinga, 1933; Spalding, 1964). In the Sea of Okhotsk *P. largha* is the only seal harmful to commercial fish (Tikhomirov, 1966). Imler and Sarber (1947) reported that fish were removed or mutilated in nets by harbour seals in the Copper River, Alaska, although this accounted for no more than 2% of the total catch, while Fisher (1952) found about

7% were damaged in the Skeena River, British Columbia. In addition, harbour seals are one of the terminal hosts of the larval nematode *Terranova (= Porrocaecum) decipiens* which infests the musculature of many species of fish, reducing marketability (Scott, 1953). Its presence in this and other species of seal, has increased the incidence of infestation in eastern Canadian groundfish (Scott and Martin, 1957; Scott and Fisher, 1958).

Harbour seals get about 90% of their fresh water from preformed water in the fish eaten, from metabolic processes and from inspired water vapour (Depocas *et al.*, 1971). Sea water is injested only incidental to swallowing food and is not used as a source of fresh water.

Internal Anatomical Characteristics

Skull

Compared to harp and ringed seals, the skull of the harbour seal (Fig. 3) is thickly boned with a broad facial area (Allen, 1880; Doutt, 1942).

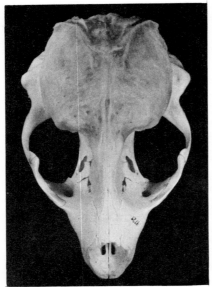 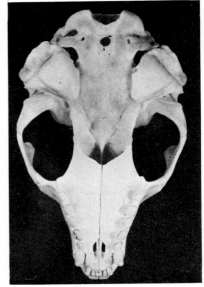

FIG. 3 (a) Dorsal and (b) ventral view of the skull of an adult female *P.v. richardsi* (BCPM 9729 Mam.).

The lower jaw is short and has a thick ramus. The postcanine teeth are crowded and obliquely set in young animals while in old animals these tend to separate and straighten, particularly in *P. largha* (Shaughnessy and Fay, 1977). No sexual differences are apparent. Only small differences in structure separate *P. largha* from *P.v. richardsi* and *P.v. stejnegeri*, and usually a combination of features must be used for species identification (Chapskii, 1967; Shaughnessy and Fay, 1977). For a detailed treatment of skull appearance the interested reader should consult Allen (1880; 1902), Doutt (1942), Belkin (1964), Chapskii (1967), Burns and Fay (1970) and Shaughnessy and Fay (1977).

TABLE 2 Selected references on the internal anatomy of harbour seals.

Organ	*P. vitulina*	*P. largha*	Reference
Integument	x		Montagna and Harrison (1957)
Musculature	x(?)		Tarasoff (1972; a literature listing)
Skeleton	x		Allen (1880)
		x	Antonyuk (1971): Tarasoff (1972— a literature listing)
Organ sizes	x		Slijper (1958)
		x	Sokolov et al. (1966); Shepeleva (1971); Kosygin and Goltsev (1971)
Digestive system	x(?)		Eastman and Coalson (1974)
		x	Sokolov et al. (1966)
Reproductive system	x		Amoroso et al. (1965); Harrison (1960); Bigg (1969b; 1973); Bigg and Fisher (1974)
Endocrine system	x		Harrison et al. (1962); Amoroso et al. (1965)
Circulatory system			
Morphology	x		Harrison and Tomlinson (1956); Tarasoff and Fisher (1970)
		x	Ivanova (1971)
Blood chemistry	x		Hubbard (1968); Lenfant (1969); Ridgway (1972); Ridgway et al. (1975)

Organ systems

An enormous body of literature exists on the macroscopic and microscopic anatomy of the various organ systems of harbour seals. For a detailed consideration of these the reader should refer to the listings of research works given in Table 2.

Behaviour

Social behaviour

Phoca vitulina tends to be solitary in water but forms small groups when hauled out (Fig. 4). Group sizes reported on Sable Island average 18 seals (range 2-125) (Boulva, 1975), in The Wash 15-500 (Harrison, 1960) while in some areas of Alaska, up to several thousand (Alaska Department of Fish and Game, 1973). Groups usually consist of mixed age and sex although Newby (1973b) and Evans and Bastian (1969) report some segregation between nursing females and weaned pups. Groups exhibit no social hierarchy like otariids. Individuals space themselves several feet apart using aggressive gestures such as head butting, snorting, growling, biting and fore-flipper waving. They sleep lightly, each frequently waking to look around, making them one of the most difficult species of seal to approach. Schusterman (1968) reported that captive harbour seals exhibit about 20 times more "looking or glancing around" than California and Steller sea lions and northern elephant seals.

The closest bond is between mother and pup during lactation. The mother keeps near to the pup and in case of danger will grab it with her foreflippers or mouth and dive to safety. Newby (1973b) found that the pup nursed for about a minute every three to four hours. At weaning, pups are abandoned.

Courtship and mating are rarely observed as they apparently take place in water. According to Venables and Venables (1957) mating seems to be preceded and followed by periods of rolling and bubble-blowing by both sexes. Johnson (1969) noted that the male generally initiated chasing, neck and flipper biting and embracing. Mating is probably promiscuous.

During the period of littoral distribution *P. largha* behaves like *P. vitulina*. However, at breeding time during the pelagic distribution, *P. largha* is monogamous with pairs forming about 10 days before pupping

FIG. 4 Typical haul-out site for *P. v. richardsi*. Note the individual variations in pelage pattern.

and lasting until mating about one month later. Pairs may defend a territory around the natal floe as family groups are separated by at least 0.25 km (Fay, 1974). Adults are located along the ice edge while immatures occur deeper in the ice floes. Following breeding, all ages congregate in large moulting patches before the ice melts and the seals move to the coast.

Diving

Pups of *P. vitulina* swim at birth. At two to three days they can dive for up to 2 min (Finch, 1966), and at 10 days for up to 8 min (Harrison and Tomlinson, 1960). In older seals the deepest simulated dive is 206 m, and the longest is 30 min (Harrison and Kooyman, 1968; Kooyman *et al.*, 1972). The usual dive, however, is probably closer to that described by Carl (1964) who observed a young seal feeding in 6-8 ft of water. In six consecutive dives it submerged for an average of 3 min 16 s with the longest of 30 dives being 4 min 15 s.

Pups of *P. largha* do not swim until after weaning, at which time they may dive to 80 m (Goltsev, 1971). From stomach contents, adults may dive to 300 m. Chugunkov (1970) reported that the species travel distances of up to 400 m underwater.

Before submerging, seals exhale, retaining only about 40% of the inspiratory lung volume and continuing to exhale periodically as the dive progresses (Harrison and Tomlinson, 1960; Kooyman *et al.*, 1972). The lung alveoli collapse during the dive, forcing air into the non-absorptive and less compressible bronchioles to reduce the chance of "the bends" (or nitrogen narcosis) occurring (Kooyman *et al.*, 1972).

Diving initiates many physiological changes to conserve oxygen. The most important is the "dive reflex" which slows circulation and redistributes blood flow. Details of these processes are described by Harrison and Tomlinson (1960), Bron *et al.* (1966), Murdaugh *et al.* (1966), Harrison and Kooyman (1968) and Kerem and Elsner (1973). Without the dive reflex drowning occurs in the same time as in land mammals.

Locomotion

Scheffer and Slipp (1944) described locomotion of *P. vitulina* on land as typically a hitching motion with foreflippers and claws thrust into the ground to pull the body along in quick caterpillar-like jerks. The hindflippers are held together and up. Tarasoff (1972) found that the hind limbs can assist in climbing by swinging from side to side.

Scheffer and Slipp (1944) noted that in water the hindflippers provide the propulsion. Each webbed foot alternatvely provides the power stroke as it expands and moves medially with perhaps some propulsion as the foot contracts and moves laterally. Occasionally both hindflippers give power strokes in a clapping fashion. By turning the hindflipper palms upwards, power strokes elevate the anterior body, but apparently palms cannot be tilted downward.

Chugunkov (1970) observed that *P. largha* swims at speeds up to 3.8 m-s.

Thermoregulation

Matsuura and Whittow (1974) found that *P. vitulina* on land overheated at 30°-35°C air temperature. However, Ohata and Whittow (1974) calculated that the seal can prevent this by hauling out on cool or damp sand. Over-heating is also prevented by increasing peripheral bloodflow through the flippers and skin (Tarasoff and Fisher, 1970; McGinnis, 1975).

Insulation is provided mainly by blubber and by allowing the skin temperature to follow closely that of the environment (Irving and Hart, 1957; Hart and Irving, 1959). The metabolic rate increases if water temperature falls below 20°C or air temperature below 2°C, with some seasonal acclimatization to these critical temperatures. Hair is of little insulative value in *P. vitulina*, although Tarasoff and Fisher (1970) feel it could be useful in air on calm days.

Phoca vitulina usually has a rectal and deep body temperature of 37-38°C although it can vary from 35 to 40 °C. Temperature drops during a dive and increases when hauled out in the sun (Scholander *et al.*, 1942; McGinnis, 1975). Peripheral tissues vary considerably more and can continue operating because they possess heterothermic enzymes (Somero and Johansen, 1970; Behrisch and Percy, 1974).

Vocalization

Harbour seals are probably the least vocal pinniped (Poulter, 1968). Pups utter a sheep-like cry. After weaning, vocalizations are limited to threatening snorts, grunts and growls. Schevill *et al.* (1963) recorded faint underwater clicks which, if used for echolocation, would probably be of value only for the last stages of catching prey. Schusterman *et al.* (1970) found that the underwater sounds were variably pulsed, of low frequency, and probably associated with threat display.

Hearing

Mohl (1968a, 1968b) concludes that *P. vitulina* hears best in water, about the same as man in air. It responds to frequencies 1-180 kHz with pitch discrimination up to 64 kHz in water while in air only 1-22.5 kHz. Peak frequency sensitivity in water is 32 kHz, while in air it is about 12 kHz. Sensitivity is about 15 db better in water than in air. The seal has directional hearing in both air and water.

Vision

Schusterman and Balliet (1970) found the visual acuity of *P. vitulina* in water to be similar to that of land mammals with sharp aerial vision. Jamieson and Fisher (1970) suggest that under low light intensities, acuity is probably better in water than in air. They point out that the cornea is flattened causing astigmatism which is eliminated in water, due to similar refractive indices, and absent in air under bright light, because of a pinhole-shaped pupil (Jamieson, 1971; Jamieson and Fisher, 1972). Under low light in air, however, the pupil is elliptical and probably does not eliminate corneal astigmatism.

The retina contains many rod photoreceptors, suggesting black and white vision, although some cones may exist, making the extent of colour vision uncertain (Jamieson and Fisher, 1971). Extensive summation of the rod photoreceptors indicates a highly sensitive retina which probably permits good vision under low lighting conditions, such as at great depths.

Vision is not essential for survival. Newby (1973b) found three healthy but blind adult females with pups in Washington.

Captivity

Harbour seals are easily kept in captivity and are frequently used as displays in zoos and aquaria and for research. They are hardy, eat most kinds of fish, can be kept in groups or alone, and quickly become relaxed under restrained conditions. Herring, mackerel, smelt and squid are commonly used as foods (Keyes, 1968). New-born pups can be hand-fed on chopped or blended fish rather than on a milk formula (Reineck, 1962; Johnson, 1969).

Breeding is not common unless adult males are raised in captivity. Wild adult males apparently lose their breeding drive in captivity, while males raised in captivity become aggressive enough to mate. Females have an annual oestrous cycle in captivity regardless of male presence (Bigg, 1973).

Hubbard (1968) gives a good account of husbandry procedures.

Reproduction

The reproductive cycle is annual and similar in both species (Bigg, 1969a; Bigg and Fisher, 1974). The birth season lasts one to two months; lactation lasts from two to six weeks followed shortly by ovulation and mating, then blastocyst implantation one and a half to three months later. Both species have regional variations in the timing of births, *P. largha* ranging from February to May and *P. vitulina* from February to September, with clinal variations between the Gulf of Alaska and Mexico (Bigg, 1969a, Scheffer, 1974; Shaughnessy and Fay, 1977). Where the distributions of the two species overlap in the North Pacific birth seasons differ by about two months. Variations in reproductive timing may have evolved to increase survival of pups in each locality (Bigg, 1973). Experimental studies with *P.v. richardsi* suggest that each population maintains its unique reproductive timing through a specific response to photoperiod and the existence of an annual endogenous reproductive rhythm (Bigg, 1973; Bigg and Fisher, 1975).

In *P.v. richardsi*, oestrus lasts one to nine weeks, with some females apparently being induced ovulators (Bigg, 1973). On the basis of sperm presence in the epididymis, males seem to be in breeding condition for up to nine months of the year (Bigg, 1969b).

Moult

Stutz (1967b) described the sequence of moult of lanugo hair in *P. vitulina* as from the face, anterior flippers and nape over the sacrum to the ventral and mid-dorsal regions and finally the lumbar mid-dorsal area. Moult of the adult-type hair in both species is annual and usually occurs between the pupping season and two or three months after. Havinga (1933) and Kosygin and Goltsev (1971) reported that immature *P.v. vitulina* and *P. largha* moult before adults, although Stutz (1967b) did not observe this in *P.v. richardsi*. Scheffer and Slipp (1944) watched a captive seal complete most of its moult in about one month, while Stutz (1967b) found that most wild seals moulted within a two month period. Although individual variations exist, shedding begins around the body openings, then along the body ventral midline, nape and sacrum and spreads to other regions (Stutz, 1967b).

Diseases

Ridgway (1972) and Ridgway *et al.* (1975) give a general treatment of diseases in marine mammals including harbour seals. Some specific examples in harbour seals are now given. Bonner (1972) reports the only case of an apparent epizootic among harbour seals, this occurring in Shetland during the 1920s; but details are lacking of its cause. The viral disease, seal pox, was recently identified and described by Wilson and Sweeney (1970) and Wilson *et al.* (1972) in wild and captive harbour seals. Ericksen (1962) at the Copenhagen Zoo, recorded cases of generalized pasteurellosis, complications of umbilical infections involving streptococci of Group G, necroses in the gingiva, keratitis of the eye, and gastroenteritis. Instances are also known of peritonitis and septicaemia involving *Clostridium* (Grafton, 1967) and of mucopurulent bronchitis and catarrhal pneumonia associated with the lung-worm *Parafilaroides grymnurus* (Van den Broek and Wensvoort, 1959). Infections around the lips and commissure of the mouth are commonly associated with *Pseudomonas* (Greenwood *et al.*, 1974). Johnston and Fung (1970) list 15 types of bacteria found around the nostrils, eyes, mouth, throat, skin, anus, vagina and prepuce of wild Pacific harbour seals. Van Pelt and Dieterich (1973) report an instance in which a hand-fed new-born contracted a generalized staphyloccal infection and toxoplasmosis. They conclude the disease, caused by the protozoan *Taxoplasma gondii,* was transmitted to the foetus across the placenta.

References

Allen, J. A. (1880). "History of North American Pinnipeds: A Monograph of Walrus, Sea Lions, Sea Bears, and Seals of North America" *US Geol. Geogr. Surv. Territ., Misc. Publ.* **12,** Government Printing Office, Washington, D.C.
Allen, J. A. (1902). The hair seals (Family Phocidae) of the North Pacific Ocean and Bering Sea. *Bull. Am. Mus. Nat. His.* **16,** 459-499.

Alaska Department of Fish and Game. (1973). "Alaska's Wildlife and Habitat", Office of the Commissioner, Subport Building, Juneau.

Amoroso, E. C., Bourne, G. H., Harrison, R. J., Matthews, L. H., Rowlands, I. W. and Sloper, J. C. (1965). Reproductive and endocrine organs of foetal, newborn, and adult seals. *J. Zool. (Lond.)* **147**, 430-486.

Antonyuk, A. A. (1971). Quantitative changes in the vertebrae in the vertrebral column of pinnipeds. *In* "Research on Marine Mammals", (eds K. K. Chapskii and E. S. Milchenko), p. 317-321. Tr. Atl. Nauchno-Issled. Inst. Rybn. Khoz. Okeanogr. 39. [In Russian] (English translation Can. Fish. and Mar. Serv. Transl. Ser. No. 3185.)

Barnes, L. G. and Mitchell, E. D. (1975). Late cenozoic North-east Pacific Phocidae. *Rapp. P. -v Réun. Cons. Int. Explor. Mer* **169**, 34-42.

Beck, B., Smith, T. G. and Mansfield, A. W. (1970). Occurrence of the harbour seal, *Phoca vitulina* (Linnaeus) in the Thlewiaza River, N.W.T. *Can. Field-Nat.* **84**, 297-300.

Behrisch, H. W. and Percy, J. A. (1974). Temperature and the regulation of enzyme activity in homeo- and heterothermic tissues of Arctic marine mammals: some regulatory properties of 6-phospho-gluconate dehydrogenase from adipose tissues of the spotted seal *Phoca vitulina*. *Comp. Biochem. Physiol., B. Comp. Biochem.* **47**, 437-443.

Belkin, A. N. (1964). A new species of seal—*Phoca insularis* sp. n.—from the Kuril Islands. *Dokl. Akad. Nauk SSSR* **158**, 1217-1219. (In Russian)

Belkin, A. N., Kosygin, G. M. and Panin, K. I. (1969). New data on the characteristics of the Island seal. *In* "Marine Mammals", (Eds V. A. Arseniev, B. A. Zenovich and K. K. Chapskii), 157-175. Science Publisher, Moscow. (In Russian.)

Bemmel, A. C. V. van. (1956). Planning a census of the harbour seal *(Phoca vitulina* L.) on the coasts of the Netherlands. *Beaufortia* **54**, 121-132.

Bigg, M. A. (1969a). Clines in the pupping season of the harbour seal, *Phoca vitulina. J. Fish. Res. Board Can.* **26**, 449-455.

Bigg, M. A. (1969b). The harbour seal in British Columbia. *Fish. Res. Board Can. Bull.* **172**, 33.

Bigg, M. A. (1973). Adaptations in the breeding of the harbour seal, *Phoca vitulina. J. Reprod. Fert. Suppl.* **19**, 131-142.

Bigg, M. A. and Fisher, H. D. (1974). The reproductive cycle of the female harbour seal off south-eastern Vancouver Island. *In* "Functional Anatomy of Marine Mammals", (Ed. R. J. Harrison), Vol. 2, 329-347. Academic Press, London and New York.

Bigg, M. A. and Fisher, H. D. (1975). Effect of photoperiod on annual reproduction in female harbour seals. *Rapp. P.-v. Réun., Cons. Int. Explor. Mer* **169**, 141-144.

Bonner, W. N. (1972). The grey seal and common seal in European waters. *Oceanogr. Mar. Biol. Annu. Rev.* **10**, 461-507.

Bonner, W. N., Vaughan, R. W. and Johnston L. (1973). The status of common seals in Shetland. *Biol. Conserv.* **5**, 185-190.

Bonner, W. N. and Whitthames, S. R. (1974). Dispersal of common seals *(Phoca vitulina)* tagged in The Wash, East Anglia. *J. Zool. (Lond.)* **174**, 528-531.

Boulva, J. (1971). Observations on a colony of whelping harbour seals, *Phoca vitulina concolor,* on Sable Island, Nova Scotia. *J. Fish. Res. Board Can.* **28**, 755-759.

Boulva, J. (1975). Temporal variations in the birth period and characteristics of newborn harbour seals. *Rapp. P.-v. Réun. Cons. Int. Explor. Mer* **169**, 405-408.

Bron, K. M., Murdaugh, H. V., Millen, J. E., Lenthall, R., Raskin, P. and Robin, E. D. (1966). Arterial constrictor response in a diving mammal. *Science (Wash., D.C.)* **152**, 540-543.

Brownell, R. L., Delong, R. L. and Schreiber, R. W. (1974). Pinniped populations at Islas de Guadalupe, San Benito, Cedros, and Natividad, Baja California, in 1968. *J. Mammal.* **55**, 469-472.

Burns, J. J. (1970). Remarks on the distribution and natural history of pagophilic pinnipeds in the Bering and Chukchi seas. *J. Mammal.* **51**, 445-454.

Burns, J. J. and Fay, F. H. (1970). Comparative morphology of the skull of the ribbon seal, *Histriophoca fasciata,* with remarks on systematics of Phocidae. *J. Zool. (Lond.)* **161**, 363-394.

Burns, J. J. and Fay, F. H. (1973) Comparative biology of Bering Sea harbor seal populations. *Proc. Alaskan Sci. Conf.* **23**, 28(Abstr.).

Burns. J. J. and Fay, F. H. (1974). New data on taxonomic relationships among North Pacific harbor seals, genus *Phoca (sensu stricto). Trans. Int. Theriol. Congr.* **1**, 99 (Abstr.).

Caldwell, D. K. and Caldwell, M. C. (1969). The harbor seal, *Phoca vitulina concolor,* in Florida. *J. Mammal* **50**, 379-380.

Caldwell, D. K., and Golley, F. B. (1965). Marine mammals from the coast of Georgia to Cape Hatteras. *J. Elisha Mitchell Sci. Soc.* **81**, 24-32.

Carl, G. C. (1964). Diving rhythm of the hair seal. *Victoria Naturalist (Vic. Nat. Hist. Soc., Victoria, B.C.)* **21**, 35-37.

Chapskii, K. K. (1967). Morphological—taxonomical nature of the *pagetoda* form of the Bering Sea *larga*. *In* "Investigations on Marine Mammals", (Eds K. K. Chapskii and M. Y. Yakovenko), 147-176. Poliarnia Pravda, Murmansk. (In Russian.) (English translation, Fish. Res. Bd. Can., Transl. Ser. No. 1108.)

Chapskii, K. K. (1969). Taxonomy of seals of the genus *Phoca sensu stricto* in the light of present craniological data. *In* "Marine Mammals", (Eds V. A. Arseniev, B. A. Zenkovich and K. K. Chapskii), 294-304. Nauka, Moscow. (In Russian.) (English translation, Can. Dept. Secr. State, Transl. Bur. No. 539.)

Chugunkov, D. I. (1970). "Pinnipeds of Kamchatka", *Priroda Moscow* **6**,12-17. (In Russian.) (English translation, Div. Foreign Fish., Washington).

Curry-Lindahl, K. (1975). Ecology and conservation of the grey seal *Halichoerus grypus*, common seal *Phoca vitulina*, and ringed seal *Pusa hispida* in the Baltic Sea. *Rapp. P. -v. Réun. Cons. Int. Explor. Mer* **169**, 527-532.

Davies, J. L. (1958). Pleistocene geography and the distribution of modern pinnipeds. *Ecology* **39**, 97-113.

Depocas, F., Hart, J. S. and Fisher, H. D. (1971). Sea water and water flux in starved and in fed harbor seals, *Phoca vitulina*. *Can. J. Physiol. Pharmacol.* **49**, 53-62.

Doutt, J. K. (1942). A review of the genus *Phoca*. *Carnegie Mus., Ann.* **29**, 61-125.

Eastman, J. T. and Coalson, R. E. (1974). The digestive system of the Weddell Seal, *Leptonychotes weddelli*—a review. *In* "Functional Anatomy of Marine Mammals", (Ed. R. J. Harrison), Vol. 2. 235-320. Academic Press, London and New York.

Eriksen, E. (1962). Diseases of seals in the Copenhagen zoo. *Nord. Veterinaermed.* 14, *Suppl. 1,* 141-149.

Evans, W. E. and Bastian, J. (1969). Marine mammal communication: social and ecological factors. *In* "The Biology of Marine Mammals", (Ed. H. T. Andersen), 425-475. Academic Press, New York and London.

Fay, F. H. (1974). The role of ice in the ecology of marine mammals of the Bering Sea. *In* "Oceanography of the Bering Sea", (Eds D. W. Hood and E. J. Kelly), 383-399. University of Alaska, Fairbanks.

Fedoseev, F. A. (1971). The distribution and numbers of seals on whelping and moulting patches in the Sea of Okhotsk. *In* "Research on Marine Mammals" (Eds K. K. Chapskii and E. S. Milchenko, 87-99. Kaliningrad. Trudy Atlant. NIRO. No. 39. (In Russian) (English translation, Fish. and Mar. Ser. No. 3185).

Finch, V. A. (1966). Maternal behavior in the harbor seal. *In* Proc. 3rd Annu. Conf. Biol. Sonar Diving Mamm. (Ed. C. E. Rice). 147-150. Stanford Res. Inst. Biol. Sonar Lab., Menlo Park, CA.

Fisher, H. D. (1952). The status of the harbour seal in British Columbia, with particular reference to the Skeena River. *Fish. Res. Board Can., Bull.* **93**, 58.

Frey, H. W. and Aplin, J. A. (1970). Sea lion census for 1969, including counts of other California pinnipeds. *Calif. Fish Game* **56**, 130-133.

Grafton T. S. (1967). Investigation of the sudden death of a female harbor seal. 47th Ann. Meeting Amer. Soc. Ichthy. and Herpetology, Aquarium Symposium Section, San Francisco. (Not seen, cited by Hubbard, 1968).

Goltsev, V. N. (1971). Feeding of the harbor seal. *Ekologiya* **2** 62-70. (In Russian.) (English translation, U.S. NOAA).

Goltsev, V. N. and Fedoseev, G. A. (1970). Dynamics of the age composition of rookeries and the replacement capacity of harbour seal populations. *Izv. Tikhookean. Nauchno-Issled. Inst. Rybn. Khoz. Okeanogr.* **71**, 309-317. (In Russian.) (English translation, Fish. Res. Bd. Canada Trans. Ser. No. 2401.)

Greenwood, A. G., Harrison, R. J. and Whitting H. W. (1974). Functional and pathological aspects of the skin of marine mammals. In "Functional Anatomy of Marine Mammals", Vol. 2. (Ed. R. J. Harrison), 73-110. Academic Press, London and New York.

Harrison, R. J. (1960). Reproduction and reproductive organs in common seals *(Phoca vitulina)* in The Wash, East Anglia. *Mammalia* **24**, 372-385.

Harrison, R. J. and Tomlinson J. D. W. (1956). Observations on the venous systems in certain Pinnipedia and Cetacea. *Proc. Zool. Soc. London* **126**, 205-233.

Harrison, R. J. and Tomlinson, J. D. W. (1960). Normal and experimental diving in the common seal *(Phoca vitulina)*. *Mammalia* **24**, 386-399.

Harrison, R. J., Rowlands, I. W., Whitting, H. W. and Young, B. A. (1962). Growth and structure of the thyroid gland in the common seal. *(Phoca vitulina). J. Anat.* **9**, 3-15.

Harrison, R. J. and Kooyman, G. L. (1968). General physiology of the Pinnipedia. In "The Behavior and Physiology of Pinnipeds", (Eds R. J. Harrison, R. C. Hubbard, R. S. Petersen, C. E. Rice, R. J. Schusterman), 211-295. Appleton-Century-Crofts, New York.

Hart, J. S. and Irving, L. (1959). The energetics of harbour seals in air and in water with special consideration of seasonal changes. *Can. J. Zool.* **37**, 447-457.

Havinga, B. (1933). Der seehund *(Phoca vitulina L.)* in den Holländischen gewässern. *Ned. Tijdschr. Ned. Dierk. Ver.* **3**, 79-111.

Hook, O. (1961). Notes on the status of seals in Iceland, June-July, 1959. *Proc. Zool. Soc. Lond.* **137**, 628-630.

Huang, Shou- jen (Ed.) (1962). Pinnipedia. In "Kexue Chubanshe. (Economically important animals of China.)", 414-417. Peking (?). (In Chinese) (English translation, Fish. Res. Board Can., Transl. Ser. No. 2328.)

Hubbard, R. C. (1968). Husbandry and laboratory care of pinnipeds. In "The Behavior and Physiology of Pinnipeds", (Eds R. J. Harrison, R. C. Hubbard, R. S. Petersen, C. E. Rice and R. J. Schusterman), 299-358. Appleton-Century-Crofts, New York.

Imler, R. H. and Sarber, H. R. (1947). Harbor seals and sea lions in Alaska. U.S. Fish. Wildl. Serv., Spec. Sci. Rep. 28.

Irving, L. and Hart, J. S. (1957). The metabolism and insulation of seals as bare-skinned mammals in cold water. *Can. J. Zool.* **35**, 497-511.

Ivanova, E. I. (1971). The morphology of the vascular system of cetaceans and pinnipeds. In "Research on Marine Mammals", (Eds K. K. Chapskii and E. S. Milchenko), 179-192. Tr. Atl. Nauchno-Issled. Inst. Rybn. Khoz. Okeanogr. 39. (In Russian.) (English translation Can. Fish. and Mar. Serv. Transl. Ser. No. 3185.)

Ivashin, M. V., Popov, L. A. and Tsapko, A. S. (1972) "Marine Mammals Handbook", (Ed. P. A. Moiseeva), Pischevaya Promyshlennost, Moscow. (In Russian.) (English translation, Fish. Res. Board Can. Transl. Ser. No. 2783.)

Jamieson, G. S. (1971). The functional significance of corneal distortion in marine mammals. *Can. J. Zool.* **49,** 421-423.

Jamieson, G. S. and Fisher, H. D. (1970). Visual discrimination in the harbour seal *Phoca vitulina,* above and below water. *Vision Res.* **10,** 1175-1180.

Jamieson, G. S. and Fisher, H. D. (1971). The retina of the harbour seal, *Phoca vitulina. Can. J. Zool.* **49,** 19-23.

Jamieson, G. S. and Fisher, H. D. (1972). The pinniped eye: a review. *In* "Functional Anatomy of Marine Mammals", (Ed R. J. Harrison), Vol. 1, 245-261. Academic Press, London and New York.

Johnston, B. W. (1969). Maintenance of harbor seals *(Phoca vitulina)* in aquarium quarters. *In* "Proc. 6th Annu. Conf. Biol. Sonar Diving Mamm.", 49-54. Stanford Res. Inst., Biol. Sonar Lab., Menlo Park, California.

Johnston, D. G. and Fung, J. (1970). Bacterial flora of wild porpoises, seals, and seal lions of Pacific coastal waters. *In* "Proc. 7th Annu. Conf. Biol. Sonar Diving Mamm.", 191-216. Stanford Res. Inst., Biol. Sonar Lab., Menlo Park, California.

Kerem, D. H. and Elsner, R. W. (1973). Cerebral tolerance to asphyxial hypoxia in the harbor seal. *Respir. Physiol.* **19,** 188-200.

Keyes, M. C. (1968). The nutrition of pinnipeds. *In* "The Behavior and Physiology of Pinnipeds", (Eds R. J. Harrison, R. C. Hubbard, R. S. Petersen, C. E. Rice and R. J. Schusterman), 359-395. Appleton-Century-Crofts, New York.

Kooyman, G. L., Schroeder, J. P., Denison, D. M., Hammond, D. D., Wright, J. J. and Bergman, W. P. (1972). Blood nitrogen tensions of seals during simulated deep dives. *Am. J. Physiol.* **233,** 1016-1020.

Kosygin, G. M. and Goltsev, V. N. (1971). Data on the morphology and ecology of the harbour seal of the Tatar Strait. *In* "Research on Marine Mammals", (Eds K. K. Chapskii and E. S. Milchenko), 238-252. Tr. Atl. Nauchno-Issled. Inst. Rybn. Khoz. Okeanogr. 39 (In Russian.) (English translation, Can. Fish, and Mar. Serv. Transl. Ser. No. 3185.)

Krog, J. and Bjarghov, R. (1973). Common seals in Svalbard. *Fauna (Oslo)* **26,** 217-218. (In Norwegian.) (English translation, T. Øritsland, Inst. Mar. Res., Oslo.)

Lenfant, C. (1969). Physiological properties of blood of marine mammals. *In* "The Biology of Marine Mammals", (Ed. H. T. Andersen), 95-116. Academic Press, New York and London.

Lockley, R. M. (1966). "Grey Seal, Common Seal", October House, New York.

McGinnis, S. M. (1975). Peripheral heat exchange in phocids. *Rapp. P.-v. Réun. Cons. Int. Explor. Mer* **169,** 481-486.

McLaren, I. A. (1966). Taxonomy of harbour seals of the western North Pacific and evolution of certain other hair seals. *J. Mammal.* **47,** 466-473.

Mansfield, A. W. (1965). The influence of pinnipeds on the fisheries of eastern Canada. *In* "A Seals Symposium", (Ed. E. A. Smith), Proc. Nat. Conserv. Consult. Comm. Grey Seals Fish. Meet. 9th Sept., 1964, 35-45. Dep. Zool. Univ. Cambridge, Edinburgh: Nat. Conserv.

Mansfield, A. W. (1967). Distribution of the harbour seal. *(Phoca vitulina Linnaeus)*, in Canadian Arctic waters. *J. Mammal.* **48,** 249-257.

Marakov, S. V. (1967). Materials on the ecology of the larga on the Commander Islands. *In* "Investigations on Marine Mammals", (Eds K. K. Chapskii and M. Ya Yakovenko), 126-136. Poliarnaia, Pravda. (In Russian.) (English translation, Fish. Res. Board Can. Transl. Ser. No. 1079.)

Matsuura, D. T. and Whittow, G. C. (1974). Evaporative heat loss in the California sea lion and harbor seal. *Comp. Biochem. Physiol., A. Comp. Physiol.* **48,** 9-20.

Mohl, B. (1968a). Hearing in seals. *In* "The Behavior and Physiology of Pinnipeds", (Eds R. J. Harrison, R. C. Hubbard, R. S. Petersen, C. E. Rice and R. J. Schusterman), 142-195. Appleton-Century-Crofts. New York.

Mohl, B. (1968b). Auditory sensitivity of the common seal in air and water. *J. Aud. Res.* **8,** 27-38.

Mohr, E. (1965). Über *Phoca vitulina largha* Pallas, 1811, und weissgeborene Seehunde. *Z. Saeugetierkd* **30,** 273-287. (In German.) (Translation, Can. Dept. Secr. State, Transl. Bur., 1973.)

Mohr, E. (1966). Determining and estimating age in the seal *Phoca vitulina* L. *Z. Jagdwiss.* **12,** 49-54. (In German.) (Translation, R. Floyd, Univ. Wash., Seattle.)

Montagna, W. and Harrison, R. J. (1957). Specializations in the skin of the seal *(Phoca vitulina). Am J. Anat.* **100,** 81-113.

Murdaugh, H. V., Robin, E. D., Millen, J. E., Drewry, W. F. and Weiss, E. W. (1966). Adaptations to diving in the harbor seal: cardiac output during diving. *Am. J. Physiol.* **210,** 176-180.

Naito, Y. (1976). The occurrence of the phocid seals along the coast of Japan and possible dispersal of pups. *Sci. Rept. Whales Res. Inst.* **28,** 175-185.

Naito, Y. and Nishiwaki, M. (1972). The growth of two species of the harbor seal in the adjacent waters of Hokkaido. *Sci. Rep. Whales Res. Inst. Tokyo* **24,** 127-144.

Naito, Y. and Nishiwaki, M. (1973). Kurile harbour seal *(Phoca kurilensis)* p. 44-49. *In* "Seals", IUCN Publ. New Ser., Suppl. Paper No. 39.

Newby, T. C. (1973a). Changes in the Washington State harbor seal population 1942-1972. *Murrelet* **54,** 4-6.

Newby, T. C. (1973b). Observations on the breeding behavior of the harbor seal in the State of Washington. *J. Mammal.* **54,** 540-543.

Ohata, C. A. and Whittow, G. C. (1974). Conductive heat loss to sand in California sea lions and a harbor seal. *Comp. Biochem. Physiol., A. Comp. Physiol.* **47,** 23-26.

Øynes, P. (1964). Sel Pa Norskekysten fra Finnmark til More. *Fisken Havet* **48**, 13. (In Norwegian; English summary).

Paulbitski, P. A. and Maguire, T. D. (1972). Tagging harbor seals in San Francisco Bay (California, USA). *In* Proc. 7th Annu. Conf. Biol. Sonar Diving Mamm., 53-72. Stanford Res. Inst., Biol. Sonar Lab., Menlo Park, CA.

Pearson, J. P. and Verts, B. J. (1970). Abundance and distribution of harbor seals and northern sea lions in Oregon. *Murrelet* **51**, 1-5.

Poulter, T. C. (1968). Marine mammals. *In* "Animal Communication", (Ed. T. Seboek), 405-465. Bloomington, Indiana University Press.

Rae, B. B. (1973). Further observations on the food of seals. *J. Zool. (Lond.)* **169**, 287-297.

Reineck, M. (1962). The rearing of abandoned sucklings of *Phoca vitulina*. *Int. Zoo. Yearb.* **4**, 293-294.

Reijnders, P. J. H. (1976). The harbour seal *(Phoca vitulina)* in the Dutch Wadden Sea: size and composition. *Neth. J. Ser. Res.* **10**, 223-235.

Ridgway, S. H. (1972). Homeostasis in the aquatic environment. "Mammals of the Sea: Biology and Medicine", (Ed. S. H. Ridgway), 590-747. Thomas, Springfield, Ill.

Ridgway, S. H., Geraci, J. R. and Medway, W. (1975). Diseases of pinnipeds. *Rapp. P.-v. Réun. Cons. Int. Explor. Mer* **169**, 327-337.

Scheffer, V. B. (1958). Seals, sea lions and walruses: a review of the Pinnipedia. Stanford University Press, Stanford, CA.

Scheffer, V. B. (1964). Hair patterns in seals (Pinnipedia). *J. Morphol.* **115**, 291-304.

Scheffer, V. B. (1974). February birth of Mexican harbor seals. *Murrelet* **55(3)**, 44.

Scheffer, V. B. and Slipp, J. W. (1944). The harbor seal in Washington State. *Am. Midl. Nat.* **32**, 373-416.

Schevill, W. E., Watkins, W. A. and Ray, G. C. (1963). Underwater sounds of pinnipeds. *Science (Wash., DC.)* **141**, 50-53.

Scholander, P. F., Irving, L. and Grinnell, S. W. (1942). On the temperature and metabolism of the seal during diving. *J. Cell. Comp. Physiol.* **19**, 67-78.

Schusterman, R. J. (1968). Experimental laboratory studies of pinniped behavior. *In* "The Behavior and Physiology of Pinnipeds", (Eds R. J. Harrison, R. C. Hubbard, R. S. Petersen, C. E. Rice and R. J. Schusterman), 87-171. Appleton-Century-Crofts, New York.

Schusterman, R. J., Balliet, R. F. and St. John, S. (1970). Vocal displays under water by the gray seal, the harbor seal, and the Steller sea lion. *Psycho. Sci.* **18**, 303-305.

Schusterman, R. J. and Balliet, R. F. (1970). Visual acuity of the harbour seal and the Steller sea lion underwater. *Nature (Lond.)* **226**, 553-564.

Scott, D. M. (1953). Experiments with the harbor seal, *Phoca vitulina*, a definitive host of a marine nematode, *Porrocaecum decipiens*. *J. Fish. Res. Board Can.* **10**, 539-547.

Scott, D. M. and Martin, W. R. (1957). Variations in the incidence of larval nematodes in Atlantic cod fillets along the Southern Canadian mainland. *J. Fish. Res. Board Can.* **14,** 975-996.

Scott, D. M. and Fisher, H. D. (1958). Incidence of the ascarid *Porrocaecum decipiens* in the stomachs of three species of seals along the southern Canadian Atlantic mainland. *J. Fish. Res. Board Can.* **15,** 495-516.

Sergeant, D. E. (1951). The status of the common seal *(Phoca vitulina L.)* on the East Anglian coast. *J. Mar. Biol. Ass. UK* **29,** 707-717.

Shaughnessy, P. D. (1975). Biochemical comparison of the harbor seals *Phoca vitulina richardi* and *P.v. largha. Rapp. P.-v. Réun. Cons. Int. Explor. Mer* **169,** 70-73.

Shaughnessy, P. D. and Fay, F. H. (1977). A review of the taxonomy and nomenclature of North Pacific Harbour seals. *J. Zool. (Lond.)* **182,** 385-419.

Shepeleva, V. K. (1971). Adaptations of seals to life in the Arctic. *In* "Morphology and Ecology of Marine Mammals, Seals, Dolphins, Porpoises", (Eds K. K. Chapskii and V. E. Sokolov), 1-58. Translated by H. Mills, Israel Progr. Sci. Transl., 1973. Halstad Press, Jerusalem.

Slijper, E. J. (1958). Organ weights and symmetry problems in porpoises and seals. *Arch. Neerl. Zool.* **13,** *suppl. 1.* 97-113.

Smith, E. A. (1965). Comments in Discussion after Session 2. *In* E. A. Smith. A Seals Symposium". Proc. Nat. Conserv. Consult. Comm. Grey Seals Fish. Meet. 9th Sept. 1964, 46. Dep. Zool. Univ. Cambridge, Edinburgh: Nat. Conserv.

Sokolov, A. S., Kosygin, G. M. and Tikhomirov, E. A. (1966). Information on the weight of internal organs in the Bering Sea Pinnipeds. *Izv. Tikhookean. Nauchno-Issled. Inst. Rybn. Khoz. Okeanogr.* **48,** 137-147. (In Russian.) (English translation, US Bureau Comm. Fish., Washington.)

Somero, G. N. and Johansen, K. (1970). Temperature effects on enzymes from homeothermic and heterothermic tissues of the harbor seal *(Phoca vitulina). Comp. Biochem. Physiol.* **34,** 131-136.

Spalding, D. J. (1964). Comparative feeding habits of the fur seal, sea lion and harbour seal on the British Columbia coast. *Fish. Res. Board Can. Bull.* **46,** 52.

Stutz, S. S. (1966). Foetal and post-partum whitecoat pelage in *Phoca vitulina. J. Fish. Res. Board Can.* **23,** 607-609.

Stutz, S. S. (1967a). Pelage patterns and population distributions in the Pacific harbour seal *(Phoca vitulina richardi). J. Fish. Res. Board Can.* **24,** 451-455.

Stutz, S. S. (1967b). Moult in the Pacific harbour seal, *Phoca vitulina richardi. J. Fish. Res. Board Can.* **24,** 435-441.

Tarasoff, F. J. (1972). Comparative aspects of the hind limbs of the river otter, sea otter and seals. *In* "Functional Anatomy of Marine Mammals", (Ed. R. J. Harrison), Vol. 1, 333-359. Academic Press, London and New York.

Tarasoff, F. J. and Fisher, H. D. (1970). Anatomy of the hind flippers of the two species of seals with reference to thermoregulation. *Can. J. Zool.* **48**, 821-829.

Tikhomirov, E. A. (1961). The distribution and migration of seals in waters of the Far East. *In* "Transactions of the Conference on ecology and hunting of marine mammals", (Eds E. H. Pavlovskii and S. E. Kleinenberg), 199-210. Akad. Nauk SSSR, Ikhtiol. Comm., Moscow. (In Russian.) (English translation, Bur. Comm. Fish., Washington.)

Tikhomirov, E. A. (1964). Distribution and biology of pinnipeds in the Bering Sea. *In* "Soviet fisheries investigations in the Northeast Pacific, Part III", (Eds P. A. Moiseev, A. G. Kagenovskii and I. V. Kisevetter), 277-285. Peschevaia Promyshlennost, Moscow. (In Russian.) (English translation, Israel Progr. Sci. Transl., 1968.)

Tikhomirov, E. A. (1966). Certain data on the distribution and biology of the harbor seal in the Sea of Okhotsk during the summer-autumn period and hunting it. *Izv. Tikhookean. Nauchno-Issled. Inst. Rybn. Khoz. Okeanogr.* **58**, 105-115. (In Russian.) (English translation, Bur. Comm. Fish., Washington.)

Tikhomirov, E. (1968). Body growth and development of reproductive organs of the North Pacific phocids. *In* "Pinnipeds of the North Pacific", (Eds V. A. Arseniev and K. I. Panin), 213-241. Tr. Vses. *Nauchno-Issled. Inst. Morsk. Rybn. Khoz. Okeanogr.* **68**. (In Russian.) (English translation, Israel Progr. Sci. Transl. 1971.)

Van den Broek, E. and Wensvoort, P. (1959). On the parasites of seals from the Dutch coastal waters and their pathogenity. *Saeugetierkd. Mitt.* **7**, 58-62. (Not seen, cited by Bonner 1972.)

Van Pelt, R. W. and Dieterich, R. A. (1973). Staphylococcal infection and toxoplasmosis in a young harbor seal. *J. Wildl. Dis.* **9**, 258-261.

Vaughan, R. W. (1971). Aerial survey of seals in The Wash. Natural Environment Research Council, Seals Res. Unit, Occas. Publ. 2.

Venables, U. M. and Venables, L. S. V. (1955). Observations on a breeding colony of the seal *Phoca vitulina* in Shetland. *Proc. Zool. Soc. Lond.* **125**, 521-532.

Venables, U. M. and Venables, L. S. V. (1957). Mating behaviour of the seal *Phoca vitulina* in Shetland. *Proc. Zool. Soc. Lond.* **128**, 387-396.

Wilson, T. M. and Sweeney, P. R. (1970). Morphological studies of seal pox virus. *J. Wildl. Dis.* **6**, 94-97.

Wilson, T. M., Dykes, R. W. and Tsai, K. S. (1972). Pox in young, captive harbor seals. *J. Am. Vet. Med. Assoc.* **161**, 611-617.

2
Ringed, Baikal and Caspian Seals

Phoca hispida Schreber, 1775; *Phoca sibirica* Gmelin, 1788 and *Phoca caspica* Gmelin, 1788

Kathryn J. Frost and Lloyd F. Lowry

Introduction

The ringed, Baikal and Caspian seals include several closely related species, subspecies and geographical races of northern phocids which are commonly united in the subgenus *Pusa* of the genus *Phoca*. As a group they are the most numerous and most widely distributed of northern pinnipeds. The unifying features of the subgenus include their small size, delicate skull and affinity for ice. The adaptability of this form of seal has resulted not only in its being the numerically dominant group in the arctic basin and several contiguous seas but also in the establishment and success of isolated populations in Lakes Baikal, Saimaa and Ladoga, and in the Baltic, *Phoca sibirica* Gmelin, 1788, and Caspian and Okhotsk Seas, *Phoca caspica* Gmelin, 1788.

Genus and Species

The ringed seal was first described as *Phoca hispida* by Schreber in 1775 based on specimens from Greenland and Labrador. The generic name is taken from the Greek word for seal while the specific name is from the Latin *hispidus* meaning barbed or bristly, referring to the coat of the adult. *Pusa* has been applied to the ringed seal group at the genus or subgeneric level since suggested by Scopoli in 1777. Although *Pusa* was elevated to generic rank by Allen (1880) and again by Scheffer (1958), we shall consider *Pusa* as a subgenus of *Phoca* as suggested by Chapskii (1955b) and Burns and Fay (1970). Such an inclusion appears justified as the seals of the subgroup *Pusa* lack a cranial character which distinguishes them from all other seals of the genus *Phoca*. Most specific and subspecific names are geographical terms referring to the locations of the populations.

As would be expected for a species with such a broad geographical range, many common names have been applied to this group of seals. Eskimos and commercial sealers, by virtue of their intimate contact with the animals, have provided an array of names which describe various sexes, age classes and conditions of seals. For example, the general term "netsiak" or "natchek" is commonly applied to ringed seals by Alaskan and Canadian Eskimos, while male ringed seals in breeding condition are distinguished as "tiggak" or "tiguk". Arctic traders commonly referred to ringed seals as "jar seals" but distinguished newly moulted young as "silver jars". Other common English names are "hair seal" and "common seal." "Akiba" is the common Soviet term for ringed seals.

The abundance of *Pusa* seals and their importance in subsistence and commercial economics have prompted and provided material for numerous scientific studies. Canadian research has been summarized in monographs by McLaren (1958) and Smith (1973). The work of Johnson *et al.* (1966) has provided a basic description of the biology of ringed seals in Alaska. Soviet investigators have published results of numerous studies concerning the taxonomy, distribution and ecology of ringed, Caspian and Baikal seals. We are indeed fortunate that the basic biology of these seals is quite well understood and documented. In addition, the distributional features of the group have spurred considerable speculation on paleoclimatology and the evolution of northern phocids (e.g. Chapskii, 1955a, Davies, 1958; McLaren, 1960; Timoshenko, 1975).

Distribution and Movements

Ringed seals have a wide general distribution in seasonally and permanently ice-covered waters of the Northern Hemisphere (Fig. 1). *Phoca hispida hispida** occurs throughout the arctic basin and ranges

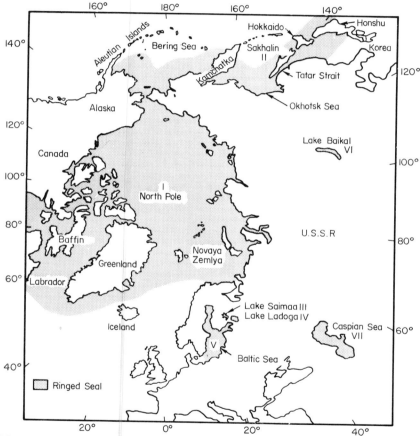

FIG. 1 Normal distribution of ringed, Baikal and Caspian seals. The centres of distribution of the commonly recognized forms are indicated as follows: I, *Phoca hispida hispida*; II, *P. h. ochotensis*; III, *P. h. saimensis*; IV, *P. h. lagodensis*; V, *P. h. botnica*; VI, *P. sibirica*; VII, *P. caspica*. (*Map adapted from Judith King, 1964*).

*see note on p. 49

seasonally into the North Atlantic, Hudson and James Bays and the Chukchi and Bering Seas. *Phoca hispida ochotensis* occurs in the Sea of Okhotsk and the northern islands of Japan. Naito (1976) recorded ringed seals as far south as Kyushu (about 33° N latitude); however, Fedoseev (1970) did not observe breeding grounds south of 45° N. *P. h. botnica* occurs in the Baltic Sea, including the Gulfs of Bothnia and Finland, while *P. h. saimensis* and *P. h. lagodensis* occur in nearby Lakes Saimaa and Ladoga. The taxonomy and inter-relationships of these last three subspecies have been analyzed by Müller-Wille (1969). Of particular zoogeographical interest are *P. sibirica* and *P. caspica*, both of which occur in complete isolation and are separated from their nearest relatives by more than 1600 km.

Within the range of each species and subspecies, regional and seasonal differences in distribution are obvious. During the pupping and breeding season these differences are largely related to differences in the quantity and characteristics of ice for which all forms show a strong affinity. McLaren (1958), Burns (1970) and Smith (1973) concluded that highest densities of breeding adults of *P. h. hispida* occur on stable landfast ice, whereas non-breeders occur in the flaw zone and the moving pack ice. In late spring, non-breeding individuals congregate along leads in the ice (Burns and Harbo, 1972). In summer all age classes and both sexes are found along the edge of the permanent ice pack and in near-shore ice remnants. This requires a migration of many hundreds of kilometers by seals which wintered in the Bering Sea. The Okhotsk Sea ringed seals whelp on pack ice in spring and spend summer and fall in the water feeding and hauling out on shore (Ognev, 1935). Similar seasonal association with ice is seen in Baikal and Caspian seals and in *P. h. botnica* of the Baltic region. Patterns of summer distribution seem mostly influenced by the distribution and movements of prey.

Although patterns of movement have been well described for most of the isolated forms, such patterns are less well known for circumarctic *P. h. hispida*. The degree of interchange between sub-populations of this subspecies is of considerable interest. The only substantial study of movements based on marking has been conducted in north-western Canada. Of over 300 seals marked at Herschel Island and Cape Parry, four recoveries have been made (T. G. Smith, personal communication). Of the recoveries, two were essentially local (Sachs Harbor, Banks Island and Holman, Victoria Island), while two indicated substantial westward movement to Point Barrow, Alaska and East Cape, Siberia.

Throughout the remainder of this chapter we shall deal primarily with the most common and well-known subspecies, *P. h. hispida*. Information on other forms will be incorporated when available and significant.

Abundance

Estimates of population size of ringed seals usually have been derived from counts of animals sighted on the ice by aerial strip sampling. Since the proportion of seals on the ice: total seals is not known, various conversion factors are applied to derive estimates of actual abundance. In practice, most investigators attempt to make counts at the time the highest proportion of animals is visible. The exact date of optimum counting varies with location, but midday during sunny calm weather in the spring or early summer seems best. Such counts, although frequently extrapolated to total population figures, are valuable in themselves for year-to-year and regional comparisons in density. Variations in density are often great even in nearby areas, due to strong habitat preference by breeding animals and a tendency for non-breeders to congregate near leads. Observed densities as high as 15 animals per km^2 were recorded in June on the fast ice along the complex coastline of Baffin Island (McLaren, 1966), while densities less than 0.1 animals per km^2 were common at the same time of year in the off-shore Beaufort Sea (Burns, unpublished).

Table 1 gives estimates of density and population size of ringed, Caspian and Baikal seals. Estimates for ringed seals are all from aerial surveys conducted at optimal times, as described above. Readers are encouraged to consult original sources for details of survey design and data analysis. The most detailed information available to us is for *P. h. hispida*. However, it seems unwise at this time to attempt an estimate of the world population of this subspecies due to the vast unsurveyed areas and the unknown relationship between observed and total numbers. In addition, data presented by Stirling *et al.* (1977) provide convincing evidence of year-to-year changes in abundance within the same geographical area. In spite of the many uncertainties, the estimate for this subspecies given by Bychkov (1971) of 2.5 million is probably a realistic minimum.

Regional variations in abundance in ringed seals are probably most pronounced in summer. Ognev (1935) cites an instance in the Sea of Okhotsk on 24-26 August 1929 when herds of thousands were seen. Conversely, we found extremely low densities of ringed seals in the off-

TABLE 1 Population estimates and densities (observed animals per km^2) of ringed, Caspian and Baikal seals.

Species or subspecies	Geographical area	Habitat surveyed	Density	Estimated population	Sources
Phoca hispida hispida	North-east Chukchi Sea	shore-fast ice	1.43-2.07	2 895	Burns and Harbo (1972)
	Alaskan Beaufort Sea	shore-fast ice	0.41-0.94	8 717	Burns and Harbo (1972)
	Canadian Beaufort Sea and Amundsen Gulf	shore-fast and off-shore ice	1974: 0.15-0.52 1975: 0.13-0.46	41 983	Stirling *et al.* (1977)
	South-west Baffin I, Frobisher Bay and north-west Quebec	shore-fast ice	2.25	121 500	McLaren (1966)
	Hudson Bay	shore-fast ice off-shore ice	2.97 0.73-0.83	455 000[a]	Smith (1975)
	James Bay	shore-fast and off-shore ice	—	61 000[a]	Smith (1975)
P. h. ochotensis	Okhotsk Sea	—	—	800 000-1 000 000	Bychkov (1971)
P. h. botnica	Baltic Sea	—	—	10 000-50 000	Scheffer (1958)
P. h. saimensis	Lake Saimaa	—	—	2 000-5 000	Scheffer (1958)
P. h. lagodensis	Lake Ladoga	—	—	5 000-10 000	Scheffer (1958)
P. caspica	Caspian Sea	—	—	470 000-650 000	Bychkov (1971)
P. sibirica	Baikal Sea	—	—	30 000-50 000	Bychkov (1971)

[a]Estimates derived from counts were doubled to account for seals in the water.

shore ice of the central Beaufort Sea in August 1977. As mentioned in the previous section, availability of food is probably responsible for such observations.

External Characteristics and Morphology

Colouration

The common name "ringed seal" refers to prominent grey-white rings found on the generally dark grey backs of adult seals (Fig. 2). These rings may be separate or somewhat fused together. The belly is usually silver, lacking dark spots (although the latter sometimes occur on pups of the year). According to Ognev (1935) the Okhotsk Sea variety of ringed seal, *P. h. ochotensis,* is light to smokey grey with whitish rings around grey centres.

Baikal seals are generally more uniform in colour. Dorsally they are dark silvery grey, shading to lighter yellowish grey ventrally. Spotted individuals are rare (Ognev, 1935; King, 1964).

The Caspian seal is irregularly spotted with brown or black against a light greyish yellow background. The spots are sometimes encircled by light coloured rings; however, the presence or absence of rings is highly variable (Ognev, 1935; King, 1964).

Ringed, Caspian and Bailkal seal pups are born with a white woolly natal coat often called lanugo. This lanugo is considerably finer and longer than that of two other northern phocids, the spotted seal and the ribbon seal. When dry the lanugo functions to keep the pup warm until it acquires an insulating layer of blubber. The white colour is probably not important for concealment from predators, since for the subspecies most exposed to predation *(P. h. hispida)* pups are born in lairs in snowdrifts and pressure ridges where they are hidden from view (Burns, 1970).

Pups begin to shed the lanugo two to three weeks after birth (Fig. 3) but are not completely shed until they are six to eight weeks old. White-coated pups have been seen in the Bering Sea in June. The coat of a newly moulted pup is finer and relatively longer than that of adults, silver on the belly (sometimes with a few dark spots) and dark grey on the back, sometimes with traces of the adult ringed pattern. These newly moulted pups are called "silver jars" by Canadian fur traders and are much sought by hunters for their pelts (Mansfield, 1967).

FIG. 2 (A) Adult ringed seal at the New York Aquarium. (Photo courtesy of G. Carleton Ray.) (B) Recently born 6.8 kg ringed seal pup in lair on moving ice in Norton Sound, Alaska, 11 April 1978. (Photo L. Lowry.)

FIG. 3 Partially moulted ring seal pup, northern Bering Sea, 10 May 1978. (Photo K. Frost.)

Growth

At birth ringed seal pups weigh about 4.5 kg and are on average 65 cm long (McLaren, 1958; Burns, 1970; Fedoseev, 1975). Our data show foetuses two to four weeks before birth to be on average 3.1 kg in weight and 54 cm in length. Fedoseev (1975) found *P. h. ochotensis* pups to be slightly smaller than Bering Sea pups, weighing 4.0 kg and measuring 60 cm on average at birth. Newborn Caspian seals weigh 1.2-2.8 kg and Baikal pups weigh 2.8-3.6 kg (Ognev, 1935).

Tikhomirov (1968) found the average weight of newly weaned ringed seal pups from the Bering and Okhotsk seas in May and early June to be 9-12 kg, or two to three times their weight at birth. By comparison, spotted seal pups (*Phoca vitulina largha*) and ribbon seal pups (*P. fasciata*) triple to quadruple their weight in a shorter nursing period. Our data from the Bering and Chukchi Seas show the following mean sizes for pups in their first year: June—14 kg, 71 cm; July—12 kg, 74 cm; August—20 kg, 79 cm; November—21 kg, 83 cm. The drop in weight from June to July reflects a natural loss of blubber which occurs after weaning, during the time in which the pup is developing proficiency in feeding. Mean standard length for one-year-old seals was 86.3 cm, or about 70% of mature adult size.

Mansfield (1967) and Pastukhov (1969a) suggested that size differences in adults from various areas may be attributable to varying ice conditions and weaning times of pups. Pups that are weaned young may grow into smaller adults. In the Canadian Arctic pups born along complex coastlines on stable ice tend to be larger than pups born along simple coastlines and on less stable ice (McLaren, 1958). Ice tends to break up later along complex coastlines, thus allowing a longer nursing period. In Lake Baikal much the same situation exists. Break-up occurs earliest in the southern part of the lake, causing weaning to take place earlier and pups to be smaller in that area (Pastukhov, 1969a). In Alaska, pups born on shore-fast ice are larger than pups born in the moving pack ice.

Growth continues throughout the first 8-10 years of life. About 86 % of final body length is attained by sexual maturity at 6-8 years (McLaren, 1958). Different parts of the body grow at about the same rate. However, flipper length seems to increase proportionately less rapidly. In all age classes there is great individual variation in length and weight. Males are on the average slightly longer than females. Average adult lengths for ringed seals vary from 121 cm in the Chukchi Sea to 128.5 cm in the Bering Sea (Fedoseev, 1975) and 135 cm in the Canadian Arctic (McLaren, 1958). McLaren suggested an extreme size of 168 cm and 113 kg. The longest of 500 ringed seals examined by

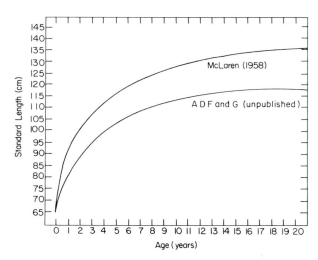

FIG. 4 Age-length relationship of Canadian and Alaskan ringed seals. Canadian data (McLaren, 1958) averages data from several populations. Alaskan data (Alaska Department of Fish and Game, unpublished) was collected in the Bering, Chukchi and Beaufort Seas from 1975 to 1977.

us was a 12-year-old male measuring 132.5 cm, and the heaviest was an 111 kg pregnant female taken in March. Mean standard length in our samples from Alaskan waters for seals older than 10 years was 114.6 cm and mean weight of both sexes was 49 kg.

Growth curves for ringed seals in Alaskan and Canadian waters are presented in Fig. 4. The difference in size of seals in these two areas is striking.

As in all the northern phocids, body weight and blubber thickness in ringed seals fluctuate markedly throughout the year. Physical body condition, as measured by weight and blubber thickness, is best during the winter months, from freeze-up time until February. Minimum body condition occurs in July-August after a prolonged period of reduced feeding associated with breeding and moulting. Johnson et al. (1966) reported winter weight gains of 31-34 % and spring-summer losses of 18-39 %. Thickness of hide and blubber at the sternum in seals we have measured decreased from an average of 5.0 cm in January through April to 2.6 cm in July.

Life History

All seals of the subgenus *Pusa* use ice as a pupping habitat. Five of the seven species or subspecies dealt with in this chapter pup almost exclusively on stable fast ice. Most construct birth lairs under the snow in which pups are born and spend the first weeks of life. Basic reproductive characteristics of the various forms are treated in Table 2. The following discussion will pertain only to *P. hispida*.

Most ringed seal pups are born in early April in birth lairs on shore-fast ice. In the western Arctic drifting but stable pack ice is also important pupping habitat. Although twinning has been reported, a single pup is by far most common. At birth, as in adulthood, the sex ratio is 1:1. Only the female participates in care of the young, nursing the pup for five to seven weeks. Although young pups can swim, they seem to avoid the water. During the nursing period, weight of the pup more than doubles and the lanugo is shed. At or around ice break-up females abandon their pups. The pups frequently remain in the vicinity of the collapsed birth lair, basking on top of the ice.

Most mating occurs in late April and early May within one month after parturition. McLaren (1958) found lactating females with pups to have new completely formed corpora lutea, indicating recent ovulation. Peak spermatogenetic activity in males occurs between

TABLE 2 Reproductive features of ringed, Caspian and Baikal seals. Sexual maturity in females is defined as age at first pregnancy.

Species or subspecies	Mean age at sexual maturity F	M	Period of parturition	Duration of nursing period	Pupping habitat	Sources
Phoca hispida hispida	6-10(7)	5-7	early April	5-7 weeks	lairs on fast ice, also drifting ice in Chukchi Sea and Arctic Ocean	McLaren (1958); Mansfield (1967); Burns (1970); Smith and Stirling (1975)
P. h. ochotensis	5-7	6-7	March-May peak in mid-April	3 weeks	drifting ice, sometimes in lairs	Fedoseev (1975)
P. h. botnica	5	3-4	March	5-6 weeks	cavities on fast ice	Curry-Lindahl (1975)
P. h. saimensis	—	—	beginning of May	—	ice	Ognev (1935)
P. h. lagodensis	—	—	—	—	ice	—
P. sibirica	3-6	4	February-March	3 months	lairs on fast ice	King (1964); Pastukhov (1969a)
P. caspica	4-6	6	mid-January-late February	4-5 weeks	fast ice, not in lairs	Ognev (1935); Fedoseev (1976)

March and mid-May (McLaren, 1958; Johnson et al., 1966) and maximum testes size is attained in April to mid-May. Implantation of the foetus is delayed for about three and a half months after fertilization. This delay is longer than for most other pinnipeds. Mean implantation date for ringed seals in Alaska is 25 August (Burns and Eley, unpublished data).

Some ringed seal females ovulate for the first time at three years of age. However, successful pregnancy does not occur until the fourth to seventh year of life. In the Canadian Arctic at Home Bay, Smith (1973) reported 41% of all four-year-olds to be pregnant. McLaren (1958), however, found no four-year-olds and only 8% of all five-year-olds to be pregnant. On the average, sexual maturity occurs at the age of six or seven years (McLaren, 1958; Mansfield, 1967; Tikhomirov, 1968; Frost, Burns and Lowry, unpublished data). Pregnancy rates of sexually mature females vary geographically. Canadian investigators report an overall pregnancy rate of 91-92% in the Baffin Island area (McLaren, 1958; Smith, 1973). In the early 1970s the rate in the Canadian Beaufort Sea dropped as low as 11% (Stirling et al., 1977). Johnson et al. (1966) reported a mean pregnancy rate of 86% for the southern Chukchi Sea. In the years from 1975-77 ringed seals from Alaskan waters had a mean pregnancy rate of 53%, with a rate of 78% for females over 10 years old. Reproductive rates appear constant from age 10 to maximum life expectancy (McLaren, 1958).

Ringed seal males mature at about the same age as females, between five and seven years of age. Testes are large at birth, decrease in size somewhat, or remain about the same during the first year, increase slowly between the first and sixth years, then increase rapidly during the sixth and seventh years. Baculum length and weight increase sharply during the seventh year (McLaren 1958; Frost, Lowry and Burns, unpublished data).

The moult in ringed seals occurs from late March until July, with the peak in June. During the moult animals haul out on the ice along cracks or leads and bask in the sun. Elevated skin temperatures may be necessary for proper regrowth of hair (Feltz and Fay, 1966). Young, reproductively immature seals moult earlier than do breeding adults. Vibe (1950) found that older animals began to haul out about 10 days later than did young ones. Early in the moult he found the maximum number of seals on the ice between 10 a.m. and 3 p.m. The amount of time spent on the ice increased as the moult season progressed.

Maximum reported age for ringed seals is 43 years (McLaren, 1958). The oldest seal in our samples ($n = 952$) was a 29-year-old male. Average lifespan is probably between 15 and 20 years. Less than

1.5% of the seals for which ages were determined by us were over 20 years of age. Sixty percent were less than seven years old.

Predators of the ringed seal include ravens, red foxes, arctic foxes, wolves, dogs, wolverines, polar bears and humans (Burns, 1970). Polar bears, arctic foxes and humans are the only significant predators. Arctic fox predation appears to be the single most important source of mortality in new-born pups in Amundsen Gulf (Smith, 1976). Smith found no indications of seals older than pups being killed by foxes. Polar bears seem to concentrate hunting in the off-shore ice, particularly in the unstable flaw zone. In the Canadian Arctic they catch mostly one- to two-year-olds from March until break-up (Smith, 1976). Of 70 ringed seals killed by polar bears in Alaskan waters, most were males older than six years of age (Eley, unpublished ms.) and few pups were taken. Best (1977) and Eley (unpublished ms.) estimated predation rates to be a minimum of one ringed seal or the equivalent thereof every 6.5 days.

Human hunters in coastal areas take ringed seals for both subsistence and commercial purposes. The Inuit Eskimos in Siberia, Alaska, Canada and Greenland have traditionally depended on the harvest of ringed seals to support nutritional and other needs. Harvest by Alaskan Eskimos from 1960 to 1970 averaged about 10000 animals annually. Harvests in recent years have been considerably lower due to economic and cultural changes in many of the villages. Caspian seals have long been subjected to intense harvesting. Annual harvest prior to 1965 averaged 70000 animals (Fedoseev, 1976). Badamshin (1960) reports annual catches as high as 227000.

Internal Characteristics and Morphology

The skull of the ringed seal is thin-walled with a relatively short nasal portion and narrow interorbital (Fig. 5). Considerable variability in skull shape and size occurs which complicates the use of cranial characters in taxonomic studies (Müller-Wille, 1969; Timoshenko, 1975). Mean condylobasal lengths reported by Timoshenko (1975) are as follows: *P. hispida*, 169.0 mm; *P. caspica*, 175.1 mm; and *P. sibirica*, 181.0 mm.

The adult dental formula of ringed seals is as in other members of the subfamily Phocinae (King, 1964):

$$I \frac{1\text{-}2\text{-}3}{0\text{-}2\text{-}3} C \frac{1}{1} PC \frac{1\text{-}2\text{-}3\text{-}4\text{-}5\text{-}0\text{-}0}{1\text{-}2\text{-}3\text{-}4\text{-}5\text{-}0\text{-}0} \times 2 = 34$$

Ringed seal pups are born with a completely erupted set of permanent teeth. The incisors, canines and first postcanines are single-rooted, while remaining postcanines are double-rooted. All postcanines are distinctly multicusped. They are offset such that when the jaws are closed small orifices remain. Water may be extruded through these orifices during the capture of small invertebrates (McLaren, 1958; Fedoseev, 1965). Dental anomalies are not uncommon.

Fay (1967) found 16 thoracic vertebrae and ribs in each of two ringed seals examined. A count of 15 is more usual among pinnipeds.

The lungs of ringed seals are lobed. The right lung has four lobes (apical, cardiac, diaphragmatic and postcardiac), whereas the left lung has only three (no postcardiac). Lobulation is similar to that in the bearded seal *Erignathus barbatus,* but distinctly different from the unlobulated lungs of ribbon seals *(Phoca fasciata)* and Caspian seals (Sokolov *et al.,* 1968).

The trachea of the ringed seal is made up of 87 (72-99) thinwalled, narrow rings. Fifteen (10-19) of these are continuous with fused ends. The remaining rings are complete with overlapping tips.

Intestinal measurements for 12 ringed seals taken at Barrow in spring 1975 and 1976 are as follows:

Small intestine: mean = 1598 cm; range = 1430-1830 cm
Large intestine: mean = 66 cm; range = 48-84 cm
Duodenum: mean = 31 cm; range = 16-41 cm

The length of the small intestine on the average was 13.8 times the standard length of the seal (mean, 116 cm).

Weights of the major organs of ringed seals are given in Table 3.

Physiology

Ringed seals have one of the highest total red cell masses recorded for pinnipeds. Mean packed cell volumes range from 55 to 63% and haemoglobin counts are about 24 g per 100 ml (Geraci and Smith, 1975; Ferren and Elsner, 1979). White blood cell counts are similar to other pinnipeds.

Oxygen capacity is about 70 ml O_2 per kg Lean Body Mass (LBM) (Ferren and Elsner, 1979). Respiratory tidal volume ranges from 1.17 to 1.85 litres (Parsons, 1977). This is larger on a body weight basis than terrestrial mammals but similar to other marine mammals. Ringed seals extract about 50% of the oxygen from the inspired air. Total blood volume is around 230 ml kg^{-1} LBM or 2960-4120 ml for an average seal (St Aubin *et al.*, 1978; Ferren and Elsner, 1979). This is

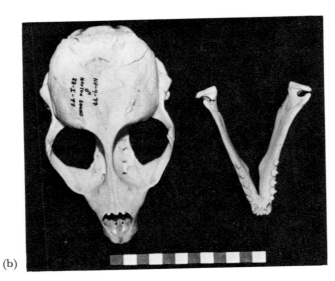

FIG. 5 Skulls of 10-year-old male *P. h. hispida*. (a) lateral view; (b) dorsal view.

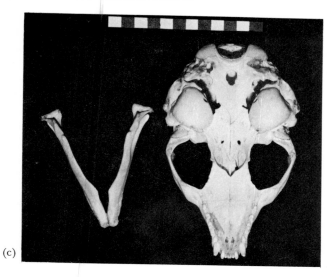

FIG. 5 (c) Ventral view.

TABLE 3 Organ weights of ringed seals and their relationship to body weight. Brain weight data were reported by Ferren and Elsner (1979) from four seals. Data for other organs are from eight specimens taken in Barrow, Alaska in spring 1975.

Organ	Mean weight (g)	Range	Mean percent total body weight	Mean percent lean body weight
Brain	154	129-180	0.78	1.35
Heart	319	255-426	0.69	1.35
Liver	916	709-1248	1.98	3.87
Kidneys	227	199-340	0.49	0.96
Lung	702	511-1078	1.51	2.96
Spleen	138	114-170	0.30	0.58

comparable to values for harbour seals and ribbon seals, but considerably greater than values for terrestrial carnivores. Blood characteristics suggest an animal adapted to deep or sustained diving.

Pastukhov (1969b) reported maximum dive time for Baikal seals in the wild to be 43 min, and maximum forced dive times in newly captured animals to be 40-68 min. He found the duration of dives to decrease as the length of time in captivity increased. Ferren and Elsner (1979) determined maximum forced dive duration to be 17 min, 51 sec for *P. hispida*. They suggested that this comparatively short diving time, despite relatively large blood volumes and oxygen capacity, may be explained by relatively large brain size (1.1-1.9% of LBM) in relation to body mass and total oxygen reserve.

Pronounced bradycardia occurs during diving. Heartbeat drops from 80-90 beats per min at the surface to 10-20 beats per min when submerged (Ferren and Elsner, 1979).

Pastukhov (1969b) found Baikal seals in captivity to consume a maximum of about 1 kg of fish at a single feeding and a total of 5-6 kg per day, or 4-6% of their body weight per day. He suggested that consumption in the wild would be at least one and a half times this rate. According to Parsons (1977), minimum energy requirements for *P. h. hispida* range from 35-55 Kcal kg^{-1} per day, with the lowest values for large animals. He reported an energy assimilation rate of 97% with about 9% lost in the urine. Basal metabolic rate ranges from about 15 to 23 Kcal kg^{-1} per day with peak metabolic rates about twice basal rate. When Baikal seals in captivity were starved for 60 days they lost 30% of their total body weight (Pastukhov, 1969b). When feeding resumed, weight gain was rapid.

Behaviour

In late summer, fall, winter and early spring, ringed seals spend most of their time in the water feeding. Foods eaten vary markedly by season and geographical area (McLaren, 1958; Fedoseev, 1965; Johnson *et al.*, 1966; Lowry *et al.*, 1980). Fishes of the cod family, pelagic amphipods, euphausiids, shrimps and other crustaceans make up the bulk of the diet. Fedoseev (1965) suggests that cusps of the postcanines are functional in filtering small organisms such as euphausiids from bites of water. Although fishes eaten by ringed seals are usually small (less than 20 cm), this is by no means always the case. Smith (1977) records an instance in which a 121 cm long ringed seal had apparently captured and was eating a 127 cm long wolf-fish *(Anarhichas sp.)*. Fishes appear to be the main food of Caspian and Baikal seals (Pastukhov, 1969b; Vorozhtsov *et al.*, 1972).

Since ringed seals feed on the most abundant organisms at at least two levels of the food chain, their food resources are shared with many other marine mammals, sea-birds and fishes. The possibility exists for direct food competition between ringed seals and other marine mammals in some areas (Lowry *et al.*, 1978).

A seasonal cycle of feeding intensity has been well documented (McLaren, 1958; Johnson *et al.*, 1966). Feeding is at a minimum during the moult period but does not cease entirely (Fedoseev, 1965). We have found that ringed seals in Alaska feed most intensively on arctic cod *(Boreogadus saida)* and large zooplankton *(Parathemisto libellula* and *Thysanoessa spp.)* (Lowry *et al.*, 1980). The question of how ringed seals locate food in the total darkness of the arctic winter has never been investigated.

During the late summer, fall, winter and early spring, ringed seals are not obvious to the casual observer. Seals can be seen swimming among the floes of the pack ice in summer, often singly but sometimes in loose groups. Observations of seals in the water show that the amount of time at the surface ranges from 12 to 130 s, while dive times range from 18 to 720 s (Mansfield, 1970). As sea ice begins to reform in fall, seals appear to be concentrated in open leads. As the ice cover thickens and becomes more extensive, holes are actively kept open by frequent use and abrasion of the ice by the claws of the front flippers (Vibe, 1950). As the time for pupping approaches, pregnant females increase the size of their holes and hollow a birth lair in the snow. Haul-out lairs are occasionally made by non-breeders and breeding males. Characteristics of lairs have been described in detail by Smith and Stirling (1975).

Although there are no direct observations of ringed seal breeding behaviour, it has been suggested that males are monogamous and mating occurs in the water. Observations of the distribution of seals during the pupping and breeding season suggest that adults may be territorial at that time of year (Burns, unpublished). Juvenile seals found frozen out of the water in spring sometimes show signs of intraspecific aggression (Smith and Memogana, 1977). Stirling (1973) has classified four types of underwater vocalizations made by ringed seals. Since all four types were heard at all times of year, it seems unlikely that any have a specific relationship to reproductive behaviour.

Although it has been suggested that birth lairs function to reduce the vulnerability of young to predators, the ease with which a lair can be detected by a trained dog suggests that the lair would be little deterrent to predators such as polar bears and arctic foxes (Smith, 1976). However, Eley (unpublished ms.) found evidence of successful kills in only 22 of 107 lairs excavated by polar bears. Smith and Stirling (1975) suggest that the lair may be important in providing a relatively warm environment in which the small pup can accumulate insulating blubber. It is interesting to note that pups of Okhotsk Sea ringed seals and Caspian seals are not born in lairs. A comparative study of meteorological conditions and pup physiology and growth in the various areas would be of great interest.

The behaviour of ringed seals basking on ice in summer does reflect a marked adaptation to their major predator, the polar bear. Stirling (1974) found that hauled out seals constantly alternated between lying flat and lifting their heads and scanning their surroundings. He found a mean looking time of seven seconds and a mean lying time of 26.3 s. Undoubtedly this behaviour explains the low success experienced by bears stalking hauled-out seals (Stirling, 1974).

Little information is available on the sensory capabilities of ringed seals. Terhune and Ronald (1976) found an upper audible frequency limit in water of 60 kHz. This value is similar to that reported for harbour seals. Our observations and those of Eskimo hunters indicate that both hearing and vision in air are well developed.

Owing to their small size and apparently wide physiological tolerances, ringed seals have been successful in captive situations. Techniques for live capture by netting have been described by Smith *et al.* (1973). Pups during nursing and shortly after weaning can be easily caught by hand. Ringed seals are quite docile and tractable in captivity and have been the subject of numerous valuable laboratory studies.

Disease

Little is known about the diseases of ringed seals. Fay (1978) has observed the following disorders: acute dermatomycosis, consolidative pneumonia, fungal dermatitis, focal necrosis of the liver and eye disorders. Thirteen serum samples tested for *Leptospira* were negative. Fay (1978) and Madin *et al.* (1976) reported negative results in tests for San Miguel Sea Lion Virus.

Organochlorine residues (DDT, DDE and PCBs) have been found

in blubber samples of Canadian Arctic ringed seals. Concentrations are in the low ppm range, and are considerably less than those reported for ringed seals in the Gulf of Bothnia or for other species of seals in eastern Canada (Addison and Smith, 1974). Residue concentrations (DDT and DDE) were found to increase with age in males; however, no such correlations existed in females.

As in many other marine mammals and birds, ringed seals contain high levels of mercury in the liver. Smith and Armstrong (1978) found concentrations to increase with age with no apparent toxic effects. Ringed seals from the Canadian Arctic had higher mercury levels than those from western Alaska.

Acknowledgements

We would like to acknowledge our many colleagues who have found the study of seals so interesting and rewarding. Without the results of their work, from which we have liberally borrowed information, this chapter would hardly have been possible. We particularly thank our close-working associates within and outside of the Alaska Department of Fish and Game for the time, effort and materials they have contributed to our studies. Our work was supported in part by the Bureau of Land Management, Outer Continental Shelf Environmental Assessment Program Contract Number 03-5-022-53 and by Federal Aid in Wildlife Restoration Projects W-17-8, W-17-9, W-17-10 and W-17-11. Mr John J. Burns has been particularly helpful in this work and in reviewing and commenting on the manuscript.

Note

We include several previously recognized subspecies in *P. h. hispida*. Scheffer (1958) suggested the inclusion of *P. h. pomororum* (Novaya Zemlya and White Sea), *P. h. birulai* (New Siberian Islands), *P. h. soperi* (west coast of Baffin Island) and *P. h. beaufortiana* (Dolphin and Union Strait, Canada). Youngman (1975) provided evidence that *P. h. beaufortiana* is indistinguishable from *P. h. hispida*. Fedoseev and Nazarenko (1970) examined *P. h. krascheninikovi* (Bering Sea) and *P. h. pomororum* and decided they are best included in *P. h. hispida*.

References

Addison, R. F. and Smith, T. G. (1974). Organochlorine residue levels in Arctic ringed seals: variation with age and sex. *Oikos* **25**, 335-337.

Allen, J. A. (1880). "History of North American Pinnipeds. A Monograph of the Walruses, Sea Lions, Sea Bears, and Seals of North America", US Geol. and Geogr. Surv. Terr., Misc. Publ. 12, Washington.

Best, R. C. (1977). Ecological aspects of polar bear predation. *In* "Proc. 1975 Predator Symp.", (Eds R. L. Phillips and C. Jonkel), 203-211. Univ of Montana, Missoula.

Badamshin, B. I. (1960). The state of the Caspian seal stock. *Zool. Zh.* **39**, 898-911. (Fish. Res. Bd. Can. Transl. Ser. No. 376. 22pp.)

Burns, J. J. (1970). Remarks on the distribution and natural history of pagophilic pinnipeds in the Bering and Chukchi Seas. *J. Mammal* **51**, 445-454.

Burns, J. J. and Fay, F. H. (1970). Comparative morphology of the skull of the ribbon seal, *Histriophoca fasciata*, with remarks on systematics of Phocidae. *J. Zool.* **161**, 363-394.

Burns, J. J. and Harbo, S. J. Jr (1972). An aerial census of ringed seals, northern coast of Alaska. *Arctic* **25**, 279-290.

Bychkov, V. A. (1971). A review of the conditions of the pinniped fauna of the USSR. *In* "Scientific Principles for the Conservation of Nature", (Ed. L. K. Shaposhnikov), 59-74. (Transl. Can. Dept. Foreign Languages No. 0929).

Chapskii, K. K. (1955a). Contribution to the problem of the history of development of Caspian and Baikal seals. *Trudy Zool. Inst. Akad. Nauk SSSR* **17**, 200-216. (Fish. Res. Bd. Can. Transl. Ser. No. 174.)

Chapskii, K. K. (1955b). An attempt at revision of the systematics and diagnostics of seals of the subfamily phocinae. *Trudy Zool. Inst. Akad Nauk SSSR* **17**, 160-201. (Fish. Res. Bd. Can. Transl. Ser. No. 114).

Curry-Lindahl, K. (1975). Ecology and conservation of the grey seal *Halichoerus grypus*, common seal *Phoca vitulina*, and ringed seal *Pusa hispida* in the Baltic Sea. *Rapp. P. -v. Réun. Cons. Int. Explor. Mer* **169**, 527-532.

Davies, J. L. (1958). Pleistocene geography and the distribution of northern pinnipeds. *Ecology* **39**, 97-113.

Eley, T. J. (1978). An analysis of polar bear predation on Alaskan ice-inhabiting pinniped populations. Alaska Dept. Fish and Game, unpubl. MS.

Fay, F. H. (1967). The number of ribs and thoracic vertebrae in pinnipeds. *J. Mammal.* **48**, 144.

Fay, F. H., Dieterich, R. A. and Schults, L. M. (1978). Mortality and morbidity of marine mammals. *In* "Environmental Assessment of the Alaskan Continental Shelf", Annual Reports of Principal Investigators for the year ending March 1978, Vol. I. 39-79. Outer Continental Shelf Environmental Assessment Program, Boulder, Colorado.

Fedoseev, G. A. (1965). Food of the ringed seal. *Izv. TINRO* **59**, 216-223.

Fedoseev, G. A. (1970). Distribution and numerical strength of seals off Sakhalin Island. *Izv. TINRO* **71**, 319-324. (Fish. Res. Bd. Can. Transl. Ser. No. 2400).

Fedoseev, G. A. (1975). Ecotypes of the ringed seal (*Pusa hispida* Schreber, 1777) and their reproductive capabilities. *Rapp. P. -v. Réun. Cons. Int. Explor. Mer* **169**, 156-160.

Fedoseev, G. A. (1976). Principal populational indicators of dynamics of numbers of seals of the family Phocidae. *Ekologiya* **5**, 62-70. (Transl. Consultants Bureau, New York. 439-446).

Fedoseev, G. A. and Nazarenko, Yu. I. (1970). On intraspecific structure of ringed seals in the Arctic. *Izv. TINRO* **71**; 301-307. (Fish. Res. Bd. Can. Transl. Ser. No. 2411).

Feltz, E. T. and Fay, F. H. (1966). Thermal requirements *in vitro* of epidermal cells from seals. *Cryobiology* **3**, 261-264.

Ferren, H. and Elsner, R. (1979). Diving physiology of the ringed seal: adaptations and implications. Proc. 29th Alaska Sci. Conf., Fairbanks, Alaska, 15-17 August 1978, 379-387.

Geraci, J. R. and Smith, T. G. (1975). Functional hematology of ringed seals (*Phoca hispida*) in the Canadian Arctic. *J. Fish. Res. Bd. Can.* **32**, 2559-2564.

Johnson, M. L., Fiscus, C. H., Ostenson, B. T. and Barbour M. L. (1966). Marine mammals. *In* "Environment of the Cape Thompson Region, Alaska", (Eds N. J. Wilimovsky and J. N. Wolfe), 897-924. U.S. Atomic Energy Commission, Oak Ridge, TE.

King, J. (1964). "Seals of the World", British Museum (Natural History), London.

Lowry, L. F., Frost, K. J. and Burns, J. J. (1978). Food of ringed seals and bowhead whales near Point Barrow, Alaska. *Can. Field-Nat.* **92(1)**, 67-70.

Lowry, L. F., Frost, K. J. and Burns, J. J. (1980). Variability in the diet of ringed seals, *Phoca hispida,* in Alaska. *Can. J. Fish. Aquat. Sci.* **37**, 2254-2261.

Madin, S. H., Smith, A. W. and Akers, T. G. (1976). Current status caliciviruses isolated from marine mammals and their relationship to caliciviruses of terrestrial mammals. *In* "Wildlife Diseases", (Ed. L. A. Page), 197-204. Plenum Press, New York.

Mansfield, A. W. (1967). Seals of arctic and eastern Canada. *Fish. Res. Bd. Can. Bull.* No. 137.

Mansfield, A. W. (1970). Population dynamics and exploitation of some Arctic seals. *In* "Antarctic Ecology", Vol. I. (Ed. M. W. Holdgate), 429-446. Academic Press, London and New York.

McLaren, I. A. (1958). The biology of the ringed seal, *Phoca hispida*, in the eastern Canadian Arctic. *Fish. Res. Bd. Can. Bull.* No. 118.

McLaren, I. A. (1960). On the origin of the Caspian and Baikal seals and the paleoclimatological implication. *Am. J. Sci.* **258**, 47-65.

McLaren, I. A. (1966). Analysis of an aerial census of ringed seals. *J. Fish. Res. Bd. Can.* **23**, 769-773.

Müller-Wille, L. L. (1969). Biometrical comparison of four populations of *Phoca hispida* Schreb. in the Baltic and White Seas and Lakes Ladoga and Saimaa. *Comm. Biólogicae* **31(3)**, 1-12.

Naito, Y. (1976). The occurrence of the phocid seals along the coast of Japan and possible dispersal of pups. *Sci. Rep. Whales Res. Inst.* **28**, 175-185.

Ognev, S. I. (1935). Mammals of the USSR and adjacent countries. Vol. III. Carnivora (Fissipedia and Pinnipedia) Moscow: Acad. Sci. USSR. (In Russian; English transl. by A. Birron and Z. S. Coles for Israel Program for Scientific Translations, 1962). Available Office of Technical Services, US Dept. Commerce, Wash. 25, D.C.

Parsons, J. L. (1977). Metabolic studies on ringed seals *(Phoca hispida)*. M.S. Thesis, University of Guelph.

Pastukhov, V. D. (1969a). Onset of sexual maturity of the female ringed seal. "Morskie Mlekopitayushchie", 127-135. 1969. (Fish. Mar. Serv. Transl. Ser. No. 1474).

Pastukhov, V. D. (1969b). Some results of observations on the Baikal seal under experimental conditions. "Morski Mlekopitayushchie", 105-110. 1969. (Fish. Mar. Serv. Transl. Ser. No. 3544.)

Scheffer, V. B. (1958). "Seals, Sea Lions and Walruses", Stanford University Press, Stanford, CA.

Smith, T. G. (1973). Population dynamics of the ringed seal in the Canadian eastern Arctic. *Fish. Res. Bd. Can. Bull.* **181**.

Smith, T. G. (1975). Ringed seals in James Bay and Hudson Bay: population estimates and catch statistics. *Arctic* **28**, 170-182.

Smith, T. G. (1976). Predation of ringed seal pups *(Phoca hispida)* by the arctic fox *(Alopex lagopus)*. *Can. J. Zool.* **54**, 1610-1616.

Smith, T. G. (1977). The wolffish. cf. *Anarhichas denticulatus*, new to the Amundsen Gulf area, Northwest Territories, and a probable prey of the ringed seal. *Can. Field-Nat.* **91**, 228.

Smith, T. G., Beck, B. and Sleno, G. A. (1973). Capture, handling and branding of ringed seals. *J. Wildl. Manage.* **37(4)**, 579-583.

Smith, T. G. and Stirling, I. (1975). The breeding habitat of the ringed seal *(Phoca hispida)*: The birth lair and associated structures. *Can. J. Zool.* **53**, 1297-1305.

Smith, T. G. and Memogana, J. (1977). Disorientation in ringed and bearded seals. *Can. Field-Nat.* **91**, 181-182.

Smith, T. G. and Armstrong, F. A. J. (1978). Mercury and selenium in ringed and bearded seal tissues from arctic Canada. *Arctic* **31**, 75-84.

Sokolov, A. S., Kosygin, G. M. and Shustov, A. P. (1968). Structure of lungs and trachea of Bering Sea pinnipeds. *In* "Pinnipeds of the North Pacific", (Eds V. A. Arseniev and K. I. Panin). 250-262. *Izv. TINRO*, Vol. 68. (Transl. Israel Program for Scientific Translations, 1971).

St Aubin, D. J., Geraci, J. R., Smith, T. G. and Smith, V. I. (1978). Blood volume determination in the ringed seal, *Phoca hispida*. *Can. J. Zool.* **56**, 1885-1887.
Stirling, I. (1973). Vocalization in the ringed seal *(Phoca hispida)*. *J. Fish. Res. Bd. Can.* **30**, 1592-1594.
Stirling, I. (1974). Mid-summer observations on the behavior of wild polar bears *(Ursus maritimus)*. *Can. J. Zool.* **52**, 1191-1198.
Stirling, I., Archibald, W. R. and DeMaster, D. (1977). Distribution and abundance of seals in the eastern Beaufort Sea. *J. Fish. Res. Bd. Can.* **34**, 976-988.
Terhune, J. M. and Ronald, K. (1976). The upper frequency limit of ringed seal hearing. *Can. J. Zool.* **54**, 1226-1229.
Tikhomirov, E. A. (1968). Body growth and development of reproductive organs of the North Pacific phocids. *In* "Pinnipeds of the North Pacific", (Eds V. A. Arseniev and K. I. Panin), 213-241. (Transl. Israel Program for Scientific Translations, 1971).
Timoshenko, Yu. K. (1975). Craniometric features of seals of the genus *Pusa. Rapp. P.-v. Réun. Cons. Int. Explor. Mer* **169**, 161-164.
Vibe, C. (1950). The marine mammals and the marine fauna in the Thule district. (Northwest Greenland) with observations on ice conditions in 1939-41. *Medd. om Gron.* **150**, 1-115.
Vorozhtsov, G. A., Rumyantsev, V. D., Sklarova, G. A. and Khuraskin, L. S. (1972). *Trudy VINRO* **89**, 19-29. (Fish. Mar. Serv. Transl. Ser. No. 3108).
Youngman, P. M. (1975). Mammals of the Yukon Territory. *Nat. Mus. Can. Publ. Zool. No. 10*.

3
Harp Seal
Phoca groenlandica Erxleben, 1777

K. Ronald and P. J. Healey

Genus and species

The harp seal was first named *Phoca groenlandica* by Fabricius in 1776. In 1777, however, Erxleben described this animal and is therefore usually credited with the name. *Pagophilus*, which means ice lover (from the Greek words "pagos" meaning ice and "philos" meaning loving), was introduced by Gray in 1850 (Scheffer, 1958). Other scientific names that have been given to this seal include *P. oceanica* (Lepechin); *Callocephalus oceanicus* (Lesson); *Phoca similunaris* (Boddaert); *P. dorsata* (Pallas); *P. mulleri* (Lesson); *Callocephalus groenlandicus* (F. Cuvier); *Phoca albicauda* (Desmarest); *P. desmarestii* (Lesson); *P. pilayi* (Lesson) (Brown, 1868). Recently *Pagophilus* was reassigned to subgeneric rank within the inclusive genus *"Phoca"* based on the decision that it, along with *Pusa, Histriophoca* and *Phoca (sensu stricto)* lacked any unique cranial characters that allow generic recognition (Burns and Fay, 1970).

The harp seal has also been given many common names by various nationalities. Sealers refer to the new-born pups as "whitecoats", partly moulted pups as "ragged jackets", the fully moulted pups as "beaters", the other immatures as "bedlamers" and the adults as Greenland seals, harp, saddle seals or saddlebacks. French Canadian names include "phoque du Groenland", "loup marin de glace", "loup marin coeur" and "brasseur" (Mansfield, 1967). The Eskimos generally call the seal "kairulik" (Mansfield, 1967) or in Pond Inlet "neitke" (Brown, 1868). Other names are saddleback (English), "svartsida" (Norwegian), "daelja", "daevok" and "aine" (Lapp), "svartsiden" (Danish), "blaudruselur" (Icelandic) and "atak" (Brown, 1868).

There are three major populations of harp seals, each with its own distinct breeding ground, one in the White Sea, one in the Greenland Sea north of Jan Mayen and a third utilizing the Gulf of St Lawrence and the ice off the east coast of Newfoundland. Although there are no obvious morphological differences between them, it is believed that the three populations rarely mix (Sergeant, 1973a).

External Characteristics and Morphology

The smoothly tapering body of the harp seal is thick and round in the thoracic region. The tail is short, slightly dorsoventrally flattened and tapers gradually until near the distal extremity. Distally, the limb becomes distinct only from the tarsus. The flippers are composed of five digits united by an interdigital web (Tarasoff et al., 1972). Adult males have 46.8 ± 0.29 labial vibrissae and 3.0 ± 0.88 eye vibrissae (Shustov and Iablokov, 1967).

Colouration and integument

As the harp seal grows it moults (Ronald et al., 1970) as well as changing its coat colour many times, and there is even variation in the coats of full-grown adults (King, 1964). The variable aspects are the darkness of the general background, the darkness of the dorsal as compared to the ventral surface, the quantity of spots, their dimensions and hue and the presence of the harp shaped dark area (Fig. 2) from which the seal gets one of its common names (Smirnov, 1924).

The harp seal pups are born with a soft, curly yellowish coat, stained by the amniotic fluids. This turns white during the first three days

FIG. 1 Front view of 10-20-year-old harp seal female moving up ice slope in the Gulf of St Lawrence. (Photograph: D. M. Lavigne.)

probably due to the crystallization of the amniotic fluids. When they are two to four weeks old this white coat is shed to reveal a short-haired coat of grey, darker dorsally, lighter on the ventral surface and marked with darker grey and black spots (King, 1964). The term "ragged jackets" is used for pups with loose white hair. There is a great variation in the shade of the ground colour and the degree of spotting but all immatures show the same general pattern (King, 1964). The dorsal areas and upper sides of the yearling seals begin to lighten while

FIG. 2 Side view of female harp seal about to enter lead in ice of Gulf of St Lawrence. (Photograph: D. M. Lavigne.)

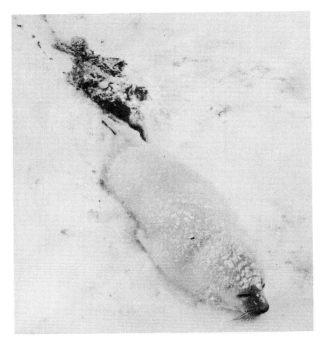

FIG. 3 Harp seal pup approximately three hours old and still attached to placental membranes. Note the amniotic fluids are crystallizing on the white-coated pup's hair. (Photograph: K. Ronald.)

the ventral surface darkens causing the spots to be less obvious. The general background of the two year coat is light ash grey and the spots are much paler than before. The three-year-old seals are a uniform light ash grey with markings becoming even less distinct. The four-year-olds or young adults may begin to show the mature adult pattern on their coats although this is not fully developed, especially in the females, for some years after sexual maturity (Smirnov, 1924). Some males as they approach maturity show a very dark coloration. This "sooty" phase is usually lost at the next moult. The adult male harp seal is usually light whitish grey with a horse-shoe-shaped black band running along the flanks and across the back. Its head, to just behind the eyes, is black. The face and "harp" of the adult female are paler and may be broken into spots (King, 1964).

The harp seal hair is made up of primary guard and secondary hairs (Tarasoff et al., 1972) and is of some insulative value, although the

FIG. 4 Group of young and old mature harp seals in ice lead. Note the fifth seal from the right who is assuming a typical ventral-dorsal swimming position. (Photograph: D. M. Lavigne.)

blubber plays an important part in this function (Frisch *et al.*, 1974). When in the water the inner part of the woolly underfur retains a stagnant layer of water, 2 mm thick, while a barrier of 14 or 15 guard hairs flattens tightly over this underfur to provide some insulation (Frisch *et al.*, 1974). The hair lengths of the coat decrease from the mid-back to the tarsus to the interdigital web (Tarasoff *et al.*, 1972).

Dimensions

The following figures represent the minimum, maximum and average lengths of the different ages of harp seals in centimetres: whitecoats—97, 108 and 103; moulting pups—90, 123 and 110; yearlings—126, 148 and 131; two-year-olds—138, 155 and 144; three-year-olds—158, 170 and 160; adult females—168, 183 and 179; and adult males—171, 190 and 183 (Smirnov, 1924). Using the formula:

$$\text{springtime fattening} = \frac{100 \times \text{maximal girth}}{\text{total length}}$$

the following maximum, minimum and medium values apply: whitecoats—81, 67 and 75; moulting pups—97, 77 and 85; yearlings—74, 65 and 69; two-year-olds—78, 63 and 69; three-year-olds—75, 62 and 67; adult females—71, 63 and 67; and adult males—73, 57 and 64 (Smirnov, 1924).

Weights

The new-born harp seal pups weigh on the average about 11.8 kg at birth but increase their weight rapidly to about 22.8 kg within four to five days. Their weight continues to increase at the same rapid rate to a maximum of 33.3 kg at weaning then declines to an average of 27.1 kg for the fully moulted pups. Adult male harp seals average 135 kg in weight and females about 119.7 kg (Sivertsen, 1941). There are seasonal weight changes in all age groups.

Distribution and Migration

Harp seals inhabit the northern Atlantic and Arctic Oceans and range from northern Russia through Spitzbergen and Jan Mayen to Greenland then southwards to Newfoundland and north and east into Baffin Bay and Hudson Bay (Fig. 5). It is found off the coasts of

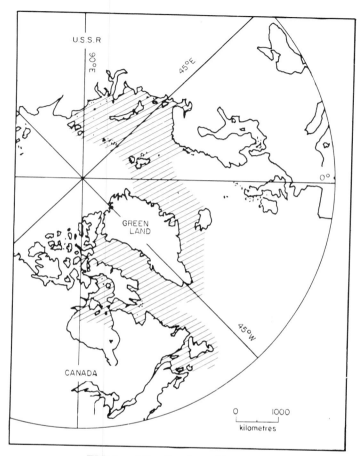

FIG 5. Distribution of harp seals.

Europe and Asia from Severnaya Zemlya and Cape Chelyuskin to northern Norway, including the Kara Sea, Novaya Zemlya, Franz Josef Land, the White Sea, Spitzbergen and Jan Mayen as well as on all the coasts of Greenland except the extreme north and on Baffin Island, Southampton Island, Labrador, the east coast of Newfoundland and the Gulf of St Lawrence (King, 1964). As a summer resident in the Arctic, it migrates as far north as Jones and Lancaster Sounds and Thule in north-west Greenland (Mansfield, 1967). The harp seal is not normally found far into Hudson Bay and rarely visits the northern coast of Iceland. Individuals are occasionally reported from Scotland and the Shetland Islands and even as far south as the Bristol Channel and the River Teign (King, 1964).

All three populations of harp seals migrate following the same general pattern of moving north in the summer and coming south in the winter and spring to breed (King, 1964). During the summer the Jan Mayen or Greenland Sea population can be found in its most northerly feeding grounds between Spitzbergen and Greenland. These seals move south to Jan Mayen in the winter and pupping occurs on the ice here in March. When the ice begins to drift south the adults mate then move to the ice north of Jan Mayen where they moult during the month of April. By mid-May this herd has returned to the northern feeding grounds (King, 1964).

The White Sea population spends the summer period of intensive feeding and growth in the area close to the ice edge in the Barents and Kara Seas. At this time, groups are found at West Spitzbergen, at the north end of Novaya Zemlya, among the many islands of Franz Josef Land and in the Straits of Severnaya Zemlya. With the approach of the cold weather and the southward advancement of the ice border, these seals travel along the west and east coasts of Novaya Zemlya in search of food. In January and early February a rapid migration occurs to the funnel and basin of the White Sea where pupping takes place. The ice drift transports the new pups to the Barents Sea where they begin to feed independently (Popov, 1966). This passive migration northward occurs during lactation and for some time after weaning and brings the pups to the areas of rich food supply by April (Popov, 1970). The adults moult from mid-April to late May then migrate north out of the White Sea (King, 1964).

During the summer, the north-west Atlantic population extends from Thule, Greenland and Jones Sound between Devon and Ellesmere Islands, south to northern Labrador waters and from Cape Farewell west to north-west and south-east Hudson Bay (Sergeant, 1965). The movement south out of the Canadian archipelago begins in late September and early October and by early November large numbers have passed Cape Chidley in northern Labrador (Mansfield, 1967) to reach the Strait of Belle Isle in late December (Sergeant, 1965). Only the adults and older immatures are involved in this migration. The remainder of the immatures move later although some stay behind in the West Greenland waters throughout the winter (Mansfield, 1967; Sergeant, 1965). In January, part of the population is thought to pass through the Strait of Belle Isle into the Gulf of St Lawrence while the remainder moves down the east coast of Newfoundland. By late February the "Gulf" seals can be found on the

FIG. 6 Aerial view of early formation of harp seal herd on the ice in the Gulf of St Lawrence. Picture taken from helicopter 100 m above the ice. (Photograph: K. Ronald.)

ice to the north and west of the Magdalen Islands while the "Front" seals are off the coast of Labrador from Belle Isle to Hamilton Inlet (Mansfield, 1967). These two groups may not be separate even though they are constant and distinct in location, numbers and breeding date (Sergeant, 1965, 1973a). In early May, after breeding, the moulting adults begin to leave the gulf and move north along the Labrador coast (Mansfield, 1967). The northward migration occurs by stages as a series of active northward movements followed by a drift southward on pack ice (Sergeant, 1970). This migration is deflected by ice towards the open-water coast of West Greenland and most of the animals arrive in south-west Greenland around mid-June (Mansfield, 1967). After weaning, the young harp seal pups undergo three successive movements: an active move away from the ice where they were born to the ice edge where they can begin feeding, a passive drift with the current and an active migration northward. These young harp seals migrate alone, separately from and later than the adults and immatures (Sergeant, 1965).

The northward migration of the harp seals is extremely accurate, especially considering the great distance that is covered. The Gulf of St Lawrence herd migrates 4000 km between 40° N, the most southerly point, and 70° N, the most northerly point (Sergeant, 1970). The harp seal's ability to accomplish this migration is partly understood. There is evidence that the seal's eye is attuned to the blue-green coloration of coastal waters (Lavigne and Ronald, 1975), or the animal may orientate into the wind (Sergeant, 1970) or perhaps its navigational capacity is innate, since young animals find their way north in the absence of older animals which have travelled the route before, and since lone individuals may make crucial decisions, involving orientation (Norris, 1966), or the animals follow the temperature profile of the moving ice edge.

Tagging experiments have shown that young animals occasionally drift to other populations. The exchange eastward from Newfoundland is about 1% while that eastward from Jan Mayen is about 10%. No westward exchange has been observed (Sergeant, 1973a).

Abundance and Life History

Harp seals are born in late February and early March. At two to four weeks of age they lose their white coat and enter the water and are deserted by their mothers. After feeding for four to six weeks at the ice border they begin a northward migration and spend the summer feeding (Sergeant, 1965). Their southward migration begins in late fall when they return to the breeding grounds. Tagging results suggest that immatures very consistently return to their place of birth (Sergeant, 1965). Until they are sexually mature harp seals do not arrive from the north until early April at which time they form moulting patches along with the adult males who have completed mating (Allen, 1975). Sexual maturity occurs around five and a half years in females (Sergeant, 1966) and eight years in males (King, 1964). In late February the females form whelping groups on the ice and give birth to their young a few days later. Breeding occurs in late March following which the adult males move onto the ice to moult. The adult females join these moulting groups around April 20-27. Following the moult the harp seals begin their northward migration again. This cycle of migrating, whelping, mating, moulting and migrating occurs yearly for the adult seals until their death in their twenties or at the most early thirties (Fisher, 1955).

At the present time the numbers of seals in the three populations are thought to be as follows: the White Sea—0.5-0.7 million, Jan Mayen—0.1 million (Lavigne, 1978), and the Gulf of St Lawrence—1.3 million (Ronald *et al.*, 1975). From 1924 to 1928 the White Sea population was composed of 20% sexually mature males, 20% sexually mature females, 40% male and female immatures and 20% pups less than one year old (Dorofeev, 1939b). During periods of heavy exploitation by man the harp seal populations show an increase in growth and maturation rates as compensation (Sergeant, 1973c). In order to ensure that the populations do not diminish further catch quotas have been determined which the sealers must adhere to. The maximum sustainable yield of whitecoat pups is determined as a percentage of the annual production of pups within the herd. There are as yet no totally accurate figures of the population but new techniques of aerial censusing (Lavigne *et al.*, 1975) combined with more classical techniques might provide such data.

Internal Characteristics and Morphology

Skull, skeleton and teeth

In the harp seal skull the nasals are long and narrow and taper gradually from the anterior to the posterior end (Fig. 7). The posterior

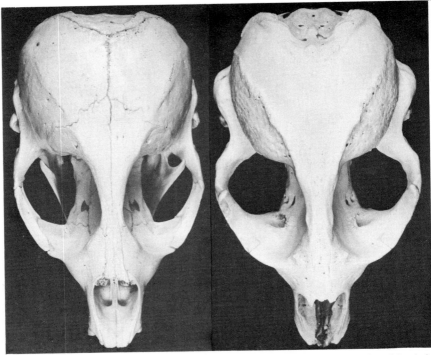

FIG. 7 Dorsal view of: (a) six-month-old male harp seal skull; (b) old adult harp seal skull

margin of the palate is round and the posterior palatine foramina lie in or anterior to the maxilla palatine suture (Doutt, 1942). The following are maximum, minimum and average measurements in mm for different aspects of the skull: total length—221, 190 and 204; width across the mastoids—123.6, 107.5 and 114.6; interorbital width—20.0, 8.3 and 11.7; length of the nasals—52.6, 35.8 and 42.1; width of the nasals at the tip—19.2, 14.6 and 16.6 (Doutt, 1942). The skulls of the adult Greenland harp seal are described as being comparatively light, thin boned and smooth with the rostral part more shortened and the breadth at the auditory bullae an average of 54.7% of the condylobasal length. The White Sea seal skulls are heavy, strong, and relatively thick boned with more conspicuous crests. The rostral part was more elongated with the breadth at the auditory bullae an average of 51.6% of the condylobasal length (Smirnov, 1924).

The vertebral formula of *P. groenlandica* is seven cervical, 15 dorsal, six lumbar, four sacral and 13 caudal segments (Murie, 1870).

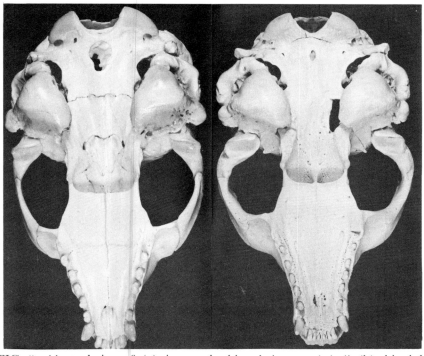

FIG. 8 Ventral view of: (a) six-month-old male harp seal skull; (b) old adult harp seal skull

At birth, harp seal pups possess milk teeth which are separated from the jaw bones and are loose in the soft tissues of the gum, but never erupt. Sometimes the pups have molars at birth but even if they don't the eruption of teeth begins immediately in the order of molars, canines, premolars and incisors and the formation is basically completed by the time the seal begins to feed independently (Beloborodov, 1975). In the adult harp seal the teeth are comparatively small with the third molariform tooth the largest. There is one large central and one small posterior cusp. The anterior end of each ramus of the mandible is narrow and pointed and slopes backward to the lower margin of the ramus. The coronoid is long, slender and pointed. The mandibular teeth are small with one large central, two small posterior and one small anterior cusp (Doutt, 1942). The dental formula (King, 1964) of *P. groenlandica* is

$$I\frac{3}{2} \ C\frac{1}{1} \ PC\frac{5}{5}$$

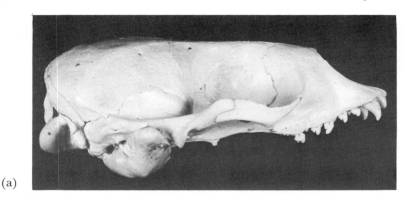

FIG. 9 Lateral view of: (a) six-month-old male harp seal skull; (b) old adult harp seal skull

Visceral anatomy and organ weights

The following is a description of the internal anatomy of a young male harp seal, 1.295 m long and weighing 18.6. kg. Its front flipper was 19 cm long, hind flipper 26.6 cm long. The viscera, including the tongue, weighed 2.325 kg, the brain 230.34 g. The heart had a well-defined bifid extremity with the cleft almost 1.2 cm deep. Its long diameter (root to apex) was 7.6 cm and its transverse diameter near the base was 9.5 cm. The right lung was without divisionary lobules while the left lung was partially divided into two lobes. Numerous and small dorsal papillae covered the terminally split tongue. The oesophagus was 40 cm long and the capacious stomach was cylindroid with a sharp pyloric bend, its long diameter being 23 cm. The small intestine from the pylorus to the caecum was 12.63 m long with an average diameter of 1.2 cm. The large intestine was 5.50 m long with a diameter of 1.9 cm at the caecal end and one of 3.1 cm at the rectum. The liver was deeply divided into seven lobes—five large elongate taper-pointed hepatic divisions and two lobules. The kidneys were compound (Murie, 1870).

The lungs of the adult harp seal are oval in outline and correspond closely to the shape of the thoracic cage. Each compact lung has three lobes and they are topographically similar although the left lung is the larger and heavier (Tarasoff and Kooyman, 1973). The harp seal heart is dorso-ventrally flattened and has a blunt rounded apex. The midsagittal line lies midway between the axis of the two carotid arteries. The interior of right ventricle follows the basic mammalian pattern. The configuration of the left ventricle, aorta and aortic valves is similar to that of terrestrial mammals. However, the valves are right, left and ventral in position in the seal rather than right, left and dorsal as in domestic animals (deKleer, 1972). Macroscopically the harp seal kidney conforms to that of other pinnipeds, being composed of many renuli which are similar to unilobar kidneys. The distinctness of the harp seal kidney lies in the fact that each renulus has a pyramid with its corresponding cortex and a calyx but neither a single branch of the renal artery nor single tributory vein. The mean dimensions of the pup kidneys are 7.6, 3.9 by 21.0 cm and the mean number of renuli 136 while the mean dimensions of the adult kidneys are 12.2, 6.2 by 2.8 cm with the same number of renuli (Dragert *et al.*, 1975).

The number of cartilaginous rings in the trachea is 42.7 ± 0.47 (Shustov and Iablokov, 1967). The length of the intestines exceeds the length of the body by as much as 10 (Murie, 1870) to 13.6 times (Shustov and Iablokov, 1967).

Weights of adult harp seal organs are: heart 0.33 ± 0.04 kg; liver 2.35 ± 0.22 kg; spleen 0.60 ± 0.16 kg (Shustov and Iablokov, 1967). The ratio of the total lung to body weight is 1.31 g per 100 g (Tarasoff and Kooyman, 1973). Mean weights and volumes of kidneys for pups are 49.6 g and 47.3 ml and for adults 173.9 g and 163.1 ml. The difference in weight and volume between left and right adult kidneys is less than 11% (Dragert *et al.*, 1975).

Central nervous system and brain weights

Little work has been done on the central nervous system and brain of phocid seals (Dykes, 1973) even less on harp seals. Some general descriptions of anatomy are available however. The bony tentorium between the cerebrum and the cerebellum is well developed and the brain is foreshortened and more spherical and convoluted than that of a land carnivore (Burne, 1909; Langworthy et al., 1938). The histological characteristics of the excito-motor cortex do not differ markedly from those of land animals. The auditory and trigeminal sensory areas are large and important (Alderson et al., 1960). The entire trigeminal nuclear complex reaches its greatest size among carnivores in *Phoca*. The cerebellum and pons are large compared with other carnivores due to the well developed paraflocculus and the relatively enlarged flocculo nodular lobe. The lobus simplex and ansiform lobule are small (Harrison and King, 1965). The weight of the adult harp seal brain is about 442 g while that of a pup is around 243 g (Sacher and Staffeldt, 1974). The anatomy of the pituitary gland of *P. groenlandica* indicates that this gland is similar in orientation to that of other types of mammals such as the ox (Leatherland and Ronald, personal communication).

Diving and swimming

Harp seals are noted for their ability to dive to great depths and stay submerged for long periods of time in their search for food. The mechanisms that allow them to do so are very complex. Harp seal skeletal muscle, in its adaptation for diving, is primarily geared for utilization of carbohydrates as the main fuel for muscular energy. The two most active locomotory muscles are the Ms psoas and longissimus dorsi, both equipped for sustained activity (George and Ronald, 1975). During a 10-18 min dive the right ventricle of the heart becomes strikingly dilated with a large and systolic volume. This is an important compensatory mechanism in the maintenance of cardiac output and the external ventricular performance (Blix and Høl, 1973). When freely swimming the dive duration of the seal ranges from 0.2 to 10 min or an average of 2.5 min (Øritsland and Ronald, 1975). During these spontaneous, short dives its heart rate drops to 41% of its surface swimming rate of 147 beats per minute. There is a level of bradycardia associated with sinus arrythmia. Many short dives show an anticipatory bradycardia occurring 2.5-7.5 s before submergence. With trained dives there is a more profound bradycardia and less anticipation (Casson and Ronald, 1975). Along with the skeletal muscle and the heart, the circulatory system also undergoes changes during a dive. When a harp seal dives the blood flow in most of the veins, draining capillary beds, drops to virtually zero indicating a complete ischaemia of many tissues (McCarter, 1973). This striking central distribution of the blood from the peripheral to the central capacitance vessels means that the blood flow in the posterior venae cavae is greatly reduced (Høl et al., 1975).

The hepatic sinus expands in apparent response to an influx of blood from the splanchnic reservoir. Blood is removed from the mesenteric veins by profound venoconstriction. At the initiation of the dive the caval sphincter appears to contract with the retained blood engorging the hepatic sinus. The blood flow leaving the brain is maintained during the dive. Therefore, during the dive of a harp seal the blood is rerouted away from the heart-brain pattern to the posterior venae cavae (McCarter, 1973).

When swimming, the harp seal uses a sculling action of the flippers as the principal means of propulsion. The flippers alternately provide power strokes as they are moved from side to side with only one flipper extended at a time. The forelimbs are used like paddles for slow forward propulsion. During inactive periods at the surface the seal uses its fore and/or hind limbs to maintain its position. Its two main poses at the surface are: the head above water and the body vertical and the prone position with the dorsal parts of the head and the mid-back exposed. The hind limbs are the primary means of propulsion with a basic pattern of side-to-side sweeps or rapid strokes in a horizontal plane with the flippers pressed together. The harp seal is capable of relatively high speeds for short periods of time in the water (Tarasoff et al., 1972). Its highest speeds are attained when it is swimming ventrodorsally.

Behaviour

Phoca groenlandica is gregarious, with only the very old males living alone or in small groups. It spends most of the year at sea and as well as being an agile and powerful swimmer, it can, when pressed, also move quickly over the ice. Its movements are of two types: either a dragging motion using the claws of the front flippers or by a humping motion. Migrating harp seals are very active, leaping and cavorting in small groups and when near the surface sometimes swimming on their backs. The harp seals use the large channels or leads in the ice to reach their breeding grounds, penetrating deep into the ice cover. These leads are also used for mounting the ice so that the haul-out sites depend on the location of these channels. They also depend on the surface of the ice and its thickness. Seals prefer rough, hummocky ice at least 0.25 m thick. When pupping, females tend to stay close to a lead and near animals with pups of the same age (Dorofeev, 1939a). The females appear to swim near their pups visiting them at intervals to suckle. The greatest number of females are seen hauled out on the ice around 1000 h and 1200 to 1500 h (Terhune, personal communication). If the females are disturbed while in the water they do not defend their young but go out to sea. On land they will either "freeze" like the pups, defend their young vigorously or re-enter the water. When it is foggy or snowing the females tend to stay in the water but on clear, windless days they like to bask in the sun. Feeding occurs two or three times a day after the female has found a comfortable place for herself and her pup (Popov, 1966).

Feeding habits

The harp seal feeds intensively in the winter and summer but eats less during the spring and autumn migrations and during spring whelping and moult (Mansfield, 1967; Sergeant, 1973b). An individual item is eaten by suction, with small fish taken in tail first. The pregnant females are partly segregated in mid-winter in the best feeding grounds, during lactation, and immediately following it are again segregated. They tend to feed on decapods at that time. In the spring all age groups are in the same geographic area and there is a stratification of feeding by size of organism and by depth (Sergeant, 1973b). The weaned young first feed in surface waters (Sergeant, 1973b) on small pelagic crustaceans such as the euphausid *Thysanoessa*, the amphipod *Anonyx* and small fish such as polar cod *Boreogadus* (King, 1964). At the intermediate depth, immature harp seals eat capelin *(Mallotus villosus)* while the moulting adults are said to dive to depths of 150-200 m to feed on herring *(Clupea harengus)*, cod *(Gadus morhua)* and other groundfish. Social feeding for the pups begins at one year with a change from Crustacea to pelagic fish (Sergeant, 1973b), this however may vary with time of day (Lavigne *et al.*, 1975).

The diet of the adult harp seal is varied, consisting chiefly of pelagic fish, especially capelin, and pelagic and benthic Crustacea (Euphausiacea, Mysidacea, Amphipoda, Decapoda) with smaller quantities of benthic fish (Sergeant, 1973b). The stomach of a seal was found to contain herring, flatfish sp., redfish *Sebastes marinus*, witch, *Glyptocephalus cynoglossus*, plaice, *Hippoglossoides platessoides* and sea mouse, *Aphrodite linnaeus* (Myers, 1959).

A recent study showed that the annual weights of food items eaten by the north-west Atlantic population of harp seals were: all organisms—

2×10^6 t; capelin—0.5×10^6 t and herring—2×10^4 t. Predation on capelin of eastern Canada occurs only in the winter when the pack ice is present (Sergeant, 1973b). The ecological efficiency of the harp seal, which is the weight of the annual increment of the population over the weight of the annual food eaten, is 0.005, a low figure (Sergeant, 1973b).

Social behaviour

Mating is believed to occur after the pups are weaned, which is about two to three weeks after their birth. Mating is thought to occur on the ice and is often preceded by fights between males in which they use their teeth and flippers (Popov, 1966). It is also believed that copulation sometimes takes place in the winter and that courtship involves fighting between male and female. The virgin females mate first followed by the whelped females (Popov, 1966). Each male is believed to mate with one (King, 1964) or more females (Merdsoy and Curtsinger, personal communication). Harp seals are at times very vocal, the many sounds they emit probably having a social and communicative function (Møhl *et al.*, 1975).

When in the pack ice, seals keep natural openings in the ice free to be used as exit and breathing holes. These holes are 60-90 cm across at the top, widening towards the base of the ice. Harp seals, unlike Antarctic seals, use communal breathing holes, with as many as 40 seals in a pod around the hole.

Sound production, hearing and vision

A total of 15 underwater phonations have been recorded for the harp seal and grouped according to frequencies, harmonic structure and duration. They include sine waves, whistles, the morse call, trills, the gull's cry, chirps, warbles, dove cooing, the frequency-shifting key, the distressed blackbird, the passerine call, the Tjok sound, grunts and squeaks (most common), knocking sounds and clicks (Møhl *et al.*, 1975). It is only in the whelping and breeding situation with its heavy concentration of animals and formation of breathing and access holes that these calls are produced. No sounds are made during moult and five of the recorded sounds have been confirmed from captive harp seals at Guelph. The only air call recorded from adult seals is produced by lactating females and sometimes by distressed captive females (Møhl *et al.*, 1975), and males. This vocalization is similar to an "URRRH" sound. The pups cry or wail like a baby and this sound is probably used as a positive clue to allow the mother to find its young on the ice (Terhune and Ronald, 1970).

Hearing plays an important role in the harp seal's life for prey location, predator avoidance and inter-animal communication (Terhune and Ronald, 1974), but is probably of little use in echolocation (Terhune and Ronald, 1975a).

The anatomy of the harp seal ear is similar in some ways to that of cetaceans and also conforms to some of the requirements for bone conduction established in terrestrial mammals. The outer ear differs from that of land animals in the loss of pinna and the modifications of the auricular muscles to close the meatus. The loss of pinna is an adaptive modification to minimize hydrodynamic resistance and is also for effective directional hearing. The presence of cavernous tissue and large, heavy ossicles are peculiarities of the seal's middle ear. These may be attributed to the seal's aquatic life or they may contribute to the general hearing mechanism in the animal. The harp seal inner ear is similar in anatomy to that of other mammals (Ramprashad et al., 1973).

The underwater hearing abilities of *P. groenlandica* are similar to those of humans in air (Terhune and Ronald, 1974). The harp seal freefield underwater audiogram, measured from 0.76 to 100 kHz, has areas of increased sensitivity at 2 and 22.9 kHz. The lowest threshold is —37 db/μbar at 22.9 kHz. Above 64 kHz the threshold increases at a rate of 40 db per octave. The seal's ear is adapted to hearing underwater as is indicated by the decrease of sensitivity in air (Terhune and Ronald, 1972). The harp seal freefield air audiogram, measured from one to 32 kHz, has its lowest threshold at 4 kHz at a level of 29 db per 0.0002 dynes-2 cm. Thus the harp seal's hearing in air is irregular and slightly insensitive with the air audiogram being generally flat. Critical ratios at 2 and 4 kHz are 10% (Terhune and Ronald, 1971). The harp seal's audiogram is similar to that of other phocids (Terhune and Ronald, 1975b).

Because seals utilize vision as a primary source of sensory information their eyes must be adapted for the two media in which they function, air and water. They must also be able to adjust to a variety of light conditions from the brightness of the sunlit snow to the underwater dimness (Lavigne et al., 1975). The harp seal has large eyes with a sensitive retina and tapetum (Nagy and Ronald, 1970) and a widely dilated pupil. These characteristics are typical of eyes adapted for nocturnal vision suggesting that the seal's eyes are adapted for dim light sensitivity (Lavigne and Ronald, 1972). The seal's rod-dominated retina is also adapted for increased sensitivity which allows for good underwater vision, especially at night. The presence of photoreceptors which function optimally in bright light, permits effective vision during the day (Lavigne et al., 1975). Thus the harp seal possesses excellent visual sensitivity in both air and water (Lavigne and Ronald, 1972).

Not much is known about the other senses of seals. The small olfactory lobes of the brain and the reduced ethmoturbinals suggest that their sense of smell is not very acute. Although taste buds are definitely present this sense is probably not very subtle, as these animals swallow their food whole. Seals vary in their awareness of touch and in their apparent liking of physical contact (Harrison and King, 1965), there is some evidence however for the use of vibrissae as contact and receptor sensors (Dykes, 1973).

Captivity

Harp seal pups can be captured by hand and placed in burlap bags to quiet them. The adults can be captured in nylon nets by cutting off access to the breathing holes. They react either by fighting or "freezing" (Ronald et al., 1970).

To mark the seals for purposes of identification cryothermic branding with liquid nitrogen can be used. This requires a four- to five-second application to shaved skin (MacPherson and Penner, 1967). When it is necessary to restrain the seals a wheeled V-shaped trough lined with a sponge mattress with canvas straps or car safety belts can be used. Moult causes considerable stress in captivity and is characterized by the sloughing off of skin as well as hair. It is often accompanied by anorexia, opacity of the eye, lethargy and irritable behaviour. The complete moult varies with the individual from 10 to 90 days (Ronald et al., 1970). It is especially important not to stress the animal at these times.

Seal pups have difficulty learning to eat in captivity and often have to be forcefed. Our preferred method consists of forcefeeding small pieces of herring four to six times a day. After the original period of one to ten days the seals will be free-feeding. The herring should be supplemented (Ronald et al., 1970).

Harp seals respond to quiet handling and will soon learn to take increasingly larger pieces of fish on their own. Throughout captivity the diet may remain mainly herring and later occasionally smelt

(Ronald et al., 1970). Only the best quality fish fit for human consumption should be used. Adult seals may consume 4-7% of their body weight daily (Geraci, 1975). While some captive seals have been known to eat up to 10% of their body weight per day (Ronald et al., 1970), others have been adequately maintained on as little as 1.5%.

Hyponatremia and the need for dietary salt supplementation occurs in captive harp seals. In fresh water, harp seals experience periods of electrolyte imbalance characterized by low plasma sodium concentrations. They can suffer mild to severe CNS disturbance and death. The cause of this condition is insufficient salt intake due to a low-salt diet in captivity with superimposed physiologic and pathologic stresses. Hyponatremia can be treated with daily sodium chloride supplements of 3 gm kg-1. This maintains the plasma electrolytes in normal or near-normal condition (Geraci, 1972).

Captive harp seals which are maintained on herring and smelt, both high in thiaminase, develop thiamine deficiencies. If the diet is known to contain thiaminase, thiamine should be supplemented. The feeding of a wide variety of fish species will help to avoid or dilute high concentrations of thiaminase which might be present in any one of the food species (Geraci, 1974).

It is often necessary to anaesthetize seals and this presents problems because of their adaptations for aquatic life. Intravenous or inhalation general anaesthesia usually results in respiratory arrest. Thiopental sodium administration via the extradural vein is a rapid and dependable means of induction, but rapid intubation and commencement of controlled ventilation are essential. Halothane/nitrous oxide induction is the safest and most controllable though slow. Halothane concentrations of 0.75-1.5% are required (McDonnell, 1972).

Reproduction

Gestation

Gestation begins in June and last 7.5 months. The gap between this time and mating, which occurs in March, is accounted for by an eleven-week delay in the implantation of the blastocyst (King, 1964). There is also evidence that the pregnant female can retain the foetus until suitable ice is available in February or even March.

Mating

The seals mate about two weeks after parturition (King, 1964). It is only at this time, in a hunted herd, that males are seen on the ice whereas males are commonly present both in the water and on the ice in a protected herd.

Maturity

Sexual maturity in females, or time of first ovulation, occurs between four and seven years of age with a mean age of 5.5 years (Sergeant, 1966). They bear their first young the following year and each successive year until they are at least 16 (King, 1964) and possibly up to 30 years of age. In captive animals at Guelph three females have reached maturity in their fifth year. In the White Sea population the age of sexual maturity for females is three to five years while in the Jan Mayen population it is three to six years with a mean age of five years. When a population approaches its maximum, sexual maturity is delayed and probably the fertility of older females is reduced. The ten-year-old female seals dominate in whelping (Sergeant, 1966). Male harp seals are sexually mature at 5.5 years (Sergeant, 1973a) but not active until 8 years (King, 1964). The maximum life span of *P. groenlandica* is over 30 with an average of well over 20 (Fisher, 1955; King, 1964). Both sexes are sexually active in their twenties (Fisher, 1955).

Breeding season

In the White Sea pupping takes place from the end of January to the beginning of April but most of the young are born between February 20 and March 5. In Jan Mayen the breeding season is slightly later. After two weeks the adults mate then move away to moult (King, 1964). In the Gulf most pups are born between February 20 and March 10; on the "Front" it is slightly later.

Birth

The pregnant females lie on the ice and the pups are born within 1.5-2.5 m of one another. The breeding grounds are preferably some distance from the margins of the large ice fields in rough, hummocky ice which gives shelter to the pups (King, 1964). The pups are mostly born at night or early in the morning.

FIG. 10   Adult female harp seal just prior to parturition. (Photograph: D. M. Lavigne.)

Lactation

Suckling lasts 10-12 days during which time the female is said to feed very little (King, 1964), although there is considerable defaecation in the water (Merdsoy and Curtsinger, personal communication). Harp seal milk is greyish-white, with a strong fish-like odour and the consistency of thick cream. It is composed of 42.6% fat, 10.4% protein, 45.3% water and 0.8% ash (Sivertsen, 1941). The high fat content causes the rapid growth in pups during lactation. The lactose and ash contents are low compared to those of other mammals. About 27% of the fatty acid content of harp seal milk consists of fatty acids of chain lengths of 20 or more, and about 69% of the total fatty acids have one or more double bonds (Cook and Baker, 1969). Both the fat and protein contents decrease during suckling to a minimum on the sixteenth day postpartum while the fatty acid composition of the fat follows no definite pattern of change during this period (Van Horn and Baker, 1971).

Size of neonates

The whitecoat seals are between 97 and 108 cm long at birth (Smirnov, 1924) and weigh an average of 11.8 kg (Sivertsen, 1941). This weight is almost all blubber (Fisher, 1955) which they will need to survive while they are learning to feed themselves. At parturition, harp seal pups lack subcutaneous blubber, the wettable, infantile fur offers poor

FIG. 11 Three-day-old whitecoat on the ice, Gulf of St Lawrence, with female in the background. (Photograph: J. M. Terhune.)

insulation and behavioural thermogenesis is not prominent. In order for them to survive a dramatic increase in heat production is necessary. Because these pups possess a layer of brown adipose tissue at birth it is thought that non-shivering thermogenesis through activated brown adipose tissue may help guard the animal from the cold (Grav *et al.*, 1974; Blix *et al.*, 1975).

Diseases

The most common symptoms of illness observed in captive harp seals are those of gastroenteritis. This consists of a general malaise with ocular opacity, anorexia and lethargy followed by emesis and diarrhoea (Ronald *et al.*, 1970). In spite of a daily supplement of 100 mg thiamine per kg food, some captive seals still show signs of a vitamin B deficiency (Blix *et al.*, 1973). This condition is invariably reversed within two days of an intramuscular injection of a vitamin B complex (Blix *et al.*, 1973). There are individual cases of captive seals developing a variety of illnesses. One seal had a gastric ulcer involving 60% of the pyloric portion of the stomach. Death resulted for another harp seal due to impaired thermoregulation associated with peritonitis and pleuritis from an infection of *Pasteurella multocida*. Polioencephalomalacia caused another seal to die. Sometimes captive animals develop dry papules on their flippers which spread until the entire flipper is bare of hair, reddened and hot to touch (Ronald et al., 1970). *Staphyloccoccal granulomas* have been known to cause skin lesions in harp seals (Wilson and Long, 1970; Ronald et al., 1970). Infectious disease caused by *Aeromonas* are found in pinnipeds. In one harp seal, aeromonads in the spleen, liver and intestines caused emaciation and inadequate feeding. Further examination showed very little content in the digestive tract, both sides of the heart dilated and the lungs oedematous and emphysematous (Dahle and Nordstoga, 1968). One captive *P. groenlandica* suddenly developed severe and persistent tetanies and was given 1.2 g magnesium sulphate intramuscularly. After one more injection the next day the seal recovered fully. Convulsions never recurred after a daily supplement of 20 mg magnesium sulphate per kg body weight was added to its diet (Blix et al., 1973).

Blood values

Haemoglobin (Ronald et al., 1969) and iron values (Vallyathan et al., 1969) are 21.80 ± 0.201 g per 100 ml blood and 55.94 ± 0.48 mg per 100 ml blood which are considerably higher than those of other mammals including man. The glucose level, 169.4 ± 2.27 mg per 100 ml blood, total lipid, 854.4 ± 20.67 mg per 100 ml blood, and total cholesterol, 242.2 ± 4.36 mg per 100 ml blood, levels are also high. The levels of activity of the three enzymes, amylase, aldolase and lactic dehydrogenase, involved in carbohydrate metabolism, are also higher than in other animals. Lactic dehydrogenase activity is particularly high. These blood properties reflect the pattern of metabolic adaptation to diving in the tissues of the harp seal, especially in its muscles. The low lipase (esterase) activity in seal blood indicates that fat is not the favoured metabolite for muscular energy (Vallyathan et al., 1969).

A high erythrocyte number in March corresponds to the natural pupping and breeding periods. This count is depressed during moult. An initial increase in erythrocyte number occurs with development. A leucocyte count of over 12 000 cells mm^{-3} indicates an abnormal condition. The mean corpuscular haemoglobin concentration (MCHC) of the harp seal averages 47.42%. The high haemoglobin level enables the seal to carry greater quantities of oxygen than terrestrial species. A pH drop in January coincides with the southward migration to the breeding grounds while a maximum in March coincides with pupping and breeding. Individuals show a lower pH immediately preceding and during part of the moult. Decreases in erythrocytes, haemoglobin and haematocrit occur during moult (Ronald et al., 1969). Erythrocyte, haematocrit and haemoglobin levels are maintained throughout life although cell volume and mean cell haemoglobin increase with age (Ronald, 1970; Geraci, 1971).

The following values were obtained from analysis of blood from harp seal pups and adults respectively: total mercury in blood (ppm)—0.16 and 0.19; and CH_3HgCl—0.20 and 0.13; selenium in ppm—1.21 and 1.45; red cells in millions per mm^3—5.16 and 4.68; mean cell volume (MCV)—109.02 and 125.96; haematocrit vol (%)—56.7 and 59.5; leucocyte differential counts (%)—neutrophils 58, lymphocytes 31, monocytes 6, and eosinophils 5; plasma sodium levels (mEq/l)—152.97 and 152.81; plasma chloride levels (mEq/l)—102.57 and 109.81; plasma potassium levels (mEq/l)—3.91 and 3.82; blood urea nitrogen (BUN) (mg%)—34.95 and 44.08; lactic dehydrogenase (I.U.)—151.29 and 118; serum glutamic pyruvic transaminase (SGPT) (I.U.)—7.81 and 38.82; serum glutamic oxaloacelic transaminase (SGOT)—71.86 and 39.3; serum cholesterol (mg%)—340.75 and 307-28; serum protein (g%)—6.49 and 7.53; serum alkaline phosphatase (I.U.)—34.5 and 10.0; and total bilirubin (mg%)—0.34 and 0.38 (Ronald and Tessaro, 1976).

Cytogenetics

The chromosome numbers of the harp seal are $2n = 32$. Their karyotypic stability is attributed to a low level of reproduction (late sexual maturity and only one pup per year), good mobility and an environment without delimited niches. Because of the above characteristics, speciation caused by chromosal rearrangement is rare (Arnason, 1972).

Parasites

The commonest of the harp seal's parasites is the nematode *Contracaecum gadi* which is a normal part of the animal's parasitofauna (Ronald *et al.*, 1970). Specimens of *Contracaecum* sp. can be found in the oesophageal, gastric and small intestinal areas of the alimentary tract. This parasite may cause disease in the form of gastric lesions (Wilson and Stockdale, 1970). Another nematode is *Terranova decipiens* (= *Phocanema decipiens* = *Porrocaecum decipiens*). The intestinal tract is a site for the infection of nematodes where they become firmly attached to the intestinal mucosa (Montreuil and Ronald, 1957). *Terranova decipiens* also occurs in the stomach. Massive infestations of these nematodes do not seem to affect the harp seal's general health. After the seal's death some worms escape through its nostrils, mouth and rectum. Infection occurs after the consumption of fish containing the larvae which are thought to reach sexual maturity in the animal's digestive tract (Myers, 1960). Flatfish are the most common source of infection of *T. decipiens* (Myers, 1957b). *Phoca groenlandica* plays some part in the dissemination of the nematode, *T. decipiens*, and has some importance as a definitive host (Ronald et al., 1969). The harp seal's internal temperature is ideal for the development of the parasite (Ronald, 1960). The harp seal is infective for the four months it spends in the Gulf. Its relative importance as a vector in the Gulf is 80.4 as compared to the harbour seal at 3.8 and the grey seal at 15.8 (Mansfield and Sergeant, 1965). The ascaroid parasites, *Contracaecum osculatum* (Rudolphi, 1802) and *Phocascaris* sp. are also found in the harp seal (Myers, 1957a). The adult worms of these three are usually free in lumen of the stomach and intestine but are sometimes attached to the mucosa (Myers, 1960). Other parasites of the harp seal are the Anisakinae, *Phocascaris phocae* (Lyster, 1940), the cestodes, *Diphyllobothrium lanceolatum, D. cordatum, D. schistochilos* (King, 1964), *Anophryocephalus anophyrs* (Smith and Threlfall, 1973), the trematodes *Orthosplanchnus arcticus* (King, 1964) and *Pseudamphistomum truncatum* (Delyamure, 1955) and the Acanthocephala, *Corynosoma strumosum.* The anoplurans *Echinophthirus groenlandicus, E. phocae* (Kellogg and Ferris, 1915) and *E. horridus* are occasionally found on harp seals but in the case of the latter not on adults (Ronald et al., 1970).

References

Alderson, A. M., Diamatopoulos, E. and Downman, C. B. B. (1960). Auditory cortex of the seal. *(Phoca vitulina). J. Anat.* **94**, 506-511.
Allen, R. (1975). A life table for harp seals in the northwest Atlantic. *In* "Biology of the Seal", (Eds K. Ronald and A. W. Mansfield), *Cons. Int. Explor. Mer, Rapp. p. -v. Réun.,* 169, Copenhagen.
Arnason, U. (1972). The role of chromosomal rearrangement in mammalian speciation with special reference to Cetacea and Pinnipedia. *Hereditas* **70**, 113-118.
Beloborodov, A. G. (1975). Postnatal dentition in the harp seal *(Pagophilus groenlandicus). In* "Biology of the Seal", (Eds K. Ronald and A. W. Mansfield), *Cons. Int. Explor. Mer, Rapp. p. -v. Réun.,* 169, Copenhagen.
Blix, A. S., Iversen, J. and Pasche, A. (1973). On the feeding and health of young hooded seals *(Cystophora cristata)* and harp seals *(Pagophilus groenlandicus)* in captivity. *Norw. J. Zool.* 21-55-58.
Blix, A. S., Høl, R. (1973). Ventricular dilation in the diving seal. *Acta Physiol. Scand.* **87**, 431-432.

Blix, A. S., Grav, H. J. and Ronald, K. (1975). Brown adipose tissue and the significance of the venous plexuses in pinnipeds. *Acta Physiol. Scand.* **94**, 133-135.
Brown, R. (1868). Notes on the history and geographical relations of the Pinnipedia frequenting the Spitzbergen and Greenland Seas. *Zool. Soc. Lond., Comm. Sci. Corresp., Proc., 1868*, 405-440.
Burne, R. H. (1909). Notes on the viscera of a walrus *(Odobenus rosmarus)*. *Zool. Soc. Lond., Proc., 1909*, 732-738.
Burns, J. J. and Fay, F. H. (1970). Comparative morphology of the skull of the ribbon seal, *Histriophoca fasciata*, with remarks on systematics of Phocidae. *J. Zool.* **161**, 363-394.
Casson, D. M. and Ronald, K. (1975). The harp seal, *Pagophilus groenlandicus* (Erxleben, 1777). 14. Cardiac arrythmias. *Comp. Biochem. Physiol.* **50A**, 307-314.
Cook, H. W. and Baker, B. E. (1969). Seal milk. 1. Harp seal *(Pagophilus groenlandicus)* milk: composition and pesticide residue content. *Can. J. Zool.* **47**, 1129-1132.
Dahle, J. K. and Nordstoga, K. (1968). Identification of aeromonads in furred animals. *Acta Vet. Scand.* **9**, 65-70.
deKleer, V. S. (1972) The anatomy of the heart and the electrocardiogram of *Pagophilus groenlandicus*. M.Sc. Thesis, University of Guelph.
Delyamure, S. L. (1955). Helminthofauna of marine mammals (ecology and phylogeny). (Ed. K. I. Skrjabin), Akad. Nauk SSSR, Moscow. Israel Program Sci. Transl., 1968.
Dorofeev, S. V. (1939a). The influence of ice conditions on the behaviour of harp seals. *Zool. Zh.* **18**, 748-761.
Dorofeev, S. V. (1939b). The relationship of age groups in seals as indicative of the condition of the stock. *In* "Sbornik Posviashchennyi Nauchnoi Deiatel 'nosti N.M. Knipovicha", (Volume in honour of scientific activity of N.M. Knipovich). Moscow.
Dorofeev, S. V. (1956). Stocks of Greenland seals and their utilization. *Rybn Kohz* **12**, 56-69. (Fish. Res. Board Can., Transl. Ser., 113, 1957).
Doutt, J. K. (1942). A review of the genus *Phoca*. *Carnegie Mus. Ann.* **29**, 61-125.
Dragert, J., Corey, S. and Ronald, K. (1975). Anatomical aspects of the kidney of the harp seal, *Pagophilus groenlandicus* (Erxleben, 1777). *In* "Biology of the seal", (Eds K. Ronald and A. W. Mansfield), *Cons. Int. Explor. Mer, Rapp, p. -v. Réun*, 169, Copenhagen.
Dykes, R. W. (1973). Characteristics of afferent fibers from the mystacial vibrissae of cats and seals. *Physiol. Can.* **4**, 176.
Fisher, H. D. (1955). Utilization of Atlantic harp seal populations. *In* "Trans. 20th North Am. Wildl. Conf. Wash., D.C.", 507-518. Wildl. Manage. Inst.
Frisch, J., Øritsland, N. A. and Krog, J. (1974). Insulation of furs in water. *Comp. Biochem. Physiol.* **47A**, 403-410.

George, J. C. and Ronald, K. (1975). Metabolic adaptation in pinniped skeletal muscle. *In* "Biology of the seal", (Eds K. Ronald and A. W. Mansfield), *Cons. Int. Explor. Mer, Rapp. p. -v. Réun,* 169. Copenhagen.

Geraci, J. R. (1971). Functional hematology of the harp seal, *Pagophilus groenlandicus. Physiol. Zool.* **44,** 162-170.

Geraci, J. R. (1972). Hyponatremia and the need for dietary salt supplement in captive pinnipeds. *Am. Vet. Med. Assoc., J.* **161,** 618-623.

Geraci, J. R. (1974). Thiamine deficiency in seals and recommendations for its prevention. *Am. Vet. Med. Assoc., J.* **165,** 801-803.

Geraci, J. R. (1975). Pinniped nutrition. *In* "Biology of the seal", (Eds K. Ronald and A. W. Mansfield), *Cons. Int. Explor. Mer, Rapp. p. -v. Réun,* 169. Copenhagen.

Grav, H. J., Blix, A. S., and Pasche, A. (1974). How do seal pups survive birth in Arctic winter? *Acta Physiol. Scand.* **92,** 427-429.

Grav, H. J. and Blix, A. S. (1975). Brown adipose tissue—a factor in the survival of harp seal pups *Vitam impendere vero. In* "Depressed Metabolism and Cold Thermogenesis", (Ed. L. Jansky), Academia, Prague.

Harrison, R. J., and King, J. E. (1965). "Marine Mammals", Hutchinson, London.

Høl, R., Blix, A. S. and Myhre, H. O. (1975). Selective redistribution of the blood volume in the diving seal *(Pagophilus groenlandicus). In* "Biology of the seal", (Eds K. Ronald and A. W. Mansfield), *Cons. Int. Explor. Mer, Rapp. p. -v. Réun,* 169. Copenhagen.

Kellogg, V. L. and Ferris, G. F. (1915). "Anoplura and Mallophaga of North American Mammals", Stanford Univ. Publ., Univ. Ser., Biol. Sci., (May, 1915).

King, J. E. (1964). "Seals of the World", British Museum (Natural History), London.

Langworthy, O. R., Hesser, F. H. and Kolb, E. C. (1938). A physiological study of the cerebral cortex of the hair seal *(Phoca vitulina). J. Comp. Neurol.* **68,** 351-369.

Lavigne, D. M. and Ronald, K. (1972). The harp seal, *Pagophilus groenlandicus* (Erxleben, 1777). 23. Spectral sensitivity. *Can. J. Zool.* **50,** 1197-1206.

Lavigne, D. M. and Øritsland, N. A. (1974). Ultraviolet photography: a new application for remote sensing of mammals. *Can. J. Zool.* **52,** 939-941.

Lavigne, D. M., Øritsland, N. A., Watts, P. and Ronald, K. (1974). Harp seal remote sensing. Can Environ., Can., Fish. Mar. Serv. Comm., Seals and Sealing, Ottawa, 115 p.

Lavigne, D. M. and Ronald, K. (1975). Pinniped visual pigments. *Comp. Biochem. Physiol.* **50B,** 325-329.

Lavigne, D. M., Bernholz, C. D. and Ronald, K. (1975). Functional aspects of the marine mammal retina. *In* "Functional Anatomy of Marine Mammals", Vol. 3, (Ed. R. J. Harrison), Academic Press, London and New York.

Lavigne, D. M. (1978). The harp seal controversy reconsidered. *Queen's Quart.* **85,** 377-388.

Lyster, L. L. (1940). Parasites of some Canadian sea mammals. *Can. J. Res., Sect. C, Bot. Sci.* **18**, 395-409.

MacPherson, J. W. and Penner, P. (1967). Animal identification. 2. Freeze branding of seals for laboratory identification. *Can. J. Comp. Med.* **31**, 275-276.

Mansfield, A. W. and Sergeant, D. E. (1965). Relative importance of seal species as vectors of codworm *(Porrocaecum decipiens)* in the Maritime Provinces. No. 10. Fish. Res. Board Can., Arct. Biol. Stn., Annu. Rep. Invest. Summ., 1964-1965.

Mansfield, A. W. (1967). Seals of Arctic and eastern Canada. *Fish. Res. Board Can., Bull.*, **137**.

McCarter, R. M. (1973). Venous circulation in *Pagophilus groenlandicus*. M.Sc. Thesis, University of Guelph, Canada.

McDonnell, W. 1972. Anesthesia of the harp seal. *J. Wildl. Dis.* **2**, 287-295.

Møhl, B., Ronald, K. and Terhune, J. (1975). Underwater calls of the harp seal, *Pagophilus groenlandicus*. In "Biology of the Seal", (Eds K. Ronald and A. W. Mansfield), *Cons. Int. Explor. Mer, Rapp. p. -v. Réun.*, 169. Copenhagen.

Montreuil, P. L. J. and Ronald, K. (1957). A preliminary note on the nematode parasites of seals in the Gulf of St. Lawrence. *Can. J. Zool.* **35**, 495.

Murie, J. (1870). On *Phoca groenlandica* (Muell.), its modes of progression and its anatomy. *Zool. Soc. Lond., Comm. Sci. Corresp., Proc., (1870)*, 604-608.

Myers, B. J. (1957a). Ascaroid parasites of harp seals *(Phoca groenlandica* Erxleben) from the Magdalen Islands, Quebec. *Can. J. Zool.* **35**, 291-292.

Myers, B. J. (1957b). Our present knowledge of *Porrocaecum decipiens* MacDonald Coll., Inst. Parasitol., Quebec, Ms.

Myers, B. J. (1959). The stomach contents of harp seals *(Phoca groenlandica*, Erxleben) from Magdalen Islands, Quebec. *Can. J. Zool.* **37**, 378.

Myers, B. J. (1960). On the morphology and life history of *Phocanema decipiens* (Krabbe, 1878) Myers, 1959 (Nemotoda: Anisakidae). *Can. J. Zool.* **38**, 331-344.

Nagy, A. R. and Ronald, K. (1970). The harp seal, *Pagophilus groenlandicus* (Erxleben, 1777). 6. Structure of retina. *Can. J. Zool.* **48**, 367-370.

Norris, K. S. (1966). Some observations on the migration and orientation of marine mammals. In "Biology Colloquium, Animal Orientation and Navigation. Proc. 27th Annu. Biol. Colloq., May 6-7, 1966", 101-125. Oregon State University Press, Corvallis.

Øritsland, N. A. and Ronald, K. (1975). Energetics in the free diving seal. *In* "Biology of the Seal", (Eds K. Ronald and A. W. Mansfield), *Cons. Int. Explor. Mer, Rapp. p. -v. Réun.*, 169. Copenhagen.

Popov, L. A. (1966). On an ice floe with the harp seals: ice drift of biologists in the White Sea. *Priroda* **9**, 93-101; (Fish. Res. Board Can., Transl. Ser., 814, 1967.)

Popov, L. A. (1970). Soviet tagging of harp and hooded seals in the north Atlantic. Fiskeridir, Skr., Ser. Havunders., 16: 1-9.

Ramprashad, F., Corey, S. and Ronald, K. (1973). Anatomy of the seal's ear *(Pagophilus groenlandicus)* (Erxleben, 1777). *In* "Functional Anatomy of Marine Mammals", Vol. 1. (Ed. R. J. Harrison), 264-305. Academic Press, London and New York.

Ronald, K. (1960). The effects of physical stimuli on the larval stage of *Terranova decipiens* (Krabbe, 1878) (Nematoda: Anisakidae). 1. Temperature. *Can. J. Zool.* **38**, 623-642.

Ronald, K., Foster, M. E., Johnson, E. (1969). The harp seal, *Pagophilus groenlandicus* (Erxleben, 1777). 2. Physical blood properties. *Can. J. Zool.* **47**, 461-468.

Ronald, K. (1970). Physical blood properties of neonatal and mature harp seals. Int. Counc. Explor. Sea, (I.C.E.S.), Counc. Meet., *Mar. Mamm. Comm.* 1970/N:5.

Ronald, K., Johnson, E., Foster, M. E. and Vander Pol, D. (1970). The harp seal, *Pagophilus groenlandicus* (Erxleben, 1777). 1. Methods of handling, moult, and diseases in captivity. *Can. J. Zool.* **48**, 1035-1040.

Ronald, K., Capstick, C. K. and Shortt, J. (1975). Effect of alternative harp seal crops on populations 1974-1993. Report to C.O.S.S., Ottawa.

Ronald, K. and Tessaro, S. V. (1976). Methyl mercury poisoning in the harp seal *(Pagophilus groenlandicus)*. Int. Counc. Explor. Sea, (I.C.E.S.), Counc. Meet., 1976/N:81.

Sacher, G. A. and Staffeldt, E. F. (1974). Relation of gestation time to brain weight for placental mammals: implications for the theory of vertebrate growth. *Am. Nat.* **108(968)**, 493-615.

Scheffer, V. B. (1958). "Seals, Sea Lions and Walruses. A Review of the Pinnipedia", Stanford University Press, Stanford, CA.

Sergeant, D. E. (1965). Migrations of harp seals *Pagophilus groenlandicus* (Erxleben) in the north-west Atlantic. *Fish. Res. Board Ca., J.* **23**, 433-464.

Sergeant, D. E. (1966). Reproductive rates of harp seals, *Pagophilus groenlandicus* (Erxleben). *Fish. Res. Board Can., J.* **23**, 757-766.

Sergeant, D. E. (1970). Migration and orientation in harp seals. *In* "Proc. 7th Annu. Conf. Biol. Sonar Diving Mamm., 23-24 Oct., 1970", 123-131. Stanford Res. Inst., Biol. Sonar Lab., Menlo Park, CA.

Sergeant, D. E. (1973a). Transatlantic migration of a harp seal, *Pagophilus groenlandicus. Fish. Res. Board Can., J.* **30**, 124-125.

Sergeant, D. E. (1973b). Feeding, growth and productivity of north-west Atlantic harp seals *(Pagophilus groenlandicus). Fish. Res. Board Can., J.* **30**, 17-29.

Sergeant, D. E. (1973c). Environment and reproduction in seals. *J. Reprod. Fertil., Suppl.* **19**, 555-561.

Shustov, A. P. and Iablokov, A. V. (1967). Comparative morphological characteristics of the harp and ribbon seals. *In* "Research on marine mammals", Issledovaniya Morskikh Mlekopitayushchie, (Eds K. K. Chapskii and M. I. Iakovenko) 51-59. Murmansk. Polyarn. Nauchno-Islled. Proektn. Inst. Morsk. Rybn. Khoz. Okeanogr., (PINRO), Tr., 21; *Fish. Res. Board Can.,* Transl. Ser.,, 1084, 1968.

Sivertsen, E. (1941). On the biology of the harp seal, *Phoca groenlandica* (Erxl.), investigations carried out in the White Sea 1925-1937. *Hvalrad. Skr.,* 26.

Smirnov, N. A. (1924). On the eastern harp seal *Phoca (Pagophoca) groenlandica.* Tromso Mus., Arsb. Nat. Avd. **47,** 3-11.

Smith, F. R. and Threlfall, W. (1973). Helminths of some mammals from Newfoundland. *Am. Midl. Nat.* **90,** 215-218.

Tarasoff, F. J., Bisaillon, A., Pierard, J., and Whitt, A. P. (1972). Locomotory patterns and external morphology of the river otter, sea otter and harp seal (Mammalia). *Can. J. Zool.* **50,** 915-929.

Tarasoff, F. J. and Kooyman, G. L. (1973). Observations on the anatomy of the respiratory system of the river otter, sea otter, and harp seal. 1. The topography, weight, and measurements of the lungs. *Can. J. Zool.* **51,** 163-170.

Terhune, J. M. and Ronald, K. (1970). The audiogram and calls of the harp seal *(Pagophilus groenlandicus)* in air. *In* "Proc. 7th Annu. Conf. Biol. Sonar Diving Mamm., 23-24 Oct., 1970", 133-143. Stanford Res. Inst., Biol. Sonar Lab., Menlo Park, CA.

Terhune, J. M. and Ronald, K. (1971). The harp seal, *Pagophilus groenlandicus* (Erxleben, 1777). 10. The air audiogram. *Can. J. Zool.* **49,** 385-390.

Terhune, J. M. and Ronald, K. (1972). The harp seal, *Pagophilus groenlandicus* (Erxleben, 1777). 3. The underwater audiogram. *Can. J. Zool.* **50,** 565-569.

Terhune, J. M. and Ronald, K. (1974). Underwater hearing of phocid seals. Int. Counc. Explor. Sea, (I.C.E.S.), Counc. Meet., *Mar. Mamm. Comm.* 1974/N:5.

Terhune, J. M. and Ronald, K. (1975a). Underwater hearing sensitivity of ringed seals *(Pusa hispida). Can. J. Zool.* **53,** 227-231.

Terhune, J. M. and Ronald, K. (1975b). Masked hearing thresholds of marine mammals. *Acoust. Soc. Am., J.* **58,** 515-516.

Vallyathan, N. V., George, J. C. and Ronald, K. (1969). The harp seal, *Pagophilus groenlandicus* (Erxleben, 1777). 5. Levels of haemoglobin, iron, certain metabolites and enzymes in the blood. *Can. J. Zool.* **47,** 1193-1197.

Van Horn, D. R. and Baker, B. E. (1971). Seal milk. 2. Harp seal *(Pagophilus groenlandicus)* milk: effects of stage of lactation on the composition of the milk. *Can. J. Zool.* **49,** 1085-1088.

Wilson, T. M. and Long, J. R. (1970). The harp seal, *Pagophilus groenlandicus,* (Erxleben, 1777). 12. Staphyloccoccal granulomas (Botryomycosis) in harp seals. *J. Wildl. Dis.* **6,** 155-159.

Wilson, T. M. and Stockdale, P. H. (1970). The harp seal, *Pagophilus groenlandicus* (Erxleben, 1777). II. *Contracaecum* sp. infestation in a harp seal. *J. Wildl. Dis.* **6,** 152-154.

4

Ribbon Seal

Phoca fasciata Zimmermann, 1783

John J. Burns

Genus and Species

The ribbon seal (Family Phocidae, genus *Phoca* Linnaeus, 1758 has long remained an animal about which little was known. This is entirely understandable in view of its distribution in the seasonally ice-covered seas of the North Pacific region and because they seldom occur near shore. There are a great many scientific papers which mention or briefly discuss this seal, but comparatively few which present much detailed information. Those indicated below are, in the opinion of this writer, significant sources of information about the distribution, natural history, systematics and management of ribbon seals.

The first available published accounts of the ribbon seal were apparently those of Pennant (1781, 1783 *in* Allen, 1880). These descriptions of "rubbon seal" were, according to Allen (1880), based entirely on information and a drawing provided to Pennant by P. Pallas, based on specimens obtained from the Kuril Islands. Pallas also published an account of this seal, as well as other species of phocids from the Asiatic coast in 1811. The ribbon seal received the binomial *Phoca fasciata* in a publication by Zimmermann (1783), based on the earlier descriptions by Pennant. According to Allen (1880) another "brief and unimportant reference" to this species was published by Siemaschko in 1851. In 1859 von Schrenck published what was then the most complete description of ribbon seals, based on specimens that he saw from the Okhotsk Sea, as well as four specimens killed near the eastern coast of the Kamchatka Peninsula, at the mouth of the Kamchatka River (Allen, 1880). In 1873, Gill placed the ribbon seal in a separate genus, *Histriophoca*, on the basis of its distinctive coloration and teeth, which were supposedly very distinct from the other members of the genus *Phoca*. Allen (1880) summarized all of the information on ribbon seals that was available at that time. Surprisingly, this information, valuable mainly because of its references to earlier writings about ribbon seals, did not advance much beyond the point of again describing pelt colour and size and recounting the previously available recorded information on distribution. Little new information had been obtained in the interval between Pennant's account (1781) and the publication of Allen's book in 1880. True (1883, 1884) commented on the osteological characters of ribbon seals.

During the first three decades of the twentieth century there was little new information obtained about ribbon seals. Commercial hunting of marine mammals in the ice-covered regions of the Okhotsk, Bering and Chukchi Seas had all but ceased. The opportunities for observation or study of this seal by persons contributing to the scientific literature were extremely limited. Smirnov (1927) and Naumov and Smirnov (1936) published an account of the systematics and geographical distribution of North Pacific phocids. Pikharev (1939) produced a note dealing specifically with the ribbon seal from the far-eastern waters of the USSR. Inukai (1940) published a note on the ribbon seal based on information from Japanese sealers active in the waters around Saghalien Island which again dealt mainly with coloration and distribution. Arseniev (1941) discussed the feeding regime of ribbon seal.

Wilke (1954) published a brief account of seals from waters around Hokkaido Island, Japan which included methods of capture, species composition of the Japanese harvest, measurements and stomach contents. Ribbon seals were included in this account. In 1955 the noted Soviet biologist, systematist and taxonomist, K. K. Chapskii, published a major paper on the systematics of seals of the subfamily Phocinae.

The tempo of ecological studies of ribbon seals accelerated in 1958 when the Soviet government established a number of laboratories for the study of marine fishes, mammals and oceanography. Investigations of ribbon seals were mainly undertaken by scientists of the Magadan branch of the Pacific Research Institute of Fisheries and Oceanography (TINRO), Soviet scientists including E. A. Tikhomirov (1961, 1964, 1966), A. P. Shustov (1965a, b, c; 1967a, b; 1970; 1972), G. A. Fedoseev (1965, 1966, 1973), Fedoseev and Goltsev (1974) as well as others, have greatly increased our knowledge about this seal.

Recent accounts of ribbon seals by American investigators include those of Burns (1969) dealing with natural history, and Burns and Fay (1970) concerned with comparative morphology. Comparative respiratory physiology of pinnipeds including the ribbon seal has been addressed by Lenfant et al. (1970). Relationships of marine mammals and ice in the Bering Sea are discussed by Burns (1970) and Fay (1974).

External Characteristics and Morphology

The most striking and obvious characteristic of the ribbon seal is its distinctive coloration and pattern of markings. Essentially all ribbon seals older than one year have a series of light bands on a darker background. There are basically four bands: one encircling the neck or head; one encircling the posterior trunk at or behind the level of the naval; and one on each side of the body, broadly encircling the foreflippers or "shoulders". Width of bands is quite variable and on some individuals they coalesce. Indistinct light bands are present on some young seals by twelve weeks after birth but they are usually not obvious until the moult at age one. After the third year of life males are reddish brown to almost black (reddish brown before the moult and almost black immediately after it) with light (approaching white) bands (Fig. 1a). These light bands increase in contrast to the dark background coloration in males up to sexual maturity. Adult females exhibit similar marking but they are much less distinct because of a considerably lighter background coloration (Fig. 1b). These changes in coloration were discussed by Burns (1969) and, in greater detail, by Naito and Oshima (1976).

At birth the pups are covered by a dense coat of near white lanugo (Fig. 3), which is normally completely shed by five weeks of age. From the time pups shed their lanugo until the following moult in March-

FIG. 1 (a) An adult male ribbon seal. June 1, 1968, central Bering Sea, 60° 20′ N, 169° 24′ W. (Photo by J. Burns.) (b) Adult male (foreground) and female ribbon seals aboard the R.V. "Alpha Helix". June 6, 1968. (Photo by J. Burns.)

April, they usually do not exhibit the distinctive coloration of older ribbon seals. They are silver-grey on the lower flanks and belly and blue-black on the upper flanks and back (Fig. 4). Based upon comparison of photographs, this colour phase resembles that of young hooded seals, *Cystophora cristata*.

At birth, ribbon seal pups weigh approximately 10.5 kg and are about 86 cm long. The average weight of weaned pups examined

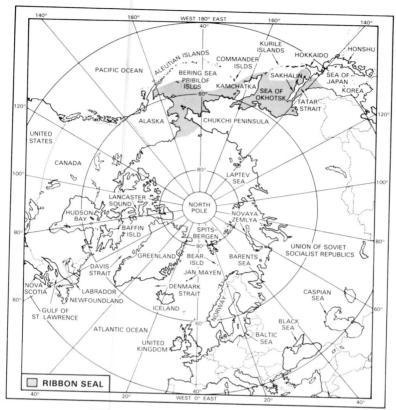

FIG. 2 Distribution of the ribbon seal.

during late May and early June was about 22 kg (15-30.9 kg, $N = 83$) and average standard length was approximately 92 cm (74.0-127.8 cm, $N = 81$). Normal increases in weight and standard length through the first six years of life are as follows: age 1, 33 kg and 106 cm; age 2, 50 kg and 130 cm; age 3, 59 kg and 139 cm; age 4, 61 kg and 144 cm; age 5, 65 kg and 148 cm; age 6, 67 kg and 148 cm. Within all age classes there was great variation in length and weight.

The rate of growth is rapid in comparison to ringed seals, *Phoca (Pusa) hispida* or bearded seals, *Erignathus barbatus*. Using the average length of 151 cm (from seals older than seven years), proportional length within age classes was as follows: weaned pups, 61%; one year, 70%; two years, 86%; three years, 92%; four years, 95%; five years, 98%; and six years, also 98%. The largest ribbon seal that I have

FIG. 3 Newborn ribbon seal pup. Central Bering Sea, April 1968. (Photo by J. Burns.)

FIG. 4 A recently weaned and moulted ribbon seal pup approximately five weeks old. Photo taken near St Lawrence Island, Alaska. May 19, 1967. (Photo by J. Burns.)

examined was a 23-year-old pregnant female taken on 28 March. This seal was 179.7 cm long, girth was 114.3 cm, blubber thickness over the sternum was 6.1 cm and it weighed 148.2 kg. She was supporting a foetus which weighed 7.3 kg.

Distribution

This distinction between total range and effective range of animals, as pointed out by Davies (1958), is pertinent to a discussion of ribbon seal distribution.

The effective range of *Histriophoca* includes the seasonally ice-covered regions of the Okhotsk, Bering and Chukchi Seas and the bays and straits contiguous with them. Total range probably includes all of the Bering Sea including the Aleutian Islands, the western north Pacific Ocean in the region of Kamchatka, the Kuril Islands and the Okhotsk Sea.

Von Schrenck (in Allen, 1880, p. 681) reported these seals from the Gulf of Tartary (Tarter Strait), the southern coast of the Okhotsk Sea and the eastern coast of the Kamchatka Peninsula. Rass *et al.* (1955, cited by Sheffer, 1958) indicated that these seals occurred all along the coast of Siberia from Bering Strait to the Kuril Islands and the shores of the Okhotsk Sea. Bobrinskii (1944) reported their occurrence in the eastern part of the East Siberian Sea.

On the American side, the effective range extends from the vicinity of Point Barrow (Bee and Hall, 1956; Morgan, personal communication) to Unalaska Island in the Aleutian Islands (Scammon, 1874).

Their range, although restricted in the vicinity of the southeastern Kamchatka Peninsula and Kuril Islands, is apparently continuous from the Bering to the Okhotsk Sea. Ribbon seals from both areas are considered to be taxonomically the same (Burns and Fay, 1970; Fedoseev, 1973).

Extreme dispersal of ribbon seals within their effective range is associated with years of unusual ice conditions. The formation of extensive ice in the Bering and Okhotsk Sea results in the occurrence of large numbers of these seals further south than they normally occur. The reverse is also true.

Inukai (personal communication) indicated that along the northern coast of Hokkaido ribbon seals occur in greatest numbers during those years of heavy ice formation. In that region, the ribbon seals drift south during the spring with ice from the central Okhotsk Sea.

Extralimital records of ribbon seals in the eastern North Pacific Ocean include a subadult male killed near Cordova, Alaska in 1962 (L. Johnson, personal communication) and an apparently debilitated male found on the beach near Morro Bay, California (Roest, 1964).

The factors affecting ribbon seal distribution are, at the present time, very poorly understood. These seals are associated with sea ice during the late winter, spring and early summer, but their distribution and activities during the late summer and fall are not well known. They are especially dependent upon the presence of sea ice on which to give birth to their pups, and for hauling out to moult. However, after the moulting period is over, in mid- to late June, they are apparently no longer dependent upon the presence of sea ice.

Concentrations of spotted seals *(Phoca vitulina largha)* and ribbon seals occur in the ice front of central Bering Sea during late winter and spring. However, by the time the ice edge is receding north through Bering Strait, there are usually only a small number of ribbon seals associated with it, although spotted seals are still numerous.

In the Okhotsk Sea there is no permanent ice pack. The various pinnipeds must become either totally pelagic during the ice-free months, or resort to coastal rookeries. Except for occasional individuals, ribbon seals do not haul out on land. Tikhomirov (1961) and Arseniev (personal communication), commenting on ribbon seals in the Okhotsk Sea, have aptly reflected my observations of these seals in the Bering Sea. The ribbon seals disappear when the seasonal sea ice melts and reappear when it forms again in winter. The question of where these seals are during the ice-free months in both the Bering and Okhotsk Seas remains to be satisfactorily answered. I have speculated (Burns, 1969; 1970) that these seals become pelagic. This conclusion was based on several different lines of evidence.

The most obvious and consistent observation is that they are common in the ice front of central Bering Sea and that they are observed less frequently near settlements farther north. Ribbon seals occur regularly in small numbers at hunting sites on St Lawrence Island, but are infrequently observed or taken by Eskimo hunters from villages in Bering Strait, even though the other species (ringed, bearded and spotted seals) occur seasonally with regularity. At the southern Chukchi Sea coastal villages of Shishmaref and Point Hope, they are only occasionally seen or taken while at Wainwright and Point Barrow in the northern Chukchi Sea they are indeed most unusual.

Ribbon seals haul out on moderately thick "clean" ice, of the kind present at the inner zone of the ice front. This unexplained propensity for clean ice, common to ribbon seals in both the Okhotsk and Bering seas has been noted by Eskimo hunters of the Bering Sea islands, by Shustov (1965c), Burns (1969) and by Naito (personal communication). Sea ice of this type seldom occurs near shore.

Compared to the other phocids of the Bering and Okhotsk Seas the behaviour of ribbon seals hauled out on the ice during late winter and spring is most unusual. They are not very wary and they rest for long periods of time without raising their heads. They commonly lie away from the edges of floes, even amidst the rough ice where their field of view is obviously greatly limited. They can be easily approached and exhibit little fear of men or boats, even at close range. These traits have made them particularly vulnerable to hunters operating from large vessels in the ice front.

I have interpreted these habits as being indicative of a species which, by virtue of its winter distribution, is not regularly subjected to harassment or predation by arctic foxes, polar bears or humans. During July through October ribbon seals do not normally occur near shore, nor do they migrate north to the summer margin of the polar ice, in significant numbers as do the ringed and bearded seals. The most likely alternative is that during the months when ice is not present they live at sea. In winter they occupy the southern edge of the ice (the front) again beyond the influence of man or quadruped predators. In most years the winter ice front occurs in the Bering Sea at or near the southern edge of the continental shelf. As long as sea ice persists in the central Bering Sea, over or near the edge of the shelf, the ribbon seals are associated with it.

As a result of observations made during the 1968 cruise of the R.V. "Alpha Helix", I speculated (Burns, 1969) that during the process of spring break up, normally prevailing oceanographic and meteorological conditions resulted in the presence of an extensive zone of sea ice which persisted in central Bering Sea, until it melted more or less in place. This disjunct band in which ribbon seals are most abundant does not normally drift north through Bering Strait. It completely melts by mid-June at which time the seals have completed the moult and live at sea.

Examination of satellite imagery available for each year since 1972 has verified the presence of this disjunct region of drifting ice which occurs comparatively close to the edge of the continental shelf.

The edge of the Bering Sea continental shelf is an extremely rich region. Several conditions contribute to high annual primary production. An indication of this high production is that the region supports one of the major commercial fisheries in the world, centered mainly on pollack *(Theragra chalcogramma)*, herring *(Clupea harengus)* and shrimp (*Pandalus* spp.), all of which are utilized for food by ribbon seals (Lowry *et al.*, 1977, 1978). Although direct evidence is lacking, it has remained attractive to speculate that during the evolution of the complex Bering Sea ecosystem, one of the ice associated phocids became adapted to this extremely productive region in which seasonal ice is present to support those vital biological events requiring a comparatively stable substrate: birth, nursing and moult, and that the species would become pelagic when the ice was not present, exploiting the shelf edge and the adjacent deep waters.

Some records of ribbon seals in Bering Sea during the summer have been recorded, including a young female taken in August 1896, 84 miles west of St Paul Island (True, 1899) and three sightings of single seals, reported by Fiscus (personal communication). These sightings were made on August 24, 30 and 31, 1963 in areas 52 miles NNE, 64 miles NE and 71 miles NE of St Paul Island. Several characteristics of this seal make it virtually impossible to observe them in the open sea, except at close range under ideal conditions.

Abundance and Life History

The preceding discussion indicated that ribbon seals seldom occur near land and are visible for counting only during the spring months when they haul out on the ice. Relatively few ribbon seals were or are taken by hunters along the coasts of Siberia or Alaska. Intensive commercial hunting was initiated in the Bering Sea in 1961 by the far-eastern sealing fleet of the USSR. From 1961 through 1967 ribbon seals were taken in the greatest numbers. The average annual Soviet harvest during that period was around 13 000 animals. Records of the Soviet catch in Bering Sea during 1966 and 1967 indicate a take of 14 582 and 12 520 ribbon seals respectively.

Bychkov (1971), citing Shustov (1965c), stated that the Bering Sea population of ribbon seals numbered 80 000-90 000 animals during the early 1960s.

The effect of this intensive commercial sealing was a noticeable reduction in the population of ribbon seals. A significant reduction in harvest was recommended (Shustov, 1967b). The Soviet harvest was reduced to 6290 ribbon seals in 1968. The quota was further reduced to 3000 in 1969 and the annual catch has remained at about that number to the present time. The harvest in Alaska has been less than 100 ribbon seals annually from 1968 to the present.

Based on aerial survey data obtained in 1969 and 1970, Shustov

(1972) indicated populations of ribbon seals numbering 140 000 animals in the Okhotsk Sea and 60 000 in Bering Sea. More recent information (Goltsev, personal communication) indicates that in Bering Sea there was an increase of 20% between 1972 and 1974. My estimates, based on aerial and shipboard surveys in eastern Bering Sea, indicate a current population of 90 000-100 000 ribbon seals in the Bering Sea.

Births occur over a period of almost five weeks extending from about 3 April to 10 May. However, the majority of births occur between 5 and 15 April. Tikhomirov (1964) indicated the period of "mass pupping" occurred between 10 and 13 April in 1962. Based on several years' observations of differences in sizes and condition of pups observed during May, there is probably some annual variation in dates of peak pupping, which may be related to prevailing weather and ice conditions.

The nursing period lasts from three to four weeks, during which time weight of the pup more than doubles (see previous discussion) and the lanugo is shed. Towards the end of lactation the pups are obese. They appear to be poor swimmers and have difficulty diving, probably due to the great amount of blubber deposited during the nursing period. Weaning appears to be abrupt, occurring when the pup is abandoned by its mother.

During the initial period of independence pups spend a great deal of time on the ice, apparently slowly achieving proficiency at diving and feeding. During this transition period, lasting an additional two to three weeks, there is a significant loss of weight from approximately 27-30 kg when weaned, to an average of 22 kg during early June. The extensive fat reserves accumulated during the nursing period are no doubt very important to the survival of young seals during this important period of adjustment. The attainment of aquatic proficiency coincides with the disappearance of sea ice during mid-June in the Bering Sea. Females continue to feed throughout the lactation period.

Most breeding occurs near or shortly after the time when females wean their pups. As with the birth period, breeding of females having previously borne pups occurs over a duration of several weeks, with a peak during the very last part of April and the first week of May. Females breeding for the first time, or having missed a successful pregnancy the previous year, probably breed outside of this peak period. My data indicate the latest date of ovulation to be 9 May. Defining sexual maturity as the age of initial conception, some females mature at two years of age, approximately 50% successfully conceive at age three and almost 100% are mature by age four (Burns, 1969).

These findings generally substantiated those of Tikhomirov (1966). However, since our respective criterion are different (age at initial pregnancy vs age at first ovulation) I found the proportion of mature females within age classes two and three to be consistently lower. After achieving sexual maturity more than 95% of the females conceive in successive years.

Based on characteristics of body size, baculum lengths, changes in testes volume and microscopic examination of testes, sexual maturity was attained in 22% of three-year-old males, 75% in age class four and 90% of five-year-olds (Burns, 1971). Male ribbon seals are in breeding condition from March to mid-June.

The moult occurs from late March through July. However, there is a considerable difference based upon age (more likely reproductive condition). Subadults complete the annual moult by early to mid-May. Adults begin moulting during the first half of May (Tikhomirov, 1964). The difference in coloration between the adults and subadults observed in May and June is quite striking in that colours of the older seals are faded. As indicated previously, completion of the moult in adult seals normally coincides with the disappearance of seasonal sea ice in the Bering Sea.

The maximum lifespan of ribbon seals may approach 30 years. The oldest ribbon seal in my samples was 23, and maximum age in the extensive samples obtained by Soviet investigators was 26 (Shustov, 1967b). Normal lifespan is probably on the order of 20 years. Less than 1.2% of the ribbon seals in a combined sample of approximately 2500 animals for which ages were determined were older than 20 (Shustov, 1967b; Burns, 1969).

Food habits

The food habits of ribbon seals remain poorly known. There are no data on the diet of seals taken in the Okhotsk and Bering Seas when they are free of ice. No samples are available from early winter and only two are available from mid-winter. Almost all of the ribbon seals examined by Soviet and American scientists have been obtained during March through June, the majority have been taken during May and June. During this period independent feeding activity is greatly reduced. This reduced feeding activity is manifested by a 50-60% decrease in thickness of the blubber and a significant loss of weight. Shustov (1965a) indicated that from the end of March to July, thickness of subcutaneous blubber overlying the sternum decreased from an average of 5.3-2.6 cm.

Adult females undergo most of their weight loss during lactation and appear to feed to a greater extent than the males during the moult. The sex ratio of moulting adults on the ice during mid-day is approximately even. During the night females are mainly in the water, some apparently feeding. Unfortunately, adult ribbon seals usually sink immediately when killed in the water during this period. Therefore, we have made little attempt to collect swimming animals in spring and early summer.

The most comprehensive studies of ribbon seal food habits were those of Arseniev (1941) and Shustov (1965b). In a sample of 1207 seals obtained by Shustov from March through July (94% taken in May and June), food remains were present in the stomachs of only 32 animals. Specimens were obtained in the course of commercial hunting activities and therefore most seals were probably killed on the ice. Shustov (1965a) indicated that of the small number of animals collected in the water the frequency of food remains in the stomachs was greater than in those collected on the ice. In the Bering Sea the food items most frequently encountered in spring were crustaceans (shrimps, crabs and mysids), followed by various fishes and in third place, cephalopods (Shustov, 1965a). Shustov notes further that although pollock, *Theragra chalcogramma* and cod, *Gadus macrocephalus* are major food items of ribbon seals in the Okhotsk Sea they were not encountered in the few seals containing food captured in the Bering Sea.

In two ribbon seal stomachs examined by Wilke (1954) from the Okhotsk Sea, he found pollock in both. One also contained remains of unidentified cephalopods. My sample of these seals from Bering Sea includes only six animals that had food remains in the stomachs. A summary of items found is included in Table 1. Recent and continuing

TABLE 1. Summary of food items in stomachs of six ribbon seals from the Bering Sea

Specimen number	Date of capture	Volume of stomach contents (ml)	Food items
S-12-67	30-IV-67	90	Fragments of shrimp, *Pandalus* sp.
S-37-67	9-V-67	1206	Remains of single fish, *Pholis* sp.
S-63-67	13-V-67	195	Milk
S-233-67	15-V-67	Trace	Digested heads of two shrimp, *Sclerocrangon* sp.
G-50-69	3-II-69	1131	Fish, *Theragra chalcogramma*
S-1-70	II-70	1150	Fish, *Boreogadus saida*

studies by Lowry *et al.* (1977, 1978) indicate that pollock, shrimps, eelpouts (family Zoareidae) and squids are all important to ribbon seals in the Bering Sea during spring.

Of the two seals collected during the winter (G-50-69 and S-1-70) both had been feeding entirely on fish—one on pollock and the other on cod.

The sum of all information on food habits of ribbon seals in the Bering Sea is small and obviously deserving of more intensified investigation.

Internal Characteristics and Morphology

Ribbons seals older than pups are more streamlined in comparison to other Bering Sea phocids. Maximum girth expressed as a proportion of length is as follows: 68% in ribbon seals, 70% in spotted seals, 73% in bearded seals and 89% in ringed seals (Burns, 1971). Is this more streamlined shape an additional indication of a seasonally pelagic existence?

Mean length of intestines is about 2070 cm, or 14.6 times greater than body length. Intestine length was found to be 14 times greater than body length in ringed seals, 16 times in bearded seals and 20 times in spotted seals (Burns, 1971).

The ribbon seal has a short, wide skull (Fig. 5). Compared to other members of the genus *Phoca (sensu lato)* the rostrum is short, the cranium is broad with reduced temporal fossae, the palate is wide with small, widely spaced teeth, the bullae are elongate and thick walled and have widely exposed petrosals. In adult specimens examined by Burns and Fay (1970), mean condylobasal length was 193.7 mm (range 181.1-206.5), mastoid width was 124.4 (112.4-135.7) and zygomatic width was 120.4 (109.6-130.8).

The dental formula is typical of the subfamily Phocinae (Scheffer, 1960):

$$I \frac{1\text{-}2\text{-}3}{0\text{-}2\text{-}3} \, C \frac{1}{1} \, PM \frac{1\text{-}2\text{-}3\text{-}4}{1\text{-}2\text{-}3\text{-}4} \, M \frac{1}{1} \times 2 = 34$$

Forty percent of our sample of 121 ribbon seal skulls exhibited deviations from this formula (Burns and Fay, 1970).

The eyes appear black in colour and are large in comparison to those of other Bering Sea phocids. Average diameter in animals older than pups was 45.2 mm. This compares with a mean diameter of 43.2 mm in ringed seals, 39 mm in bearded seals and 38.5 mm in spotted seals.

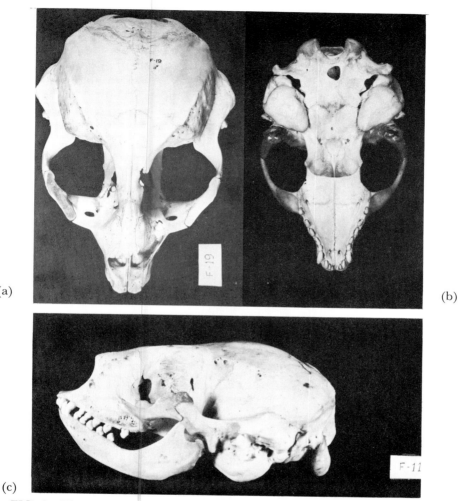

FIG. 5 Three views of ribbon seal skull: (a) dorsal, (b) ventral and (c) lateral.

Fay (1967) reported the number of ribs and thoracic vertebrae in ribbon seals to be 15.

The trachea of the ribbon seal is distinctive. On the average there are 46 tracheal rings, range was 41-50 (Sokolov et al., 1968; Burns, unpublished). The anterior part of the trachea is composed mostly of complete rings. The middle part consists of incomplete rings of which the ends either meet or are connected by a narrow fold of thin membrane. The posterior rings are only partially formed and broadly connected by a tissue membrane on the ventral surface. A slit-like opening, located approximately three-quarters of the distance from the anterior end, is connected to an air sac. This structure occurs in both males and females, although the air sac (Fig. 6) is considerably more developed in adult males (Burns, 1969). Studies by Abe *et al.* (1977) indicate that although the tracheal slit is present in adult females the air sac is poorly developed or in some cases absent.

The air sac overlies the ribs on the right side of the body, extending posteriorly. In adult males the inflated air sac apparently extends across most of the ribs below the level of the foreflipper. Function of this interesting structure is unknown, but it may assist in providing

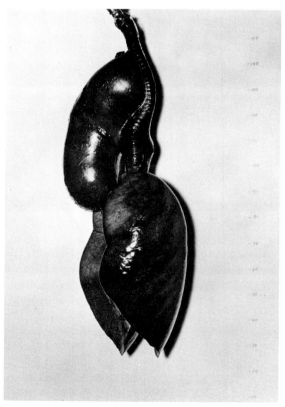

FIG. 6 Excised and inflated trachea, air sac and lungs of a three-year-old male ribbon seal (Spec. No. GS-50-69). (Photo by J. Burns.)

additional buoyancy to what I believe is a seasonally pelagic animal, and/or be involved in the production of sounds such as recorded by Ray (personal communication) and reported by Watkins and Ray (1977).

Physiology

Respiratory physiology of Bering Sea pinnipeds was investigated by Lenfant *et al.* (1970). Their experiments were conducted on seals taken aboard the research vessel "Alpha Helix" within a few hours after capture. Results of their work provide evidence of a greater diving ability in ribbon seals than in spotted seals. Total oxygen storage capacity of ribbon seals, including the lungs, blood and tissues was 78.7 ml kg^{-1}. In spotted seals it was 65.2 ml kg^{-1}. Blood characteristics contributing to this high oxygen storage capacity in ribbon seals included a mean haemoglobin concentration of 24.48 g %, a mean haematocrit of 66.6% and a mean cellular haemoglobin content of 55.1 pg. Respiratory properties of the blood indicated an oxygen capacity of 34.2 volume % and tissue myoglobin concentration was found to be 8.07 g %.

A mean haematocrit of 66.6% for ribbon seals and 58.2% for spotted seals, as reported by Lenfant *et al.* (1970), can be compared with values of 64% for ringed seals, 56.4% for bearded seals and 47.6% for harbour seals, *Phoca vitulina richardsi* (Burns, unpublished). The difference in observed values between the closely related spotted and harbour seals is noteworthy. Spotted seals are more pelagic than the coastal-dwelling harbour seals.

Behaviour

Some aspects of ribbon seal behaviour have already been discussed. These include their unusual tolerance of boats and humans, frequency of resting on ice at considerable distances from water and, compared to other Bering Sea phocids, the comparatively long duration of head-down resting periods.

Ribbon seals are solitary, except that they aggregate in favourable areas. Although several seals are frequently seen on a large ice floe they are usually quite distant from each other.

Throughout the nursing period the pups are left unattended by the females for long periods of time. In fact, most pups I found on the ice throughout the daylight hours were unattended. Mothers will seldom defend their pups from human intruders although they will occasionally remain on an ice floe or watch from the safety of the water. Adult males are conspicuously absent during the birth and early nursing period and apparently only seek out the females as they approach oestrus. Although current information is inadequate, it appears that breeding is polygamous. These observations are in marked contrast to the monogamous and solicitous spotted seal, which also inhabits the ice front (Burns et al., 1972).

Ribbon seals have a very flexible neck which, when extended, appears much longer than that of the spotted, ringed or bearded seals. When aroused on the ice they frequently remain stationary, holding their heads very high while looking for the source of disturbance. Their eyesight in air seems to be poor. This probably accounts for the comparative ease with which they can be approached.

When disturbed, their movement over the ice is surprisingly rapid. Over short distances they can move as fast as a man can sprint. Unlike most phocids which move by wriggling forward, the ribbon seal alternately extends the foreflippers. The neck is extended, close to the ice, and is moved from side to side. The pelvis is also moved from side to side and the hind flippers are held together above the ice. Except for the position of the neck and head, locomotion of ribbon seals appears to be accomplished in a manner similar to that of crabeater seals, *Lobodon carcinophagus* (Kooyman, this volume 230).

References

Abe, H., Hasegawa, Y. and Wada, K. (1977). A note on the air sac of ribbon seal. *Sci. Repts Whales Res. Inst.* **291,** 129-135.

Allen, J. A. (1880). "History of North American Pinnipeds", U. S. Geol. and Geog. Surv. Terr., Washington. Misc. Publ. 785 pp.

Arseniev, V. A. (1941). The feeding of the ribbon seal *(Histriophoca fasciata Zimm.)*. *Izv. TINRO.* **20,** 121-127. (Orig. not seen).

Bee, J. W. and Hall, E. R. (1956). Mammals of northern Alaska. Univ. Kansas. Mus. Nat. Hist. Misc. Publ. No. 8. 309 pp.

Bobrinskii, N. A. (1944). The Order Pinnipedia. *In* "Key to the Mammals of the USSR", (Eds N. A. Bobrinskii, B. A. Kuznetsov and A. P. Kuzakin), 162-168. State Publ. Office, Moscow.

Burns, J. J. (1969). "Marine Mammal Investigations", Alaska Dept. Fish and Game, Juneau. 25 pp.

Burns, J. J. (1970). Remarks on the distribution and natural history of pagophilic pinnipeds in the Bering and Chukchi Seas. *J. Mammal.* **51**, 445-454.

Burns, J. J. (1971). Biology of the ribbon seal, *Phoca (Histriophoca) fasciata*, in the Bering Sea (Abstract). Proc. 22nd Alaska Sci. Conf. p. 135.

Burns, J. J. and Fay, F. H. (1970). Comparative morphology of the skull of the ribbon seal, *Histriophoca fasciata*, with remarks on systematics of Phocidae. *J. Zool. (London)* **161**, 363-394.

Burns, J. J., Ray, G. C., Fay, F. H. and Shaughnessy, P. D. (1972). Adoption of a strange pup by the ice-inhabiting harbor seal, *Phoca vitulina largha*. *J. Mammal.* **53**, 594-598.

Bychkov, V. A. (1971). A review of the conditions of the pinniped fauna of the USSR. *In* "Principles for the Conservation of Nature" (Ed. L. K. Shaposhnikov), 59-74. Min. Agri. SSSR. (Transl. Bureau Foreign Language Division, Canada, No. 0929, 1972.)

Chapskii, K. K. (1955). An attempt at revision of the systematics and diagnostics of seals of the subfamily Phocinea. *Trudy Zool. Inst. Akad Nauk SSSR* **17**, 160-201. (Transl. Fish. Res. Bd. Can., No. 114.)

Davies, J. L. (1958). Pleistocene geography and the distribution of northern pinnipeds. *Ecology* **39**, 97-113.

Fay, F. H. (1967). The number of ribs and thoracic vertebrae in pinnipeds. *J. Mammal.* **48(1)**, 144.

Fay, F. H. (1974). The role of ice in the ecology of marine mammals of the Bering Sea. *In* "Oceanography of the Bering Sea", (Eds D. W. Hood and E. J. Kelly), 393-399. Inst. Mar. Sci., Univ. Alaska, Fairbanks.

Fedoseev, G. A. (1965). A note on the ecology of seal reproduction in the northern sea of Okhotsk. *Izv. TINRO* **57**, 212-216.

Fedoseev, G. A. (1966). Aerial observations of marine mammals in the Bering and Chukchi seas. *Izv. TINRO* **58**, 173-177.

Fedoseev, G. A. (1973). The morpho-ecological characteristics of ribbon seal populations and the grounds for protection of its stocks. *Izv. TINRO* **86**, 148-177.

Fedoseev, G. A. and Goltsev, V. N. (1974). Some morphological adaptations of the Okhotsk Sea seals and their ecological conditionality. Zool. Invest. of Siberia and the Far East. Acad. Sci. SSSR. Far Eastern Sci. Center. Vladivostok. 63-69.

Gill, T. (1873). The ribbon seal of Alaska. *Am. Nat.* **7**, 178-179.

Inukai, T. (1940). A preliminary note on the ribbon seal, *Histriophoca fasciata* (Zimm.) Gill, from the waters of Saghalien. *J. Fac. Sci. Hokkaido Imperial Univ.* Series 6, **7(3)**, 299-303.

Lenfant, C., Johansen, K. and Torrance, J. D. (1970). Gas transport and oxygen storage capacity in some pinnipeds and the sea otter. *Respiration Physiol.* **9**, 227-286.

Lowry, L. F., Frost, K. J. and Burns, J. J. (1977). Trophic relationships among ice-inhabiting phocid seals. *Ann. Rept Outer Continental Shelf Environmental Assessment Program*, contract 03-5-022-53. 59 pp.

Lowry, L. F., Frost, K. J. and Burns, J. J. (1978). Trophic relationships among ice inhabiting phocid seals. *Ann. Rept Outer Continental Shelf Environmental Assessment program*, contract 03-5-022-53. 68 pp.

Naito, Y. and Oshima, M. (1976). The variation in the development of pelage of the ribbon seal with reference to the systematics. *Sci. Repts Whales Res. Inst.* **28,** 187-197.

Naumov, S. P. and Smirnov, N. A. (1936). Data on the systematics and geographical distribution of phocidae in the northern part of the Pacific Ocean. *Izv. TINRO* **3,** 161-178 (Orig. not seen).

Pallas, P. S. (1811). "Seals", Vol. 1, 99-119. *In* Russo-Asiatic Zoological Writings, recording of all the animals observed in the vast Russian empire and adjacent seas . . . Office of the Imperial Academy of Sciences, St. Petersburg. 3 vols. (Partial transl. by L. L. Renner, 1972).

Pikharev, G. A. (1939). The striped seal. *News of the Far East Div. of Acad. Sci., USSR.* **33,** 191-196. (Orig. not seen).

Roest, A. I. (1964). A ribbon seal from California. *J. Mammal.* **45,** 416-420.

Scammon, C. M. (1874). The Marine Mammals of the Northwestern Coast of North America. John H. Carmany and Co., San Francisco.

Scheffer, V. B. (1958). "Seals, Sea Lions and Walruses: A Review of the Pinnipedia", Stanford University Press, Stanford. CA. 179 pp.

Scheffer, V. B. (1960). Dentition of the ribbon seal. *Proc. Zool. Soc. London.* **135,** 579-585.

Shustov, A. P. (1965a). The food of the ribbon seal in Bering Sea. *Izv. TINRO* **59,** 178-183.

Shustov, A. P. (1965b). Some biological features and reproductive rates of the ribbon seal *(Histriophoca fasciata)* in Bering Sea. *Izv. TINRO* **59,** 183-192.

Shustov, A. P. (1965c). Distribution of the ribbon seal *(Histriophoca fasciata)* in the Bering Sea. *In* "Marine Mammals", (Eds E. H. Pavlovskii, B. A. Zenkovich, S. E. Kleinenberg and K. K. Chapskii), 118-121. Akad Nauk SSSR, Ikhtiol. Comm., Moscow.

Shustov, A. P. (1967a). Toward the question of rational exploitation of pinniped stocks of the Bering Sea. *Problems of the North* **11,** 182-185.

Shustov, A. P. (1967b). The effect of sealing on the state of the population of Bering Sea ribbon seals. *Izv. TINRO* **59,** 173-178.

Shustov, A. P. (1970). Geographic variability of the ribbon seal *(Histriophoca fasciata). Izv. TINRO* **70,** 138-148.

Shustov, A. P. (1972). On the condition of the stocks and the distribution of true seals and walruses in the North Pacific (Abstract). Fifth All-Union Conf. on Studies of Marine Mammals, Makhachkala, USSR, Vol. 1, Part 1, 146-147. (Transl. by F. H. Fay, 1975.)

Smirnov, H. (1927). Diagnostical remarks about some seals (Phocidae) of the Northern Hemisphere. *Tromso Mus. Arsh.* **48(5),** 1-23.

Sokolov, A. S., Kosygin, G. M., Shustov, A. P. (1968). Structure of the lungs and trachea of Bering Sea pinnipeds. *Izv. TINRO* **62,** 252-263.

Tikhomirov, E. A. (1961). Distribution and migration of seals in waters of the Far East. *In* "Transactions of the Conference on Ecology and Hunting of Marine Mammals", (Eds E. H. Pavlovskii and S. K. Kleinenberg), 199-210. Akad. Nauk SSSR, Ikhtiol. Comm., Moscow.

Tikhomirov, E. A. (1964). Distribution and biology of pinnipeds in the Bering Sea. *Izv. TINRO* **52**, 272-280.

Tikhomirov, E. A. (1966). On the reproduction of seals belonging to the family Phocidae in the North Pacific. *Zool. Zhur.* **45**, 275-281.

True, F. W. (1883). The osteological characters of the genus *Histriophoca*. *Am. Nat.* **17**, 798.

True, F. W. (1884). On the skeleton of *Phoca (Histriophoca) fasciata*, Zimmerman. *Proc. U.S. Nat. Mus.* **6**, 417-426.

True, F. W. (1899). The mammals of the Pribilof Islands. *In* "The Fur Seals and Fur Seal Islands of the North Pacific Ocean", (Ed. D. S. Jordan), Part 3. 345-354. U.S. Government Printing Office, Washington, D.C.

Wilke, F. (1954). Seals of northern Hokkaido. *J. Mammal.* **35**, 218-224.

Watkins, W. A. and Ray, G. C. (1977). Underwater sounds from ribbon seal, *Phoca (Historiophoca) fasciata. Fishery Bull.* **75**, 450-453.

Zimmermann, E. A. W. Von (1783). Geogr. Ges. Mensch. Allegem. Verbreit. Thiere. vol. 3, p. 277. (Orig. not seen).

5

Grey Seal

Halichoerus grypus Fabricius, 1791

W. Nigel Bonner

Genus and Species

Fabricius, in his memoir on the Seals of Greenland (1791), was first to describe the grey seal, which he named "Krumsnudede sael", or hook-nosed seal. This was latinized as *Phoca grypus*. Fabricius' account was probably based on that of Olavsen of the Út-selur or Wetran-selur of Iceland, which "produces its young on land at the time of the withering of the grass in November" (quoted in Allen, 1880). Nilsson (1820) redescribed the same species as *Halichoerus griseus,* and while the genus *Halichoerus* (Greek = "sea pig") has been retained, it is recognized that Fabricius' *"grypus"* has priority for the trivial epithet. Allen (1880, p. 683) discusses the use of *Pusa* Scopoli 1777 as a synonym for *Halichoerus*, but rejects this, showing that *Pusa* was properly used of the ringed seal. *Halichoerus grypus* is a member of Family Phocidae, and subfamily Phocinae.

There is no type specimen.

Halichoerus is a monotypic genus and within the single species no subspecies have been described, though both geographical and seasonal breeding isolation occur.

Amongst English-speaking people, *Halichoerus grypus* is generally known as the grey (gray) seal, no doubt a reflection of Nilsson's name *"griseus"*, though the term "horse-head" has been used in Canada and "Atlantic seal" in England. Names in languages other than English are: "havert" (Norwegian); "út-selur" (Icelandic); "grasel" (Swedish); "grasael" (Danish); "grijze zeehond" (Dutch); "kegelrobben" (German); "phoque gris" (French); "seriy tyulen" (= grey seal) or "dlinnomordiy tyulen" (= long-nosed seal) (Russian).

External Characteristics and Morphology

The grey seal is the largest of the Phocinae. (For sizes, see Table 1.) The adult male is about half as big again as the female, and some individual bulls may be three times the weight of a small adult cow. Amongst other phocids, only in the elephant seals is there a greater

TABLE 1 Dimensions of a sample of grey seals taken during the breeding season at the Farne Islands, Northumberland

		Range	$\bar{x}$	SD	n
Adult males					
Nose-to-tail length		195-230 cm	207.3	9.5	25
Total weight		170-310 kg	233.0	37.6	25
Adult females					
Nose-to-tail length		165-195 cm	179.6	7.4	25
Total weight		105-186 kg	154.6	24.1	25
Pups					
Length at birth		90-105 cm			
Weights[a]					
New-born	males	12.3-19.9 kg	15.8 kg		25
	females	10.9-19.5 kg	14.8 kg		26
Moulted	males	17.2-51.3 kg	39.8 kg		26
	females	19.5-47.7 kg	39.6 kg		27

[a] Data from Boyd and Campbell (1971)

degree of sexual dimorphism. Bartholomew (1970) has drawn attention to the correlation of sexual dimorphism with land-breeding and polygyny.

Besides the difference in size, there are differences in shape between males and females. The shoulders of an adult male grey seal are very massive, while the skin in this region and over the chest is much scarred and thrown into heavy folds and wrinkles. Head-shape also is characteristic. The snout of an adult male is elongated, with a convex outline above the wide and heavy muzzle. This produces a "Roman nose" or equine appearance, the origin of Fabricius' term "hook nosed" (*grypus* in Latin), and the Canadian colloquial name "horse-head". In the female the profile is flatter and the muzzle more slender (Fig. 1).

Although the term "grey" is accurate for most grey seals there is much individual variation. Some bulls are almost completely black, while some cows are predominantly creamy white with only a few scattered black markings on the back. There are sexual differences in both colour and pattern (Hewer and Backhouse, 1959). Apart from the generally darker tone of the back, which shades into a lighter-coloured belly (more noticeable in the female) there are two tones of colour pattern, a lighter and a darker. In males the darker tone is more extensive, forming a continuous background with lighter patches or reticulation; in females the lighter tone is continuous with the darker forming spots or blotches, usually more densely distributed on the back. Immature males and females have a much less pronounced pattern than the adults and a yearling grey seal near the moult may be almost uniformly fawn in colour.

Sometimes there is a reddish or orange tinge on the head, neck, belly and flippers, but this is not universally present and it is not known whether it is genetically determined (Backhouse and Hewer, 1960; Hook, 1960).

Pups are born in a creamy white and rather silky fur, occasionally with some pigmentation in the tips of the hairs, giving a smokey tinge to the fur. The white coat is shed after two to three weeks and is replaced by a second coat which nearly matches the adult pattern. The sexual differences in pattern are already distinguishable at this stage. Partial, or complete, pre-natal moult has been recorded (Davies, 1949). At the Farne Islands, and less frequently at Orkney, the Hebrides, North Rona and Pembrokeshire, a small number of pups moult into a very dark melanistic pelage. These dark pups are generally, but not invariably, males. It is not known whether they remain dark-coated throughout their lives.

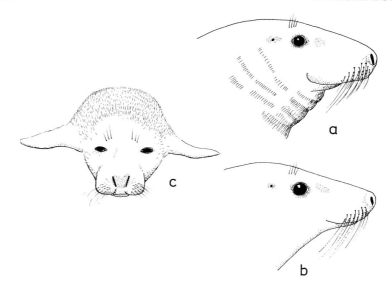

FIG. 1 (a) Profile of mature male grey seal; (b) profile of mature female grey seal; (c) head-on view of a new-born pup: note position of nostrils and heavy muzzle; (from photographs). (d) Adult female grey seal with new-born pup (Photograph: Sea Mammal Research Unit, Cambridge).

Five claws are conspicuous on the fore limb. They are long and slender and overhang the tips of the digits by 2-3 cm.

The nostrils are almost parallel slits, separated below by a conspicuous gap (Fig. 1(c)). In the Common seal, *Phoca vitulina,* the nostrils are inclined and nearly joined at the base (Wynne Edwards, 1954) and this characteristic can be used to distinguish the two species where both occur together and when only a front-view of the head of the seal is to be seen out of the water.

Distribution

The distribution of grey seals has been reviewed by Smith (1966) and Bonner (1972). Mansfield and Beck (1977) have provided an up-to-date account of their distribution and status in the western Atlantic.

There are three distinct populations centred on the Baltic Sea, the eastern North Atlantic and western North Atlantic (Davies, 1957). These are separated both geographically and physiologically, by differences in breeding season. Probably the three groups were united in the last interglacial period, when there would have been a continuous population from the Labrador coast across Greenland and

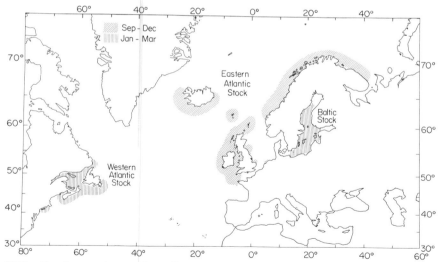

FIG. 2 Approximate breeding distribution of grey seals. Outside the breeding season the animals, particularly newly weaned pups, may stray much further (after Bonner, 1972).

Iceland to North Norway. In the last glacial period, about 100 000 years ago, the southward extension of the ice separated the western Atlantic group from grey seals on the western seaboard of France and the Iberian Peninsula. With the retreat of the ice, the eastern Atlantic group spread northward into its present range. The isolation of the Baltic Sea, as the Ancylus Lake, about 9000 years ago, separated the Baltic population from the other grey seals of the eastern Atlantic. By this time, differences in the breeding season had already developed and did not allow the two groups to mix when the Kattegat opened, connecting the Baltic with the North Sea.

Davies (1957) thinks the differences in behaviour, development and morphology are significantly greater in the eastern Atlantic group than in the other two, but concludes that there are insufficient grounds to warrant sub-specific distinction. However, further examination of material might allow such distinction to be made (Hewer, 1974).

Another view holds that in early historical times, the grey seal was a much commoner animal than it is now, and migration could have occurred to and from the American continent by way of South Greenland, and also between Iceland and western Europe, so that there was no discontinuity of distribution (van Bree, 1972).

It is scarcely possible in the light of present knowledge to choose between these theories. Wide dispersal, but no migration, is characteristic of grey seals, but we do not know enough about the genetic composition of the various stocks to speculate about the dynamics of gene-flow between them.

The western Atlantic group

These seals are found around the coasts of Newfoundland and Nova Scotia and in the Gulf of St Lawrence. The limits of distribution are Cape Chidley in the north of Labrador and Nantucket Island, RI, in the south (Mansfield, 1967). Breeding colonies occur at the Magdalan Islands (Deadman Island), Amet Island in Northumberland Strait, Sable Island (a large and rapidly expanding colony), Point Michaud on the east coast of Cape Breton Island, and on the fast-ice of Northumberland Strait and George Bay. During late spring and summer there is a dispersion away from the breeding areas, and large concentrations of grey seals are seen at Miquelon Island, south of Newfoundland, and in, for example, the Miramichi estuary.

Grey seals have probably occurred as stragglers in Greenland, but have not been reliably reported in the last 100 years (Kapel, quoted in Mansfield and Beck, 1977).

The eastern Atlantic group

Grey seals occur rather sparsely in Iceland, particularly on the west coast (Hook, 1961; Arnlaugsson, 1973). They are much more abundant in the Faroe Islands. In Norway breeding colonies are found on islands off the coast of Møre and northwards towards North Cape, the greatest number occurring in Nordland and Trondelag (Øynes, 1964). Small numbers of grey seals found on the Murman coast near the mouth of the White Sea (Karpovich et al., 1967) represent the most easterly extension of the range.

The majority of the eastern Atlantic Grey seals breed around the coasts of the British Isles (Bonner, 1976). The largest group is found in the Outer Hebrides, particularly on the islands of the Monach group and Gasker. Smaller breeding groups are found on Shillay, Coppay, Causamul and Deasker. The Inner Hebrides support a much smaller stock of seals, but some colonies such as those at Treshnish and around Oronsay have figured in the literature (e.g. Hewer and Backhouse, 1960). The isolated islet of North Rona, which lies about 65 km north-west of Cape Wrath (the north-western extremity of Scotland) supports a large and relatively stable stock of grey seals. The Orkney archipelago has another large population. Although some grey seals breed in the Pentland Firth, most are found on the northern isles, particularly the Greenholms and Little Linga. In Shetland the grey seal is far less abundant than in Orkney, but it is widespread with concentrations on the coast to the west of Ronas Hill and on the isle of Fetlar. The Farne Islands, off the Northumberland coast, are an important breeding site for grey seals. They are, in fact, the only breeding site on the east coast of the Scottish and English mainland, apart from very minor sites in Caithness and on Scroby Sands, off Norfolk. Grey seals occur in small numbers at a large number of scattered sites in the south-west of England and Wales. Summers (1974) has described their distribution in Cornwall and Scilly and Anderson (1977) has given details for Wales. There is a small breeding colony on the southern tip of the Isle of Man (Farmer and Roberts, 1970). Grey seals are found sporadically round the coast of Ireland (Lockley, 1966). Most occur on the west coast, but some are found on the Saltee Islands in the south and at Lambay Island near Dublin. In France a few grey seals are found in Brittany (Roux, 1957) and occasionally individuals or small groups are found off the continental European North Sea coast, as in the Wattenmeer of Schleswig-Holstein.

The Baltic group

The grey seal is now much less common in the Baltic Sea than it was quite recently and it is likely that its present range is smaller. It is to be found over the whole of the Gulf of Bothnia along the south-western shore of the Gulf of Finland and south to the Gulf of Danzig with a coastal distribution westwards as far as the border between Poland and Germany; on the west side of the Baltic the grey seal is found as far south as Oland (Hook, 1964). On the Swedish coast the range extends to Malmo in the extreme south of Sweden and with a further extension up the west Swedish coast, though some of these seals are of British (or Norwegian) origin (Curry-Lindahl, 1965). Wolk (1969) has given an account of the distribution in the southern Baltic. Grey seals bred in Danish waters in historical times (Møhl, 1970), but are now extinct.

Migration and Dispersal

Grey seals are not known to make definite and regular migratory movements. However, marking schemes have shown wide dispersal of weaned pups from their natal rookeries around the British Coasts.

Remarkable distances are covered by pups. A moulted pup marked at Sable Island on 5 February 1973 was recovered at Barnegat Light, New Jersey, on 2 March, a distance of 1280 km in 25 days, or an average minimum rate of travel of about 50 km per day (Mansfield and Beck, 1977). Less far, but speedier, was a pup marked on the Isle of May, in the Firth of Forth, Scotland, and recovered nine days later at Karmøy, Norway, a distance of 580 km, or an average minimum speed of about 65 km per day (Hickling, 1962).

Movements of adult seals are less well attested. Probably most seals return to breed at the rookery of their birth, but the establishment of new rookeries is clear evidence that this is not always the case. The observation of seals branded as pups at North Rona breeding on the Monach Isles (Harwood *et al.*, 1976) was the first published record of intercolony movement by adult seals, though records of pups branded at the Basque Islands in 1964 being seen as breeding adults on Sable Island in 1968, 1969, 1971 and 1973 are given in Mansfield and Beck (1977).

Major seasonal movements occur when seals congregate for the breeding season, and disperse when it is over, but these do not appear to be directional in nature.

GREY SEAL

Abundance and Life History

In the last 50 years several populations of grey seals have shown dramatic increases. Some of these increases have been the result of better investigative methods revealing previously unknown seal colonies, as in parts of Canada, but others reflect real changes in abundance. Thus the pup production at the Farne Islands increased from 751 in 1956 to 2010 in 1971 (Bonner, 1975a) and at the Monach Isles in the Outer Hebrides pup production rose from about 50 in 1961 to 1400 in 1974 (Bonner, 1976). There has been a four-fold increase in pup production in Canada in the ten years since 1966 (Mansfield and Beck, 1977). Figure 3 shows the changes at the Farne Islands and Sable Island.

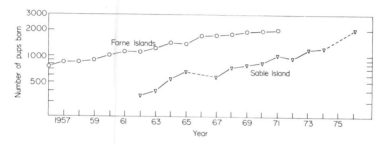

FIG. 3 Annual pup production of grey seals at the Farne Islands and Sable Island. After 1971 the numbers of pups born at the Farne Islands were affected by controlled killing of breeding females (data from Bonner, 1976 and Mansfield and Beck, 1977).

With such major changes taking place, it is difficult to provide firm figures for the various stocks, particularly for some key areas such as the Faroe Islands and the Baltic Sea, where no adequate surveys have been made. However, most of these populations (about 80%) in Canada and the United Kingdom are well known and reliable figures are available for the various times at which the surveys were made.

Table 2 shows the approximate status of stocks, from which a world population of about 120 000 grey seals is deduced, of which about 25% are found in Canada and slightly more than 55% in the United Kingdom.

TABLE 2. Status of grey seal stocks

	Pup production (thousands)	All-age population (thousands)	Notes
1. *Western Atlantic Stock*			
Gulf of St Lawrence	3.8	17.86	Mansfield and Beck (1977)
Eastern shore (Cape Breton Island to Halifax)	0.6	2.82	(All-age totals obtained by adding pups and older seals totals)
Sable Island	2	9.4	
Total		30	
2. *Eastern Atlantic Stock*			
Iceland		8-10	Einarsson (personal communication)
Faroes		3	Smith (1966) (no recent estimate)
United Kingdom			Summers (1978)
SW Britain	0.8-0.9	3.0	
Inner Hebrides	1.2	5.0	
Outer Hebrides and North Rona	9.4	35.0	
Shetland	1.0	3.5	
Orkney	4.5	14.0	
Farne Islands	1.425	8.5	
Subtotal		69.0	
Ireland		2	Lockley (1966)
Norway		2-3	Øynes (1964), Summers *et al.* (1978)
USSR		1-2	Karpovich *et al.* (1967) (adjusted)
Total		85-89	
3. *Baltic Stock*		5	Smith (1966) (no recent estimate)
World Total		120-124	

Population structure

Age-determination in grey seals is based on the study of incremental layers in the cementum of the canine teeth (Hewer, 1964). Maximum ages for wild grey seals are: females 46 (Shetland—Bonner, 1971a); 44 (Canada—Mansfield *et al.*, 1977); males, 26 (Farne Islands—Platt *et* al., 1975) 30 (Canada—Mansfield et al., 1977). A male died at age 41 after living in captivity at Skansen zoo (Mohr, 1952). Most females become pregnant for the first time at age four or five and bear their first pup a year later (Table 3). Males do not contribute to the breeding population till age eight and even then form only a small proportion of the breeding stock, the bulk of the breeding bulls being age 12-18 (Platt et al., 1975). Sex ratio at birth is near to unity (Hewer, 1964; Mansfield and Beck, 1977), but the ratio is not constant through the breeding season, more males being born early and more females later at the Farne Islands (Coulson and Hickling, 1961). At North Rona more males than females are born, but because of differential mortality, more female pups leave the rookery for the sea (Boyd and Campbell, 1971).

Based on ages at sexual maturity, a sex ratio of unity and arbitrary pregnancy and mortality rates, Hewer (1964) produced life tables for the grey seal. These were designed for a population with zero growth and thus represented an artificial concept, since all investigated grey seal colonies show a state of change (generally an increase). Life tables based on survivorship, reproductive rate and a sex ratio of 52.2% males were prepared from random samples of grey seals collected in Canadian waters. These show that a cohort of 1000 pups is produced by a total population of 3688 male and female seals and gives a better idea of population structure than Hewer's (1964) equivalent figure, 3941.

Such ratios can be used to calculate grey seal total population from counts of pups. Hewer (1964) suggested using a multiplier of 3.5 with pup counts (since the ratio varied from 3.1 at the beginning of the season to 3.9 at the end); Mansfield and Beck (1977) used 3.7 without further adjustment. However, the multiplier to be used will vary with the rate of increase of the population which is most likely to be determined by variations in adult and juvenile survival; for most populations a value of 3.5-4.5 will be appropriate (Harwood and Prime, 1978). Total pup production can be calculated from a single pup count, provided the count is classified into morphological age classes, enabling the form and timing of the birth-rate curve and estimates of pup mortality rates to be determined (Radford et al., 1978).

Habitat

The feeding range of grey seals, and particularly the extent to which they travel out from the coast, is not well known. It seems likely it is a general dispersion from the breeding areas. Outside the breeding

TABLE 3 Factors affecting grey seal populations

Location	Adult female mortality	First year mortality	Pregnancy rate Age[a]					Sample size	Population change	Authors
			3	4	5	6	7			
UK	6.6%	60%	0	0	80%	80%	80%	—	Assumed static	Hewer (1964)
Canada	13.5%[b]	49% F 55% M	16	71	89	85	85	76	?	Mansfield and Beck (1977)
UK	6.5%	34%	0	17	60	90	90	1011	Increasing 7%	Harwood and Prime (1978)
Faroe	6%	—	5	19	59	86	87	109	?	Prime (1978)

[a]The ages are the ages at which the seal *becomes* pregnant
[b]Increasing after age 25

season grey seals are most often sighted off rocky or cliffy shores, or around small islands, often where currents run strongly. However, sightings in estuaries, particularly when salmon are present, are also frequent.

Grey seals most characteristically breed on rocky shores, or on small offshore inlets; a very small proportion breed in sea-caves. Hewer (1960) and Bonner (1972) have described the principal breeding grounds. Breeding on sandy beaches was regarded as exceptional until the large grey seal colony at Sable Island, a narrow sand bar 40 km long by 1.5 km wide, east of Nova Scotia and south of Labrador, was discovered. However, the now-extinct Danish breeding colony at Anholt (Bynch in Møhl, 1970) seems to have been situated on sand, as is the small extant colony at Scroby Sands, Norfolk. Ice-breeding grey seals are found in the Baltic (Hook and Johnels, 1972; Curry-Lindahl, 1975) and in the Gulf of St Lawrence (Mansfield and Beck, 1977).

Feeding habits

Grey seals are mostly coastal in their feeding habits. They feed principally on a wide variety of fish, but smaller quantities of crustacea and molluscs (chiefly squid and octopus) are taken. The food taken by grey seals has been described by Rae (1968 and 1973) for Scottish waters, and by Mansfield and Beck (1977) for Canadian waters. In Scotland, salmon *(Salmo salar)* was the most important prey item, being found in 26% of stomachs with recognizable food remains. Rae found that the stomach contents of seals taken both in and away from salmon nets were similar thus the high proportion of salmon was not the result of biassed sampling. Cod *(Gadus morhua)* was the next most important item, occurring in 21% of the stomachs (all gadoids together occurred in 47% of stomach). Herring *(Clupea harengus)* were found in only 1.7% of stomachs. Mansfield and Beck (1977) present their data for all stomachs examined, but when values are transformed for percentage occurrence in stomachs containing recognizable food items, it is found that the principal food items for Canadian seals are herring (24.5%) with cod (17.9%), pleuronectids (15.3%) and skates (14.8%) also being important. In Canada the most important species all the year round are the benthic skates and pleuronectids, but herring, cod, squid and mackerel become of great importance when they begin their inshore migrations and this is reflected in the final order of importance. Salmon form a small (4.1%) part of the diet of Canadian seals; conversely, pleuronectids are unimportant (4.5%) in Scottish grey seals (Table 4).

TABLE 4 Percentage occurrence of various items in grey seal stomachs containing recognizable food remains. Data from Rae (1968) and Mansfield and Beck (1977)

	Scottish seals (sample size 176)	Canadian seals (sample size 196)
Salmon *(Salmo salar)*	26.1	2.6
Cod *(Gadus morhua)*	21.6	17.9[a]
Herring *(Clupea harengus)*	1.7	24.5
"Flounders" (Pleuronectidae)	4.5	15.3
Skates *(Raia* spp.)	2.8	14.8
Other fish	32.4	49.0[b]
Cephalopoda	11.9	8.6
Crustacea	13.1	1.0

[a] Given as *Gadus* spp in Mansfield and Beck (1977)
[b] This total contains 22.9% of unidentified fish

There is no evidence that the young of the year feed differently from older seals. Small prey items such as sand-eels (*Ammodytes* sp.) are eaten by large and small seals alike.

Grey seals fast during the breeding season, the cows for about 3 weeks, and the bulls for much longer; they fast or feed at a reduced rate during the moult. As a result, they may be said to feed for the equivalent of three-quarters of the year. From observations on the average weight of food in the stomach and the inference that the seals eat only once daily, it has been calculated that females ate about 4% and males 4.5% of body-weight per day; there is little difference in proportional food intake between adults and young. This is equivalent to a daily intake of 5.7 kg food per "average seal", slightly less than previous estimates (Mansfield and Beck, 1977). Grey seals can on occasion eat much larger meals; a bull grey seal has been recorded whose stomach contained 10 kg of salmon flesh and bones (Rae in Bonner, 1972).

Relation to fisheries

Grey seals cause appreciable damage to fisheries where they occur in large numbers (Rae, 1960; Bonner, 1972; Söderberg, 1975; Mansfield and Beck, 1977; Parrish and Shearer, 1977). The damage tends to be most severe where set nets or other forms of fixed gear are used.

Seals cause potential damage to fisheries by eating fish. Canadian grey seals are calculated to consume 47 000 tonnes of fish a year; by the same calculation the consumption of fish by British grey seals would amount to about 180 000 tonnes per year. Another estimate, based on an average food consumption of 15 lb (6.80 kg) of which two-thirds is fish, throughout the year, is 112 000 tons, of which about half would be marketable fish worth (at 1974 prices) about £13-£17 million (Parrish and Shearer, 1977). Obvious damage is caused when seals take fish from nets, lines or traps. Gilled salmon and cod are particularly vulnerable. Scottish salmon fishermen are sometimes troubled by individual seals (often reputed to be old bulls) who raid the nets. Of all salmonids taken by fixed nets on the Scottish east coast 20.3% were killed by seals, mostly grey seals (Rae and Shearer, 1965). Nets set for schooling fish in the spring runs of mackerel in Canada can be emptied of fish by grey seals. Once the seals have located a net, it must be moved to another site, or the seals will continue to take all the fish from it (Mansfield and Beck, 1977). Grey seals have been claimed also to strip fish, usually cod, from hooks during line-hauling. Seals may enter traps set for lobsters. It seems that this is probably to eat the bait, as the use of baits salted for at least a week reduces the damage. It is doubtful if grey seals eat many lobsters. Although seals are often seen with lobsters in their mouths, Mansfield and Beck (1977) believe that the seals were trying to shake off the lobsters which had seized their noses or lips. Grey seals are known to eat octopus, which are predators on trapped lobsters in Orkney.

Grey seals can cause consequential damage to fisheries by acting as vectors of the cod-worm *Phocanema decipiens* (see Diseases and Parasites).

Predators

There appear to be no major natural predators of grey seals. Sharks of various species or killer whales *(Orcinus orca)* may take occasional seals, but it is unlikely that either of these significantly affect grey seal population. Man is, and has been since prehistoric times, the principal predator, though in recent years human predation has lessened (see Conservation).

Diseases and Parasites

Most reports of diseases in grey seals in the wild refer to pups. In these cases starvation, caused by separation from the mother, is probably the primary factor, pathological states supervening as the pup's resistance is lowered by starvation (Bonner, 1970, 1972). However, pneumonia was a primary epizootic in grey seal pups at St Kilda (Gallacher and Waters, 1964). Infected eyes are of common occurrence. Conjunctivitis might be caused by the pup's habit of smearing virulent pus from the nose over the eye when scratching (Gallacher and Waters, 1964). *Neisseria* sp., *Staphylococcus albus, Pseudomonas aeruginosa,* and a variety of yeasts have been cultured from infected eyes (Appleby, 1964; Bonner, 1970). Haemolytic and non-haemolytic staphylococci and α and β haemolytic streptococci, *Corynebacterium* sp. and *Neisseria* sp. have been isolated from both infected and healthy eyes of Farne Island seal pups (Bonner, 1972). Septicaemia associated with *Pasturella multicida* and *P. haemolytica,* and *Escherichia coli* has been found in starved Cornish seal pups (Bonner, 1970).

There are few accounts of adult pathology. Scheffer (1958), quotes Flemming writing in 1828, who reported an epizootic of seals (unspecified, but probably grey seals) in the north of Scotland. Uterine tumours (leiomyomata and endometrial carcinomata) have been found in an aged grey seal (Mawdesley-Thomas and Bonner, 1971). Young and Lowe (1969) found stomach lesions caused by repeated penetration by nematode larvae (see below). There is a high incidence of dermal lesions associated with *Corynebacterium phocae* in grey seals from the Dee Estuary (Anderson *et al.*, 1974). The lesions were associated with high stomach nematode burdens. Similar lesions have been noted in *Phoca vitulina* in the United Kingdom (Anderson *et al.*, 1974), the Waddensee (van Haaften, 1962) and the Wattenmeer (Drescher, 1978).

Lice, *Echinophthirius horridus,* are common on grey seals. They do not appear to cause any pathological symptoms but very high lice infestations are usually associated with poor nutritional status. The nasal mite *Halarachne halichoeri* is commonly found in the nose and upper respiratory tract.

Numerous species of nematodes, acanthocephala, cestodes and trematodes have been reported from the gut, lungs, liver and kidneys (Bonner, 1972). Only exceptionally do they seem to cause pathological symptoms. One nematode, the anisakine *Phocanema* (= *Terranova* = *Porrocaecum) decipiens,* is of economic importance. This is the codworm, whose larvae live in cod and related gadoid fishes, while the adults mature in the seal's stomach when infected fish are eaten. The appearance of the larvae in fillets makes the fish unsaleable, unless the larvae are removed manually, an expensive process. *Phocanema* occurs in other species of seals, but is most abundant in grey seals. Control action against grey seals in Canada and the United Kingdom is based in part on the damage to fisheries caused by the spread of this parasite.

Internal Characteristics and Morphology

The skull

The skull of the grey seal is characterized by its long, high and wide snout. Proportionately, no other phocid, apart from the elephant seals, shows such a development of the anterior part of the skull. King (1972) has discussed many of the features of the skull. The enlargement of the snout is associated with an elevation of the fronto-nasal area, with a consequent increase in the length of the premaxillae, which ascend to reach the nasals. The narial openings are slightly key-hole shaped, a tendency carried further in *Cystophora*, a genus which also has an enlarged snout. The maxillae are swollen, increasing the width of the snout. The palate is strongly arched and its posterior border rounded. Despite the increased length of the maxilla the proportional length of the tooth-row is rather low. The teeth themselves are stout and simple. The dental formula is:

$$I\frac{3}{2} \; C\frac{1}{1} \; PC\frac{5}{5} \; (\text{Often}\frac{6}{5})$$

The cheek teeth are large and strong, nearly circular in cross-section and each with a single conical cusp. Secondary cusps are insignificant, usually only on the 5th upper and 4th and 5th lower postcanines.

General accounts of the anatomy of the grey seal date from the nineteenth century (Nehring, 1883; Hepburn, 1896; and Broch, 1914). The reproductive system conforms to the general phocid pattern (Harrison *et al.*, 1952; Amoroso *et al.*, 1965). There is a remarkable hypertrophy of the testis, prostate and ovary in late foetal and newborn grey seals, possibly mediated by placental gonadotrophins. Grey seal milk has the following composition (Amoroso *et al.*, 1950):

Total solids	Solids not fat	Fat	Lactose	Protein	Ash
67.7%	14.5%	53.2%	2.6%	11.2%	0.7%

Some endocrine organs (thyroid, adrenal and pituitary) are described by Amoroso *et al.*, (1965). The forelimb is used extensively for locomotion on land by grey seals; its function and anatomy are discussed by Backhouse (1961). Embryology and foetal growth are described by Hewer and Backhouse (1968). The somatic chromosomes of the grey seal have been studied in cultures of lung tissue by Árnason

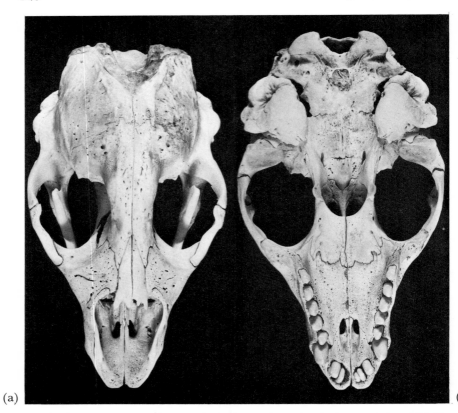
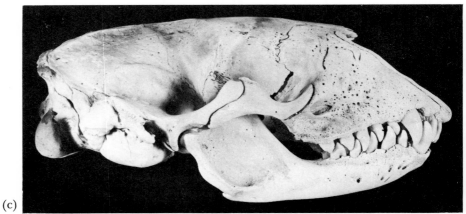

FIG. 4 Skull of male grey seal, *Halichoerus grypus*. Craniobasal length: 249 mm. (a) Dorsal view; (b) ventral view; (c) lateral view.

TABLE 5 Viscera lengths and weights of three grey seals from Scroby Sands, Norfolk (original data)

Specimen no.	Standard length (cm)	Total weight (kg)	Digestive tract lengths (cm)						Viscera weights (g)					
			Oesoph-agus	Fundus of stomach	Pyloric limb of stomach	Small gut	Large gut	Total	Liver	Kidneys	Lungs	Heart	Spleen
HG 22, M 24 April 1968 Age 4 months	109	32	45	17	8.5	1700	52	1823	—	307	—	280	155
HG 23, M 18 May 1968 Age 5 yr 5 months	134	41	52	24	13	1830	57	1976	2405	435	1050	390	200
HG 24, F (pregnant) 20 May 1968 Age 10 yr 5 months	167	82	71	23	15	2050	64	2223	2690	495	1100	645	197

(1970). The chromosome number is $2n = 32$. Árnason concluded the karyotypes of *Halichoerus*, *Phoca* and *Pusa* (which are similar) had evolved from a primitive karyotype with $2n = 34$, found in *Erignathus* and *Leptonychotes*, by fusion between two chromosome pairs. Electrophoretic investigation of blood protein systems in grey seals have been reported by Bonner and Fogden (1971), McDermid and Bonner (1975) and Heath (1978). Only small numbers have been sampled, however.

Some dimensions of viscera are given in Table 5.

Reproduction

The annual cycle

No other phocid shows such a wide range in the timing of the breeding season as does the grey seal. Not only do the three principal stocks differ, but there is much variation within them. Figure 5 shows in diagrammatic form the timing of the main events for a typical grey seal colony, such as that at Orkney in the eastern Atlantic stock. Aggregation at the breeding sites is followed by births, which occur in the autumn, with a peak in October. The dates of first and last birth vary considerably, even at a single colony, and Coulson and Hickling (1964) suggest using the standard deviation of the mean date of pupping to indicate the spread of the season, thus avoiding errors inherent in using the time interval between births, as well as in comparison of large and small groups.

Birth is followed after about three weeks by copulation, at the end of the short lactation period, soon after which the cows leave the breeding grounds. After impregnation the blastocyst remains free in the lumen of the uterus for a mean period of 102 days (Hewer and Backhouse, 1968). Attachment of the embryo takes place on about 21 February (range 10 February to 4 March). For most of this period the pregnant cow is at sea feeding, but shortly before implantation the females haul out to moult, the peak of moulting occurring at the beginning of February. It seems likely that implantation is consequent on the completion of the moult.

The male moult is later than that of the female, with a peak in the middle of March.

There is some evidence of aggregation during the moulting period and the seals spend long periods hauled out ashore, although they do

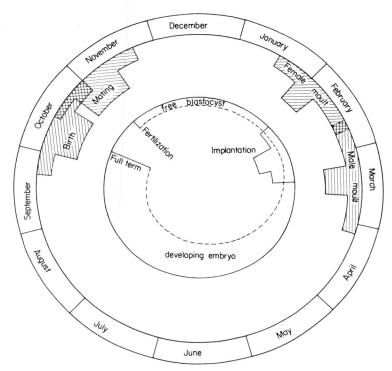

FIG. 5 Diagram of the main events in the annual cycle of grey seals from a colony such as that at Orkney in the eastern Atlantic stock.

not become as wholly terrestrial as they often are during the breeding season. Breeding sites are not usually used for moulting.

Although the situation at Orkney is typical for most British grey seals, births occur slightly earlier in the south-west (Summers, 1974; Anderson, 1977), and markedly later in the Farne Islands, from the fourth week of October to the third week of December (Coulson and Hickling, 1964), with most occurring in the first fortnight of November. Bonner and Hickling (1971) give details of the spread of birth. The very small breeding colony at Scroby Sands off the Norfolk coast produces a dozen or so pups, usually in the last days of December or early in January. It is tempting to view the situation in Great Britain as representing a cline, with the earliest breeding in the Scilly Isles, and the latest at Scroby Sands. If indeed there is such a cline, there is very marked discontinuity between the stocks at Orkney and the Farne Islands.

Grey seals at Iceland and Norway have a peak of pupping in October.

The Western Atlantic stock of grey seals breeds mostly from mid-January to mid-February on islands or on fast-ice near the shore. The moult occurs in May when schools of cod, mackerel and herring move in-shore (Mansfield and Beck, 1977).

In the Baltic grey seals breed on the ice in late February or early March (Curry-Lindahl, 1975); there seems to be little variation with latitude. Historical records indicate a January breeding season for the now-extinct Danish grey seals at Anholt (Bynch in Møhl, 1970).

Davies (1957) put forward three hypotheses to explain how the differences in breeding season, and consequent reproductive isolation, had come about. The first suggested that the eastern Atlantic grey seals represented the original environment and breeding season and that the seals of the western Atlantic and Baltic had become adapted to seas that were prone to freeze in winter by producing their pups at the end, rather than the beginning, of winter. A second explanation suggests that grey seals were originally animals of the pack ice and that the eastern Atlantic stock are seals which have colonized a new habitat and altered their breeding season. The third hypothesis suggests all the groups have deviated from an original breeding season in late spring or early summer, as in most other pinnipeds.

It is difficult to choose between these hypotheses, and in particular to explain the change from spring to autumn (or winter) breeding, which does not occur in other seals. It is possible that this feature is a comparatively recent one, resulting from prehistoric human predation on this rather vulnerable terrestrially breeding seal.

Breeding behaviour

Breeding behaviour has been described by many authors; important accounts are Hewer (1957 and 1960), Coulson and Hickling (1964), Cameron (1967 and 1969), and Anderson *et al.* (1975). Bonner (1972) provides a brief survey.

The difference in timing has already been commented on. It is likely that there are other differences in behaviour during the breeding season, partly as a result of different substrates (e.g. ice in the Baltic and Gulf of St Lawrence, rock in Scotland, sand at Sable Island), and partly as a result of differing evolution in the various stocks. The grey seal is polygynous (though the number of cows controlled by a single bull varies widely), which implies a greater degree of competition between the males than in monogamous species. However, agonistic behaviour and territoriality are not as highly developed as in the other highly social land-breeding phocids—the elephant seals.

Hewer (1960) has described various types of breeding habitat in the eastern Atlantic. The topography of the breeding site determines the behaviour of the seals. There is sometimes a pre-breeding assembly near the breeding grounds (Hewer, 1957; Hickling, 1962; and Cameron, 1969). The season starts with the birth of the first pups; soon after this the bulls arrive and take up position. Where there are only a small number of cows in an isolated cove, it is usual for the bulls to station themselves in the sea at the approach to the beach (Hewer, 1957; Fogden, 1971), but where the cows extend inland far from the beach, as at the larger and more crowded colonies (e.g. North Rona, the Farne Islands) the bulls station themselves ashore (Hewer, 1957; Anderson et al., 1975). Hewer (1957) suggested that the bulls with previous breeding experience adopted the most advantageous breeding positions, with little challenge from the other bulls. He describes the behaviour which establishes and maintains territory and comments on a behaviour at the end of contests in which one contestant rolls over sideways. This behaviour, which Hewer regarded as being associated with territoriality, was seen regularly only in the Outer Hebrides, though it occurred rarely or occasionally at North Rona or the Farnes (Hickling, 1962). It is clear that there is no physically defined territory in the grey seal, as is noted in fur seals (Peterson, 1968). The area dominated by a bull changes from day to day. At North Rona bulls often lie within three or four metres of each other, suggesting that the tolerance a bull will show to the nearness of other bulls might depend on the density of cows. Boundary displays are absent and fighting minimal, so that successful males conserve energy needed for a prolonged stay on the breeding ground. Bulls do not investigate individual cows, but once oestrous cows are present on the rookery, they will approach any cow and attempt to copulate, often being stimulated to activity when a cow changes position. Thus by constant sexual attention to the cows, once oestrous animals are present, a bull ensures that no cow in his area of influence is overlooked, and decreases the chance of its being mated by another bull. For its reproductive strategy the grey seal bull uses sexual activity, rather than territorial fighting or boundary displays (Anderson et al., 1975).

Birth takes place soon after a cow's arrival on shore. A cow may spend some time looking for a suitable place to pup, and perhaps hauling out several times to do so. On crowded colonies, where the cows pup away from the sea, they often choose a site near a stream or fresh-water pool. Birth is a speedy process as in all seals. Immediately after the expulsion of the pup, the cow turns and sniffs at it (Burton *et al.*, 1975). Smelling continues at intervals and the cow vocalizes to the pup. In this way a bond is formed between the cow and her pup (Fogden, 1971). The cow may defend the placenta, which is delivered 5-40 min after birth, against gulls (Burton *et al.*, 1975).

Cows may remain with their pups ashore throughout lactation, or may return to the sea between feeds, depending on the ease of access to and from the sea. Fogden (1971) has described mother-young behaviour at crowded and uncrowded beaches. Disturbance is often caused by human activity (boats, hunters, observers), but may be a consequence of the increased activity associated with dense concentrations of seals. In an undisturbed situation the pup lies where it was last suckled, usually asleep. A cow returning to feed her pup comes back to the place where she left it and calls. The pup responds to its mother's voice and the cow, recognizing her pup's voice, approaches it and identifies it finally by smell, before feeding it. This simple system may break down on crowded beaches, and there are accounts of deserted, starving pups (Bonner and Hickling, 1971) and of cows suckling pups not their own (Smith, 1968; Fogden, 1968, 1971). This is probably the consequence of the mother-pup bond being broken, or never having been established. Those pups on a disturbed beach which sucked from most cows did not thrive as well and spent less total time feeding than those which sucked from one cow only (Fogden, 1971).

Cows with pups are very aggressive to each other (and to bulls) and on a disturbed beach pups may become involved with cow fights, with serious or sometimes fatal consequences.

The average duration of "primary" sucking episodes (those where the pup sucks continuously, alternating infrequently between the two nipples) is 6.1 ± 0.4 min (Fogden, 1971). Evidence of frequency of feeding is sparse, but it seems usual for the pup to be fed at intervals of five to six hours, occasionally more often.

Grey seal milk contains 67% solids, with over 53% fat (Amoroso *et al.*, 1950) and the pups grow very rapidly, an average of 1.8 kg per day for the Farne Islands (Coulson, 1959) and 1.3 kg per day (males) and 1.4 kg per day (females) at North Rona (Boyd and Campbell, 1971). Lactation lasts about 16-21 days (Davies, 1949; Coulson, 1959) after which the pups may remain on shore for about 14 days, before taking to the sea and feeding independently.

Cows come into oestrus at the end of lactation. Precopulative behaviour has been described (Hickling, 1962; Anderson *et al.*, 1975), but this does not seem to be usual or well developed. Copulation can take place either in the water or on land. The bull lies at the side of the cow with a flipper over the cow's back, often grasping the cow's neck with its teeth. Copulation lasts from 15 to 45 min (Hewer, 1957). An oestrous cow is usually mated a number of times, often by several bulls. Soon after copulation, the cow will leave the breeding area and does not return until the next season.

Cows spend an average of 18 days ashore and this is also the average time spent by bulls (Anderson et al., 1975) but some bulls spend much longer. These authors found that the length of stay ashore by bulls is positively correlated with the number of copulations they achieve. The three most active of the 31 bulls observed at North Rona accounted for 35.6% of all copulations, thus a form of hierarchy exists among the breeding bulls. The average ratio of cows to breeding bulls at North Rona was 7.5:1.

The above account has referred mostly to conditions in the eastern Atlantic stock. Less information is available about grey seals in the western Atlantic or Baltic. In the western Atlantic, land-breeding grey seal cows tend to stay on land longer after birth (Cameron, 1967) and the ratio of cows to bulls is lower than in the eastern Atlantic stock. In the Baltic, the grey seals breeds mainly on drift ice or fast ice near open water. Only rarely do Baltic grey seals breed on the rocky skerries that fringe the coast (Curry-Lindahl, 1975). The seals are polygamous with loosely organized harems (Curry-Lindahl, 1965) but sometimes pups are found up to 1.6 km or so apart, indicating a monogamous relationship (Hook and Johnels, 1972).

Management and Conservation

Because of their fish-eating habits, and their role in the transmission of cod-worm, grey seals are generally unpopular with fishermen and there have been a number of official and unofficial attempts to control their number. In a few places grey seals, usually moulted pups, are used as a source of furs, but they are generally of little commerical interest, most of the skins collected being taken in the course of fishery-inspired control operations against the seals. Lately, a reappraisal of attitudes to wildlife, and particularly marine mammals, has led to the adoption of conservation policies. However, these are often combined with management plans to control seal numbers in countries with large seal stocks.

In Canada bounties are paid on grey seals older than one year, or on pups after February away from the pupping grounds. Government agents have killed large numbers of pups in the Gulf of St Lawrence and on the Basque Islands (average 800 per year) in an attempt to control the population (Mansfield and Beck, 1977). Grey seals, like other marine mammals, are totally protected in the USA.

In the UK, grey seals are protected from September to January but licences are issued to take seals in the close season to control numbers (Bonner, 1971b, 1975b). Some licences are related directly to fishery damage (e.g. those issued in Orkney), while others, at the Farne Islands, have been related to general environmental management (Bonner and Hickling, 1974). There is no protection for grey seals in Northern Ireland, the Isle of Man or the Channel Isles. Grey seals have been protected in the Republic of Ireland since 1976, but licences can be issued to kill for the protection of fisheries.

France has protected grey seals since 1961 and Denmark since 1967. In Norway, grey seals have been protected, except from December to April from Møre north. However, seals may be killed to protect salmon nets.

Neither Iceland nor Faroes appear to have either control or management of their grey seal stocks.

In the Baltic, Sweden has protected grey seals, except in the vicinity of salmon nets, since 1974; Finland has a close season from March to May and the USSR provided total protection in 1970.

Pollution and Grey Seals

In common with other marine mammals, grey seals acquire pollutant burdens from their environment: Holden (1978) has reviewed the situation. Attention has concentrated on organochlorine compounds, particularly synthetic insecticides such as DDT and dieldrin, and the industrial polychlorinated biphenyls (PCB). Very large concentrations of DDT residues up to 420 mg kg^{-1} fresh weight) and PCB (up to 140 mg kg^{-1} fresh weight) have been found in grey seals from the Baltic (Olsson et al., 1974). Smaller quantities are found elsewhere (Table 6), though a high concentration of PCB was found in seals from the east of England (Heppleston, 1973). The distribution of organochlorines in the tissues of grey seals is related largely, but not entirely, to their lipid content (Holden, 1975). Concentrations in brain lipids are consistently much lower than in lipids from other tissues.

Mercury concentrations have been studied in grey seal livers (Table

TABLE 6 Mean organochlorine concentration (mg per kg fresh weight) in blubber of adult Grey seals (modified from Holden, 1978)

Area	No	Dieldrin	ΣDDT	PCB	Reference
N. Baltic	18	—	420	140	Olsson et al. (1974)
S. Baltic	15	—	210	100	Olsson et al. (1974)
E. England	4	0.46	15.5	152	Heppleston (1973)
E. Scotland	20	0.24	12.4	35	Holden (1975)
Orkney	7	0.14	8.3	26.6	Holden (1978)
Shetland	8	0.14	8.9	11.1	Holden (1978)
N.W. Scotland	6	0.16	7.4	14.4	Heppleston (1973)
Gulf of St Lawrence	5	0.25	45.4	27	Holden (1972)

7). High concentrations have been found in east Scotland and Canada (387 mg kg^{-1} fresh weight of mercury in the liver of a Canadian grey seal—Sergeant and Armstrong, 1973). High values in liver may be related to age, rather than to incidence of local pollution (Holden, 1975). Only about 5% of the mercury is present as the methyl compound (Sergeant and Armstrong, 1973); it is possible that, like other seals, grey seals can demethylate organic mercury.

Grey seals have not been shown to suffer any toxic effects from the levels of pollutants found, though such effects have been deduced in other species (Helle et al., 1976).

Grey seals are frequently exposed to contamination by floating or beached petroleum residues. Davis and Anderson (1976) found in Wales that oiled pups had lower peak weights than unoiled pups. They concluded that disturbance caused by attempting to clean oiled pups was likely to prove more damaging than the oil itself.

TABLE 7 Mercury concentrations (mg per kg fresh weight) in liver of adult Grey seals (modified from Holden, 1978)

Area	No	Range	References
E. England	15[a]	30 (mean)	Heppleston and French (1973)
E. England	1	175	Holden (1975)
E. Scotland	10[a]	10-720	Holden (1975)
E. Scotland	8[a]	1-428	Holden (1975)
N.W. Scotland	6	113 (mean)	Heppleston and French (1973)
E. Canada	7	24-387	Sergeant and Armstrong (1973)

[a]includes some common seals, *Phoca vitulina*

Captivity and Physiological Studies

Seals, especially those taken at three to four weeks of age (just after weaning), adapt readily to captivity. They have been kept in zoos in North America, Iceland, and in Europe. Grey seals have bred successfully in several zoos.

Scholander (1940) employed the grey seal in his classic studies of diving physiology. Trained grey seals made dives to 225 m depths (Scronce and Ridgway, unpublished).

Schevill et al. (1963) have studied the sounds produced by grey seals and Ridgway and Joyce (1975) have investigated hearing in the species. In air the peak sensitivity was at about 4 kHz with greatly response as high as 150 kHz. It is the only marine mammals in which 15-30 kHz range with much reduced sensitivity past 60 kHz, but some response as high as 150 kHz. It is the only marine mammal in which electrophysiological and behavioural studies have been conducted on sleep. Grey seals can sleep soundly underwater on the bottom, while bottling at the surface and while hauled out on land (Ridgway, et al., 1975). Some blood values in grey seals have been reported by Greenwood et al. (1971).

Acknowledgements

I am much indebted to my former colleagues in the Seals Research Division, Institute for Marine Environmental Research (now the Sea Mammal Research Unit) for the great amount of help they provided in the preparation of this chapter.

References

Allen, J. A. (1880). "History of the North American pinnipeds: A Monograph of the Walruses, Sea Lions, Sea Bears and Seals of North America", Miscellaneous Publications No. 12. U.S. Geological and Geographical Survey of the Territories. Government Printing Office, Washington, D.C.

Amoroso, E. C., Bourne, G. H., Harrison, R. J., Harrison Matthews, L., Rowlands, I. W. and Sloper, J. C. (1965). Reproductive and endocrine organs of foetal, newborn and adult seals. *J. Zool. (London)* **147**, 430-486.

Amoroso, E. C., Goffin, A., Halley, G., Harrison Matthews, L., and Matthews, D. J. (1950). Lactation in the grey seal. Proc. Physiol. Soc. 4th Nov. *J. Physiol. (London)* **113**.

Anderson, S. S. (1977). The grey seal in Wales. *Nature in Wales* **15(3)**, 114-123.

Anderson, S. S., Bonner, W. N., Baker, J. R. and Richards, R. (1974). Grey seals, *Halichoerus grypus*, of the Dee Estuary and observations on a characteristic skin lesion in British seals. *J. Zool. (London)* **174**, 429-440.

Anderson, S. S., Burton, R. W. and Summers, C. F. (1975). Behaviour of grey seals *(Halichoerus grypus)* during a breeding season at North Rona. *J. Zool. (London)* **177**, 179-195.

Appleby, E. C. (1964). Observations on wild grey seals in Britain. *Tijdschr. Diergeneesk.* **89** Suppl. 1. 201-203.

Árnason, U. (1970). The karyotype of the grey seal *(Halichoerus grypus)*. *Heriditas* **64**, 237-242.

Arnlaugsson, T. (1973). Selir vid Island. Rannsoknastofnun fiskidnadarins. 25 pp. Mimeo.

Backhouse, K. M. (1961). Locomotion of seals with particular reference to the forelimb. *Symp. Zool. Soc. Lond.* **5**, 59-75.

Backhouse, K. M. and Hewer, H. R. (1960). Unusual colouring in the grey seal, *Halichoerus grypus* (Fabricius). *Proc. Zool. Soc. Lond.* **134(3)**, 497-499.

Bartholomew, G. A. (1970). A model for the evolution of pinniped polygyny. *Evolution* **24**, 546-559.

Bonner, W. N. (1970). Seal deaths in Cornwall: Autumn 1969. Nat. Environment Res. Council, Publications Series C, No. 1.

Bonner, W. N. (1971a). An aged grey seal *(Halichoerus grypus)*. *J. Zool. (London)* **164**, 261-262.

Bonner, W. N. (1971b). Legislation on the seals in the British Isles. *Salmon Net* **7**, 30-33.

Bonner, W. N. (1972). The grey seal and common seal in European waters. *Oceanogr. Mar. Biol. Ann. Rev.* **10**, 461-507.

Bonner, W. N. (1975a). Population of grey seals at the Farne Islands, north-east England. *Rapp. p. -v. Réun. Cons. Int. Explor. Mer* **169**, 366-370.

Bonner, W. N. (1975b). International legislation and the protection of seals. *In Proc. Symp. The Seal in the Baltic.* National Swedish Environment Protection Board (SNV PM 591), Solna, 12-29.

Bonner, W. N. (1976). The stocks of grey seals *(Halichoerus grypus)* and common seals *(Phoca vitulina)* in Great Britain. The Natural Environment Res. Council, Publications Series C, No. 16.

Bonner, W. N. (1979). Grey seal. *In* "Mammals in the Seas" 90-94 FAO Fish. Ser., (5) Vol. 2.

Bonner, W. N. and Fogden, S. C. L. (1971). A note on blood typing seals. *Rapp. P. -v. Réun. Cons. Int. Explor. Mer* **161**, 139-141.

Bonner, W. N. and Hickling, G. (1971). The grey seals of the Farne Islands: report for the period October 1969-July 1971. *Trans. Nat. Hist. Soc. Northumb.* **17(4)**, 141-162.

Bonner, W. N. and Hickling, G. (1974). The grey seals of the Farne Islands: 1971-1973. *Trans. Nat. Hist. Soc. Northumb.* **42(2)**, 65-84.

Boyd, J. M. and Campbell, R. N. (1971). The grey seal *(Halichoerus grypus)* at North Rona, 1959-1968. *J. Zool. (London)* **164**, 469-512.

Bree, P. J. H. van (1972). On a luxation of the atlas joint and consecutive ankylosis in a grey seal, *Halichoerus grypus* (Fabricius, 1791), with notes on other grey seals from the Netherlands. *Zool. Meded.* **47**, 331-336.

Broch, H. (1914). Bemerkungen ueber anatomische verhaeltmisse der Kegelrobbe. *Anat. Anz.* **45**, 548-560 (Pt 1), **46**, 194-200 (Pt 2).

Burton, R. W., Anderson, S. S., Summers, C. F. (1975). Perinatal activities in the grey seal *(Halichoerus grypus). J. Zool. (London)* **177 (2)**, 197-201.

Cameron, A. W. (1967). Breeding behaviour in a colony of western Atlantic gray seals. *Can. J. Zool.* **45 (2)**, 161-174.

Cameron, A. W. (1969). The behaviour of adult grey seals *(Halichoerus grypus)* in early stages of the breeding season. *Can. J. Zool.* **47 (2)**, 229-234.

Coulson, J. C. (1959). The growth of grey seal calves on the Farne Islands, Northumberland. *Trans. Nat. Hist. Soc. Northumb.* **13**, 86-100.

Coulson, J. C. and Hickling, G. (1961). Variation in the secondary sex-ratio of the grey seal, *Halichoerus grypus* (Fabricius) during the breeding season. *Nature (London)* **190**, 281.

Coulson, J. C. and Hickling, G. (1964). The breeding biology of the grey seal, *Halichoerus grypus* (Fab.), on the Farne Islands, Northumberland. *J. Anim. Ecol.* **33**, 485-512.

Curry-Lindahl, K. (1965). The plight of the grey seal in the Baltic. *Oryx* **8**, 38-44.

Curry-Lindahl, K. (1975). Ecology and conservation of the grey seal, *Halichoerus grypus,* common seal, *Phoca vitulina,* and ringed seal, *Pusa hispida* in the Baltic Sea. *Rapp. P. -v. Réun. Cons. Int. Explor. Mer* **169**, 527-532.

Davies, J. L. (1949). Observations on the grey seal *(Halichoerus grypus)* at Ramsey Island, Pembrokeshire. *Proc. Zool. Soc. Lond.* **119**, 673-692.

Davies, J. L. (1957). The geography of the gray seal. *J. Mammal.* **38**, 297-310.

Davis, J. E. and Anderson, S. S. (1976). Effects of oil pollution on breeding grey seals. *Mar. Pollut. Bull.* **7 (6)**, 115-118.

Drescher, J. E. (1978). Hautkrankheiten beim Seahund, *Phoca vitulina* Linne, 1758, in der Nordsee. *Saugtier. Mitt.* **26 (1)**, 50-59.

Einarsson, S. T. (1977). Seals in Icelandic waters. ICES C.M. 1977/N:19 (mimeo).

Fabricius, O. (1791). Udfarlig besknivelse over de Gronlandske saele. Skrivter af Naturhistorie-Selskabet. *Kjøbenhavn. Bd. 1. Hft. 2.* 73-170.

Farmer, A. S. and Roberts, D. (1970). On strandings of the grey seal, *Halichoerus grypus* (Fabricius), in the south of the Isle of Man. *Rep. Mar. Biol. Stn. Port Erin* **83**, 43-50.

Fogden, S. C. L. (1968). Suckling behaviour in the grey seal *(Halichoerus grypus)* and the northern elephant seal *(Mirounga angustirostris). J. Zool. (London)* **154**, 415-420.

Fogden, S. C. L. (1971). Mother-young behaviour at grey seal breeding beaches. *J. Zool. (London)* **164**, 61-62.

Gallacher, J. B. and Waters, W. E. (1964). Pneumonia in grey seal pups at St. Kilda. *Proc. Zool. Soc. Lond.* **142**, 177-180.

Greenwood, A. G., Ridgway, S. H. and Harrison, R. J. (1971). Blood values in young grey seals. *J. Am. Vet. Med. Assoc.* **159**, 571-574.

Haaften, J. L. van (1962). Diseases of seals in Dutch coastal waters. *Nord. Vet. Med.* **14**, Suppl. 1. 138-140.

Harrison, R. J., Matthews, L. H. and Roberts, J. M. (1952). Reproduction in some Pinnipedia. *Trans. Zool. Soc. Lond.* **27 (5)**, 437-541.

Harwood, J., Anderson, S. S., and Curry, M. G. (1976). Branded grey seals *(Halichoerus grypus)* at the Monach Isles, Outer Hebrides. *J. Zool. (London)* **180**, 506-508.

Harwood, J. and Prime, J. H. (1978). Some factors affecting the size of British grey seal populations. *J. Appl. Ecol.* **15**, 401-411.

Heath, C. (1978). Serum protein analysis of grey seals. *Mammal Rev.* **8**, 47-51.

Helle, E., Olsson, M. and Jensen, S. (1976). PCB levels correlated with pathological changes in seal uteri. *Ambio* **5 (5-6)**, 261-263.

Hepburn, D. (1896). The grey seal. Observations on its external appearance and visceral anatomy. *J. Anat. Physiol.* **30**, 413-488.

Heppleston, P. B. (1973). Organochlorines in British grey seals. *Mar. Pollut. Bull.* **4**, 44-45.

Heppleston, P. B. and French, M. C. (1973). Mercury and other metals in British seals. *Nature (London)* **243**, 302-304.

Hewer, H. R. (1957). A Hebridean breeding colony of grey seals, *Halichoerus grypus* (Fab.), with comparative notes on the grey seals of Ramsey Island, Pembrokeshire. *Proc. Zool. Soc. Lond.* **128**, 23-64.

Hewer, H. R. (1960). Behaviour of the grey seal *(Halichoerus grypus* Fab.) in the breeding season. *Mammalia (Paris)* **24 (3)**, 400-421.

Hewer, H. R. (1964). The determination of age, sexual maturity, longevity and a life-table in the grey seal *(Halichoerus grypus)*. *Proc. Zool. Soc. Lond.* **142**, 593-624.

Hewer, H. R. (1974). "British Seals", Collins New Naturalist, London.

Hewer, H. R. and Backhouse, K. M. (1959). Field identification of bulls and cows of the grey seal, *Halichoerus grypus* (Fab.). *Proc. Zool. Soc. Lond.* **132 (4)**, 641-645.

Hewer, H. R. and Backhouse, K. M. (1960). A preliminary account of a colony of grey seals *Halichoerus grypus* (Fab.) in the southern Inner Hebrides. *Proc. Zool. Soc. Lond.* **134 (2)**, 157-195.

Hewer, H. R. and Backhouse, K. M. (1968). Embryology and foetal growth of the grey seal, *Halichoerus grypus*. *J. Zool. (London)* **155**, 507-533.

Hickling, G. (1962). "Grey Seals and the Farne Islands", Routledge and Kegan Paul, London.

Holden, A. V. (1972) Monitoring organochlorine contamination of the marine environment by the analysis of residues in seals. *In* "Marine Pollution and Sea Life," (Ed. M. Ruivo), 266-272. Fishing News Books Ltd., West Byfleet, England.

Holden, A. V. (1975). The accumulation of oceanic commitments in marine mammals. *Rapp. P. -v. Réun. Cons. Int. Explor. Mer* **169**, 353-361.

Holden, A. V. (1978). Pollutants and seals—a review. *Mammal Rev.* **8**, 53-66.

Hook, O. (1960). Some observations on the dates of pupping, and the incidence of partial rust and orange coloration in grey seal cows, *Halichoerus grypus* (Fabricius), on Lunga, Treshnish Isles, Argyll. *Proc. Zool. Soc. Lond.* **134**, 495-497.

Hook, O. (1961). Notes on the status of seals in Iceland. June-July 1959. *Proc. Zool. Soc. Lond.* **137**, 628-630.

Hook, O. (1964). The biology of the grey seal: the distribution and breeding of the grey seal in the Baltic. *In* "A Seals Symposium. Nature Conservancy Consultative Committee on Grey Seals and Fisheries", (Ed. E. A. Smith), 75-84. Nature Conservancy, Edinburgh.

Hook, O. and Johnels, A. G. (1972). The breeding and distribution of the grey seal *(Halichoerus grypus*, Fab.) in the Baltic Sea, with observations on other seals of the area. *Proc. R. Soc. Lond. B.* **182**, 37-58.

Karpovich, V. N., Kokhanov, V. D. and Tatarinkova, I. P. (1967). The grey seal in the Murman coast. *Trudy PINRO* **21**, 117-125.

King, J. E. (1972). Observations on phocid skulls. *In* "Functional Anatomy of Marine Mammals", Vol. 1, (Ed. Harrison, R. J.), 81-115. Academic Press, New York and London.

Lockley, R. M. (1966). The distribution of grey and common seals on the coasts of Ireland. *Irish Naturalists J.* **15 (5)**, 136-143.

Mansfield, A. W., (1967). Seals of Arctic and eastern Canada. *Fish. Res. Board Can., Bull. 137,* (second ed).

Mansfield, A. W. and Beck, B. (1977). The grey seal in eastern Canada. Dept. Fish. Env. Fisheries and Marine Service Tech. Rep. 704.

Mawdesley-Thomas, L. E. and Bonner, W. N. (1971). Uterine tumours in a grey seal *(Halichoerus grypus). J. Path.* **103**, 205-208.

McDermid, E. M. and Bonner, W. N. (1975). Red cell and serum protein systems of grey seals and harbour seals. *Comp. Biochem. Physiol.* **50 B**, 97-101.

Møhl, V. (1970). Fangstdyrene ved de danske strand—den zoologiske baggrund for harpunerne. KUML. Aabog for Jysk Arkaelogisk Selskab, 297-329. København.

Mohr, Erna (1952). Die Robben der Europäischen Gewässer. Frankfurt am Main, Paul Schops Monogr. Wildsäugetiere bd. 12.

Nehring, A. (1883). Uber Gebiss und Skelet von *Halichoerus grypus. Zool. Anz.* **6**, 610-615.

Nilsson, S. (1820). Däggande djuren. *In* "Skandinavisk fauna, en Handbok for Jägere och Zoologer", Vol. 1. Lund. 1820-42.

Olsson, M., Johnels, A. G., Vaz, R. (1974). DDT and PCB levels in seals along the east and west coasts of Sweden. Report SNV PM 420. Statens Naturvardsverk, Stockholm.

Øynes, P. (1964). Sel på norskekysten fra Finnmark til Møre. *Fiskets Gang* No. 48, 694-707.

Parrish, B. B. and Shearer, W. M. (1977). Effects of seals on fisheries. ICES C.M. 1977/M:14 (mimeo).

Peterson, R. S. (1968). Social behavior in pinnipeds with particular reference to the Northern fur seal. *In* "The Behavior and Physiology of Pinnipeds", (Eds. R. J. Harrison, R. C. Hubbard, R. S. Peterson, C. E. Rice and R. J. Schusterman), 3-53. Appleton-Century-Crofts, New York.

Platt, N. E., Prime, J. H. and Witthames, S. R. (1975). The age of the grey seal at the Farne Islands. *Trans. Nat. Hist. Soc. Northumb.* **42 (4)**, 99-106.

Prime, J. H. (1978). Analysis of a sample of grey seal teeth from the Faroe Islands. ICES. C.M. 1978/N:5. (mimeo).

Radford, P. J., Summers, C. F. and Young, K. M. (1978). A statistical procedure for estimating grey seal pup production from a single census. *Mammal Rev.* **8**, 35-42.

Rae, B. B. (1960). Seals and Scottish fisheries. *Mar. Res. 1960* **(2)**.

Rae, B. B. (1968). The food of seals in Scottish waters. *Mar. Res. 1968 (2)*.

Rae, B. B. (1973). Further observations on the food of seals. *J. Zool. (London)* **169**, 287-297.

Rae, B. B. and Shearer, M. W. (1965). Seal damage to salmon fisheries. *Mar. Res. 1965* **(2)**.

Ridgway, S. H. and Joyce, P. L. (1975). Studies on seal brain by radiotelemetry, *Rapp. P. -v. Réun. Cons. Int. Explor. Mer* **169**, 81-91.

Ridgway, S. H., Harrison, R. J. and Joyce, P. L. (1975). Sleep and cardiac rhythm in the grey seal. *Science* **187**, 553-555.

Roux, F. (1957). Sur la présence de phoques à l'Ile d'Ouessant. *Penn ar Bed* **11**, 13-18.

Scheffer, V. B. (1958). "Seals, Sea Lions and Walruses", Stanford University Press, Stanford, CA.

Schevill, W. E., Watkins, W. A. and Ray, C. (1963). Underwater sounds of pinnipeds, *Science (N. Y.)* **141**, 50-53.

Scholander, P. F. (1940). Experimental investigations on the respiratory function in diving mammals and birds. *Hvalrådets Skrifter* **22**, 1-131.

Scopoli, G. A. (1777). Introductio ad historiam naturalem, sistens genera lapidum, plantarum, et animalium . . . Prague.

Sergeant, D. E. and Armstrong, F. A. J. (1973). Mercury in seals from eastern Canada. *J. Fish. Res. Board Can.* **30**, 843-846.

Smith, E. A. (1966). A review of the world's grey seal population. *J. Zool. (London)* **150**, 463-489.

Smith, E. A. (1968). Adoptive suckling in the grey seal. *Nature (London)* **217**, 5130, 762.

Söderberg, S. (1975). Feeding habits and commercial damage of seals in the Baltic. *In* "Proc. Symp. The Seal in the Baltic", 66-75. National Swedish Environment Protection Board. (SNV PM 591). Solna.

Summers, C. F. (1974). The grey seal *(Halichoerus grypus)* in Cornwall and the Isles of Scilly. *Biol. Conserv.* **6 (4),** 285-291.

Summers, C. F. (1978). Trends in the size of British grey seal population. *J. Appl. Ecol.* **15,** 395-400.

Summers, C. F., Bonner, W. N. and Haaften, J. L. van (1978). Changes in the seal population of the North Sea. *Rapp. P. -v. Réun. Cons. Int. Explor. Mer* **172,** 278-285.

Wolk, K. (1969). Factors affecting seal population levels in the Southern Baltic Sea. *Saugetierkundliche Mitteilungen. Jhg. 17 Heft 2* 155-158.

Wynne-Edwards, V. C. (1954). Field identification of the common and grey seals. *Scottish Nat.* **66,** 192.

Young, P. C. and Lowe, D. (1969). Larval nematodes from fish of the sub-family Anisakinae and gastro-intestinal lesions in mammals. *J. Comp. Pathol.* **79,** 301-313.

6

Bearded Seal
Erignathus barbatus Erxleben, 1777

John J. Burns

Genus and Species

Two subspecies of the circumpolar boreoarctic bearded seal are currently recognized, *Erignathus barbatus barbatus* (Erxleben, 1777) which is reported to occur from the Laptev Sea westward to the Hudson Bay region, and *E. b. nauticus* (Pallas, 1811) in the remaining region from the Canadian Arctic westward to the Laptev Sea (King, 1964). They are both contained in Family Phocidae (Gray, 1825). The distributional limits for each have never been established, although Manning (1974) stated that one boundary is probably somewhere in the central Canadian Arctic and the other in the central palearctic. Tavrovskii (1971), also indicating that the eastern boundary of these two forms is not known, stated that it may be in the eastern part of the East Siberian Sea.

This question is still very much open to further investigation. Zoologists familiar with bearded seals and the habitats they occur in are hard pressed to visualize the barriers to interchange in the boundary areas currently recognized. The possibility exists that there is a pan-Arctic *Rassenkreis*. However, the appropriate specimen material required adequately to investigate this possibility and to describe some of the local forms which may occur within it is still lacking.

Common names

Some common names of *E. barbatus* include: "bearded seal" (English), "square flipper" (English derivation of Norwegian sealers' term, also used in Canadian Maritime provinces), "morski zaits" (Russian, meaning sea hare, commonly used in the western portion of the USSR), "laktak" (from the Kamtschatdal term, now generally used throughout the Soviet far-east), "mukluk" (Upik Eskimo term used in southwest Alaska, St Lawrence Island and southern Chukchi Peninsula) and "oogruk" (Inupiat Eskimo term, a close approximation of which is used by these Eskimos from the northern Chukchi Peninsula east to Greenland).

External Characteristics and Morphology

The scientific name is descriptive of two important features of bearded seals. As indicated by King (1964), the generic name, *Erignathus*, is from Greek and refers to the deep jaw. The specific name, *barbatus*, is derived from Latin and refers to the relatively long and numerous moustachial vibrissae. Obviously, the English common name, bearded seal, is aptly applied. The Norwegian sealers' term, square flipper, is in reference to the shape of the distal portion of the foreflippers, on which the third digit is slightly longer than the others, giving the appendage a blunt, squared appearance. This is in marked contrast to the other northern phocids in which the first digit is the longest.

Bearded seals are the largest of the northern phocids (Figs 1-3). However, they have a disproportionately small head. This characteristic was noted by Allen (1880) who wrote, "I find that the lower jaw of a very old male *P. vitulina* just fits an adult skull of *Erignathus barbatus*, except that the latter is slightly longer".

The general features referred to thus far are illustrated in Fig. 1.

These seals have four retractable teats (the other northern phocids have two) which are evenly spaced, two on each side of the midline near the navel.

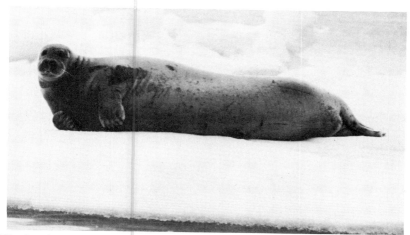

FIG. 1 An adult male bearded seal, northern Bering Sea, May 1978. (Photo courtesy of L. Schults.)

Coloration

The colour of bearded seals is variable. Most adults are basically light to dark grey, slightly darker down the middle of the back. Other individuals vary from tawn-brown to dark brown, also being darkest on the dorsal surface. The sexes are similarly coloured. Some younger animals, especially weaned pups, have faint irregularly shaped blotches which are unevenly distributed on the entire body. In general, however, these seals have none of the distinct and diagnostic colour

FIG. 2. New-born bearded seal pup, 9 April 1971, south-eastern Bering Sea.

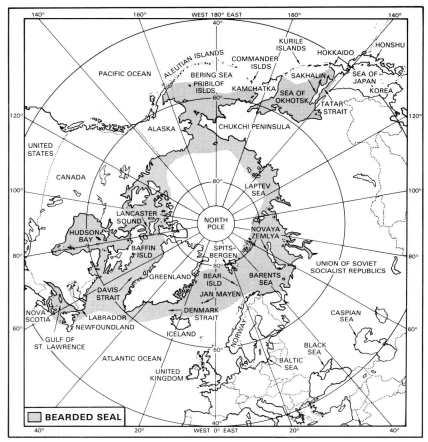

FIG. 3 Distribution of the bearded seal, *Erignathus barbatus* (Erxleben).

patterns like the spots, rings or bands found in other species. The hair is relatively short and straight. On many animals the face and foreflippers have a rust or reddish brown colour.

Term foetuses and new-born pups have dark (usually brown), dense slightly wavy hair with light coloration on the facial region and one to four broad, transverse light bands on the crown and back (Fig. 2). By the time these pups are weaned the pelage resembles that of older seals. Prior to Chapskii's (1938) excellent study of this species, there was considerable confusion in the literature about the pelage coloration of foetal and new-born bearded seals.

Age and Growth

Tooth wear in bearded seals is rapid and in the majority of animals older than about eight years teeth are missing or worn down to the gum line. Therefore, in several previous studies teeth were not examined to determine age (i.e. Chapskii, 1938; McLaren, 1958; Johnson et al., 1966; Burns, 1967). In most of these earlier studies, growth ridges on claws of the foreflippers were used, as described by Plekanov (1933). However, Benjaminsen (1973) found that the roots of the upper canine teeth remained throughout life in most bearded seals. The remaining portions of the canines could be sectioned and the age determined on the basis of annuli in the cementum layer. In our continuing studies of bearded seals in the Bering Sea we have adopted a combination of the two methods. Initial determination of age is based on claw annuli. If the determined age appears doubtful because annuli extend out to the tip of the claw, the roots of the upper canine teeth are sectioned. In most animals, aged 0-8 years, results of both methods were in close agreement.

Nonetheless, the change in procedure has resulted in minor difficulties when comparing growth curves from studies conducted prior to Benjaminsen's. In order to expand upon results of our previous studies, and to compare them with other findings, all seals 10 years and older are considered as a single age cohort designated as 10 +. This does not appear to be a problem as physical maturity is obtained by age 10 (McLaren, 1958; Burns, 1967; Benjaminsen, 1973).

Unless otherwise indicated, the following discussion is based on seals from the Bering-Chukchi Sea population.

Average length and weight of these seals at birth, determined from 13 full-term fetuses, was 131.3 cm and 33.6 kg. At age 10 and older, they average 233 cm and 229 kg (Burns, 1967, and unpublished). The age:length relationship for Bering-Chukchi seals is in close agreement with that reported by McLaren (1958) for the eastern Canadian Arctic. Benjaminsen (1973) found that the lengths in his samples from the area of Svalbard and the Barents Sea were significantly smaller than those from eastern Canadian or Alaskan waters. The reported age length relationships are shown in Fig. 4.

There is a consistent but slight difference in size between males and females from the Bering-Chukchi Sea. Johnson et al. (1966) reported that in Alaskan waters adult females were 4% longer than males. Burns (1967) indicated that, on the average, females were 3.1% longer than males. Continuing studies in this region further confirm this

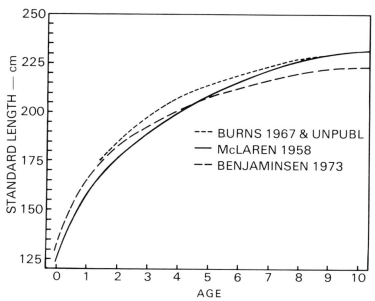

FIG. 4 Age:length relationship in Atlantic (McLaren, 1958; Benjaminsen, 1973) and Pacific bearded seals.

difference (Burns, unpublished). A difference in length between the sexes was not apparent in the eastern Canadian Arctic (McLaren, 1968). In the eastern North Atlantic region Benjaminsen (1973) indicated a slightly longer average length for adult females than for males, but the difference ($\bar{x}$ = 226.6 cm vs $\bar{x}$ = 222.5 cm) was not statistically significant. Chapskii (1938) also noted that females from the Kara and Barents Sea were slightly longer than males, but was of the opinion that the difference was not significant.

For purposes of describing growth, my data from males and females were combined (N = 143) and the average length of seals 10 years and older, 233 cm, was used as the mean length of adults. Length and proportion of adult length for the younger age classes were as follows: new-born pups, 131.3 cm, 56% of adult length; age 1, 165 cm, 71%; age 2, 185 cm, 79%; age 3, 195 cm, 84%; age 4, 209 cm, 90%; age 5, 214 cm, 92%; age 6, 219 cm, 94%; age 7, 222 cm, 95%; age 8, 228 cm, 98%; and age 9, 203 cm, 99%.

The maximum recorded standard lengths of bearded seals from the Bering-Chukchi region were 243 cm for a female (Burns, 1967) and 233 cm for a male (Johnson *et al.*, 1966). These values compare with a maximum standard length of 252 cm (sex of animal unknown) reported for the eastern North Atlantic region by Benjaminsen (1973) and 253 cm for a female seal taken in the eastern Canadian Arctic (McLaren, 1958). The maximum and mean standard lengths for mature males and females from the Kara and Barents Sea reported by Chapskii (1938) was 238 cm ($\bar{x}$ = 215.4 cm) and 242 cm ($\bar{x}$ = 209.3 cm), respectively.

Although differences in length between the sexes were slight, the differences in maximum weights were marked. Maximum reported weights in the Bering-Chukchi Sea were 262 kg for a male (Burns, 1967) and 360.5 kg for a female (Johnson et al., 1966). In our recent studies the heaviest seal examined was a 316 kg, pregnant female taken on 29 March 1977. She was supporting a 32.3 kg foetus. Differences in average weights of males and females (disregarding females supporting large foetuses) are similar to the differences in length. The length:weight relationship in 106 seals is illustrated in Fig. 5

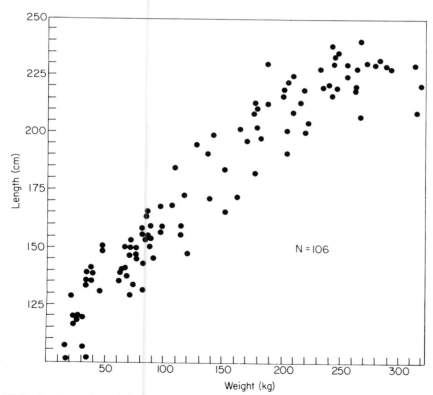

FIG. 5 Length:weight relationship in 106 Pacific bearded seals from the Bering and Chukchi Seas.

The robust body form of bearded seals can be appreciated by a comparison of standard length and girth immediately behind the foreflippers. In a series of 40 seals four years and older, mean girth was between 71-83% of standard length. This relationship changes seasonally, lowest values occurring during summer and highest values during late fall through early spring. Lactation in females and reduced feeding activity associated with the moult in both sexes contribute to these seasonal differences.

Seasonal changes in thickness of the blubber layer which account for the changes in girth are marked. Thickness of the blubber layer varies from an average of about 7.2 cm in late fall through early spring to about 4.4 cm during summer. Similarly, the hide and blubber layer account for an average of about 39% of total body weight when these seals are fattest, and about 29% when they are lean.

Distribution and movements

Bearded seals, by virtue of their association with sea ice and there dependence on benthos for food, are restricted to relatively shallow water areas within their circumpolar range (Fig. 3). It is difficult to summarize their distributions and movements in all of the different geographical regions within which they occur. In some regions, the extent of shallow waters and of suitable sea ice conditions are such that although bearded seals are present they occur in very low numbers. As an example, in the Laptev and East Siberian Seas bearded seals are well known to local peoples. They occur near the shores of the New Siberian Islands and on the sea ice. However, they are rare throughout the region (Tavrovskii, 1971).

The following comments relate specifically to bearded seals in the Bering and Chukchi Seas. This region probably comprises the largest area of continuous habitat for these seals in the world.

A shallow intercontinental shelf (the Bering-Chukchi Platform) underlies approximately half of the Bering Sea and is contiguous, through the Bering Strait, with the shelf underlying all of the Chukchi Sea. This shelf region is usually covered by sea ice during late winter and spring and is mostly ice free by late summer and fall. The southern margin of multi-year ice which persists through the melting period is usually near the northern limit of the shelf, extending completely across the Chukchi Sea.

Bearded seals can and do make and maintain breathing holes in relatively thin ice. They avoid regions of continuous, thick shore-fast ice (unlike ringed seals, *Phoca hispida*) and are not common in regions of unbroken, heavy, drifting ice. These seals utilize areas of shallow water where the ice is in constant motion, producing leads, polynya and other openings. In the Chukchi Sea during winter and spring such conditions occur as extensive "flaw zones" which occur where heavy drifting ice, influenced by winds and currents, interacts with coastal features. The extent of such favourable winter habitat in the Chukchi Sea is relatively limited. In comparison, this combination of suitable ice conditions and water depths occur over a much broader area in the more temperate Bering Sea. Most of the population moves south through the Bering Strait in late fall-early winter and "winters" in the Bering Sea. During the winter and early spring bearded seals are widely, though not uniformly, distributed throughout the drifting ice of this sea. Highest densities occur in the northern part of the ice-covered shelf. Relationships of marine mammals to sea ice in the region under discussion are described by Burns (1970) and Fay (1974).

As implied by the remarks above, bearded seals move great distances during the year, mostly maintaining an association with ice. These movements are directly related to the seasonal advance and retreat (as well as the growth and degeneration) of the seasonal sea ice cover. During winter most of them are in the Bering Sea; in summer along the wide, fragmented margin of multi-year ice. The northward spring migration through the Bering Strait, occurring from mid-April through June, is more marked and noticeable than the southward movements in late fall through winter.

Bearded seals do not resort to coastal hauling areas in the Bering-Chukchi region. This is probably because they are able to maintain their association with ice on a year-round basis, in water depths suitable for bottom feeding.

In other parts of their circumpolar range they regularly come ashore during summer. This has been recorded to occur in the Okhotsk Sea (Tikhomirov, 1961; Fedoseev, personal communication), the White Sea (Heptner, 1976) and the Laptev Sea (Tavrovskii, 1971). In these regions sea ice either melts in summer, or recedes beyond the limits of shallow water.

Some bearded seals, particularly juveniles, occur in the open sea of the Bering-Chukchi region during summer. They also enter small bays and ascend some rivers. In Imurak Basin on the south coast of the Seward Peninsula, pups were occasionally taken in nets at the time of fall freeze-up. This basin is separated from a larger bay by a long, narrow channel. There are several reports of seals becoming trapped by freeze-up of this brackish body of water and seeking escape by travelling over ice or land. In two reported instances, pups were tracked down in the snow and were dead when found (Kugzruk, personal communication). A similar occurrence in the Canadian Arctic was noted by Smith and Memogana (1977).

Abundance and Life History

In the Bering-Chukchi population, sexual maturity occurs mainly at ages five and six in females and six and seven in males. First ovulations occur in some three-year-old females.

The pupping period is long, extending from mid-March through the first week of May. Pups are born on the ice. The main period of births occurs during the last third of April in the Bering Strait region. This is approximately 10 days to two weeks later than the peak period indicated by Tikhomirov (1966) for the southern Bering Sea. In the Okhotsk Sea, the peak period of births is also in early to mid-April (Tikhomirov, 1966; Fedoseev, personal communication).

The late April peak of pupping in the Bering Strait is in agreement with findings of Johnson *et al.* (1966) for the central Chukchi Sea, Chapskii (1938) and Potelov (1975) for the Kara and Barents Sea (late April and the first days of May) and McLaren (1958) for the eastern Canadian Arctic. As indicated by Sleptsov (1943), Mohr (1952) and McLaren (1958), timing of births may be related to latitude.

The nursing period is comparatively short, lasting 12-18 days. During this time the average weight of pups increases from 33 kg at birth to 85 kg at initial independence. Weaning is abrupt, the pups simply being deserted. Bearded seal pups can swim as soon as they are born. Some independent feeding apparently occurs during the last stages of the nursing period, based on the presence of milk and small amounts of other foods found in a few stomachs. Newly weaned pups are active feeders, unlike ribbon seal *(Phoca fasciata)* and spotted seal *(Phoca vitulina largha)* pups which remain on the ice and apparently utilize the fat reserves accumulated during the nursing period.

The main breeding period coincides with the end of lactation. Most breeding occurs during the first three weeks of May. Some females apparently breed while they are still lactating. Male bearded seals are in breeding condition for a longer period of time than are the females, from mid-March through mid-June.

In the Bering and Okhotsk Seas, breeding is annual (Tikhomirov, 1966; Burns, 1967). The incidence of pregnancy in adult females from Alaskan waters has remained surprisingly constant at approximately

85% for samples collected from 1962 through 1978 (Burns, 1967; Burns and Eley, 1978). Earlier studies by Chapskii (1938), Sleptsov (1943) and McLaren (1958) indicated that females become pregnant only every other year. If this is the case it may reflect differences in quality and/or carrying capacity of bearded seal habitat in their study areas.

Implantation occurs mainly during mid-July to early August, after a delay of approximately two months. The prenatal growth rate is shown in Fig. 6. The entire period of pregnancy is approximately 11 months.

Benjaminsen (1973) indicated that bearded seals live to an age of about 31 years. In our samples, collected between 1975 and 1977 ($N = 448$), the oldest age animal was 25. Our data are obtained primarily from animals killed by coastal-based Eskimo hunters. At some

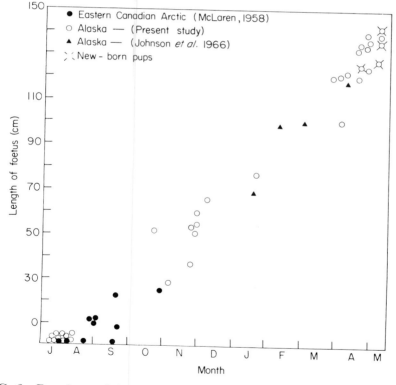

FIG. 6 Foetal growth in Pacific bearded seals from the Bering and Chukchi Seas.

locations these samples are significantly biased towards younger seals, which may occur in larger numbers close to shore. Thus, as an example, at one location in the south-eastern Bering Sea 65% of the bearded seals taken in the spring harvest were pups. In all samples combined, pups accounted for 30% of the catch. Only 2% of our combined samples were seals older than age 20.

Our samples also consistently show more females than males in all age groups entering the harvest. Of 426 seals taken in 1975-77, for which sex was determined, 56% were females. A preponderance of females was also reported by Johnson et al. (1966).

More effort will be required to relate composition and age structure of the catch made by coastal-based hunters to that which may actually exist in the population.

No detailed studies of the moult in bearded seals have been undertaken. Statements about the moult are based mainly on observations of when the hair is loose enough to be pulled out or when it is obviously shedding. The more important annual changes in structure and condition of dermal and epidermal skin layers remain to be investigated.

Based on general observations, moult in these seals occurs mainly from April through August, with a peak in May-June. This peak, as expected, coincides with the period of maximum hauling out. The relationship of moult to different ambient temperature regimes has been studied, *in vitro*, by Feltz and Fay (1966). Optimal ambient thermal conditions which facilitate the growth of new skin cells probably occur during May and June.

Bearded seals having loose or shedding hair or bare patches of skin occur in all months of the year. This suggests that condition of the hair covering may not reflect seasonal changes in the skin as accurately as it seems to in other species such as spotted and ringed seals.

The epidermal skin layer of bearded seals is black.

Although there is a peak in hauling out of bearded seals, these seals rest on the ice in all months of the year. Their tolerance to high winds and low temperatures is less than that reported for walruses *(Odobenus rosmarus)* (Fay and Ray, 1968) but significantly greater than that of the other (smaller) Bering Sea phocids. The differences among species may be a function of the great differences in size (body volume to surface area as well as total size). Bearded seals have been seen resting on the ice on calm days in February in temperatures below —30 °C.

Major predators of bearded seals in the Bering and Chukchi Seas are polar bears *(Ursus maritimus)* and man. The extent of predation by polar bears is unknown. It is probably greatest during late summer and fall when both species are concentrated along the margin of multi-year ice and less so in winter when most bearded seals are in areas south of the main distribution of bears. The main prey of polar bears is ringed seals.

During several recent years, unusually heavy and extensive ice cover in the southern Chukchi and northern Bering Seas resulted in an abundance of bears in these regions. At St Lawrence Island (63° 30′ N) polar bears usually occur during winter in very low numbers. However, during the winters of 1975 through 1977 bears were unusually numerous and record catches were taken by the island residents. Bears were also present as far south as Nunivak Island (60 °N). These major incursions by bears into the Bering Sea undoubtedly increased predation on several marine mammal species including bearded seals and walruses. It appears that several behavioural attributes of bearded seals have evolved in the face of continuous predation by polar bears.

Hunting of bearded seals by man throughout most of their circumpolar range has occurred for several thousand years. This seal is important to local subsistence economies as a dependable and significant source of food and necessary by-products. Boats, lines, clothing and other items are made from their durable skins. Formerly, implements were made from their bones, rain gear and translucent windows from their intestines, fuel from their oil, and waterproofing compounds and dyes from their blood, etc.

Although the methods used by coastal residents to hunt these seals and some of the traditional uses of by-products have changed, the major uses of meat and hides remain the same. Harvests of bearded seals in the Bering-Chukchi region are of two kinds, coastal-based subsistence hunting in Alaska and Siberia and pelagic commercial sealing by Soviet hunters. Combined American and Soviet harvests in the Bering-Chukchi region during recent years are indicated below.

The estimated population size of the Bering-Chukchi population is 300 000 animals (Burns, unpublished). Bychkov (1971) citing Chapskii (1966) presented estimates of 300 000 animals for the North Atlantic (including the North, White, Barents, Kara and Laptev Seas) and 450 000 for the North Pacific (all regions). No estimates were given for the size of bearded seal populations in the Canadian Arctic. In the eastern Beaufort Sea, Stirling et al. (1977) conducted aerial surveys in 1974 and 1975. In 1974, density of these seals in strata of their study area ranged from 0.0047 to 0.0351 per km^2. Abundance appeared to be significantly lower in 1975.

Year	American harvest	Soviet harvest	Total annual harvest
1966	1 242	6 230	7 472
1967	1 300	7 009	8 309
1968	1 050	4 577	5 627
1969	1 772	1 986	3 758
1970	1 759	2 533	4 292
1971	1 754	1 490	3 244
1972	1 353	1 428	2 781
1973	1 500	1 293	2 793
1974	1 600	1 256	2 856
1975	1 200	1 220	2 420
1976	2 125	1 644	3 769
1977	4 750	1 204	5 954

Behaviour

Except for mother-pup pairs and breeding animals, bearded seals tend to be solitary. They do not form herds, although loose aggregations are occasionally seen, especially in April-June. At this time certain features of the coastline and ocean currents produce eddies or gyres which entrain drifting ice and hold it after the main body has drifted away or disintegrated. Under these conditions the density of bearded seals in a small area may become quite high. Mostly, single animals will haul out on adjacent ice floes. Occasionally, several animals will haul out around the perimeter of a single large floe, each well spaced from its neighbour and each facing in a different direction.

They always rest with their head within inches or at most a few feet of a hole or crack through which they escape when disturbed. When sufficiently alarmed, they bolt into the water, raising and propelling their bodies by the simultaneous movement of both foreflippers. This leaping, forward movement of alarmed animals escaping from the ice is the behavioural basis for the Russian common name of sea hare (Kaitkulov, personal communication). I have never seen them move away from water when disturbed, as will spotted and ribbon seals.

The capability of pups to swim shortly after birth, the habit of resting on the ice immediately adjacent to a route of escape and the explosive escape response of alarmed animals are thought to be adaptations which have evolved in the face of continuous predation by bears over a significant portion of their range.

The senses of sight, hearing and smell are difficult to evaluate as the responses of these seals to disturbance are so varied. On a warm, calm spring day when they are basking, they often show little concern for a low-flying aircraft, or the close presence of men or boats. With care it is sometimes possible to crawl across the ice and touch one. This is in marked contrast to responses of these seals in winter when the slightest sound-producing movement of a man on the ice will cause a basking seal to flee, or entice a swimming seal to surface several hundred yards from the sound source. An undisturbed bearded seal will characteristically swim with its head and back above the water.

After watching bearded seals over many years and in a wide variety of ice and weather conditions, one must conclude that when alert they have good senses of sight and hearing and a fair sense of smell.

The mother-pup bond is strong during the early nursing period. Females often remained near pups which were being marked, occasionally coming out on the ice in an attempt to frighten the intruder. Similarly, they will remain in the water near a pup killed by hunters on the ice, or assist a wounded pup in its attempts to escape.

Young animals have been observed in a variety of different situations, engaged in what appeared to be play. In one instance, two small bearded seals kept up a constant chase which involved a repetition of riding swells into the beach, active rolling, mock fighting with the foreflippers and tail-chasing back out through the surf. This activity continued for 37 minutes. In another situation, two small seals moved along with a small outboard-powered boat and engaged in active tail-chasing, rolling, jumping partially clear of the water, slapping the water surface with their foreflippers and swimming from one side to the other immediately beneath the boat.

Formerly, hunters using kayaks in the open sea were especially successful in harpooning these seals during fall. On moderately windy days the seals could be approached very closely. It is thought that seals remaining at the surface for long periods of time were sleeping and could not hear an approaching kayak over the noises of wind and waves. Unlike swimming seals which expose both head and back, these "sleeping" seals are vertical in the water (Koezuma, personal demonstration).

Mothers and pups frequently nose and gently scratch each other. Mutual nosing was part of every encounter observed in which a female rejoined her pup on the ice. It was similar to that reported by Burns *et al.* (1972) for spotted seals.

Use of foreflippers is probably important in social interactions among adults. Adults often bear deep, parallel scratch marks or scars on the posterior third of the body, particularly the abdomen. These marks are usually attributed to polar bears. However, although the distance between the parallel marks is about the same as that between the claws of a small- to medium-sized bear, it is also similar to that between the claws of a flexed foreflipper. Additionally, it is difficult to imagine that a significant number of bearded seals struck on the abdomen by bears could escape the strong curved claws relatively unscathed.

In the only killing of a bearded seal by a bear which I observed, the seal's head was crushed by a single blow as it lay on the ice. The bear then grasped the seal's neck in its mouth and pulled it away from the edge of the ice floe. Tooth marks have not been found on seals having scratch marks.

Bearded seals are very vocal and produce a distinctive song. The long, musical underwater sounds are well known to Eskimo hunters who, in the days when kayaks were extensively used for spring hunting, tried to locate animals by listening for them. Although parts of the song are audible at close range in air, it is easily heard by placing a paddle in the water and pressing an ear against the butt of the handle. The Alaskan Eskimo terms "aveloouk" (Upik Eskimo) and "ayuktuk" (Inupik Eskimo) specifically refer to bearded seals singing in the water.

The production of sounds by bearded seals has been referred to by several earlier workers. Chapskii (1938) summarized the available information up to the time of his studies. He and others, notably Dubrovskii (1937), associated these sounds with the mating period (Dubrovskii referred to the nuptial whistle) and stated that, "it cannot be explained otherwise than being a sound expression of sexual excitement." Some of Chapskii's informants apparently indicated that both males and females made these sounds.

In the Bering Sea, the bearded seal song can be heard from March through July. There is definitely a marked peak in "singing" during April and May.

Analysis of these songs shows them to be complex, long, oscillating frequency-modulated warbles that may be longer than a minute in duration, followed by a short unmodulated low-frequency moan. These sounds are stereotyped and repetitive (Ray *et al.*, 1969). Once located, the singing animals are easy to approach, apparently being inattentive to minor disturbances. They dive slowly, apparently in a loose spiral (judged by the rhythmic changes in strength of the sounds), release bubbles (which signal their general location to the seal hunters) and surface in the centre of their area of activity. This behaviour is repeated many times.

All of the "singing" bearded seals taken by Eskimo hunters in my presence were adult males. It was suggested that the song is produced by mature males during the spring breeding season and that it is a proclamation of territory or of breeding condition or both (Ray et al., 1969). Future studies will probably show that females also produce these songs. This is suggested by the number of songs that have been heard simultaneously at some locations. It seems that most of the seals observed had to be participating in the singing and it does not seem probable that they would all be males.

Food habits

As indicated previously, the distribution of bearded seals is directly related to the distribution of seasonally ice-covered shallow waters. They are benthic feeders utilizing primarily epibenthos. Organisms of the infauna and some schooling, demersal fishes are also consumed. Feeding depths of up to 200 m are reported by Kosygin (1971). Chapskii (1938) indicates feeding depths of up to 50 m. Based on observed distribution of these seals in the southeastern Bering Sea they are restricted to waters of 130 m or less.

The total array of food items consumed by bearded seals is quite large. In the Okhotsk Sea, Fedoseev and Buktiyarov (1972) reported that their diet included 41 different species of benthic and nektonic invertebrates. However, in all areas where food habits have been investigated, a relatively few species comprise the bulk of the diet. The important food items vary in relation to composition of benthic communities which occur throughout this seal's range.

Chapskii (1938) indicated that two groups of bottom organisms constitute the basic foods in the Kara and Barents Seas. These were crustaceans and molluscs. Shrimps were the most important of the crustaceans while both gastropods and bivalves were the important molluscs. Other foods included a variety of worms (notably *Priapulus*), cod and other demersal fishes. Although fishes were not a major item, there was a regularity to their occurrence in seal stomachs. The echiuroid *(Echiurus echiurus)* was reported as probably being taken accidentally. Feeding activity was found to be reduced but not stopped during the spring moult of these seals.

Pikharev (1940) and Fedoseev and Bukhtiyarov (1972) investigated the food habits of bearded seals in the Okhotsk Sea. In the former study the major kinds of prey items, listed in order of importance, were: crabs (mainly *Hyas coarctatus aleuticas*); echiuroids; crangonid shrimps; lamellibranchs; pandalid shrimps; gastropods; holothuroidians; and cephalopods. In the latter study crabs and shrimps accounted for 87% of the diet in the northern part of the Okhotsk Sea. In the southern part, around Sakhalin Islands, clams comprised 40% of the diet; gastropods, 12%; and "worms", 23%.

Food habits of these seals in the Bering-Chukchi regions have been reported by Kenyon (1962), Kosygin (1966, 1971), Johnson *et al.* (1966), Burns (1967) and Lowry *et al.* (1977, 1978). Summarizing these studies, it was found that in the southern, ice-covered regions of the Bering Sea the snow crab *Chionocetes opilio* was the most important prey. Other significant prey included the crab *Hyas coarctatus*, shrimps, gastropods and octopus. In the northern Bering Sea, *Hyas coarctatus* was most important, followed by the snow crab, clams (mainly *Serripes groenlandicus*) and shrimps. Major food items in the Chukchi Sea were similar to those reported for the northern Bering Sea with the notable exception that snow crabs were of only minor importance. This crab probably reaches the limits of its northern distribution in the Chukchi Sea.

Lowry *et al.* (1977, 1978), working in the Beaufort Sea, indicated that the major food items in that region included the crab *Hyas coarctatus*, the shrimp *Sabinea septemcarinata* and arctic cod *Boreogadus saida*.

Significant amounts of sand or gravel are reported in the stomachs of seals from all areas where studies have been conducted.

In the northern Bering and Chukchi Seas utilization of clams appears to be seasonal, occurring mainly in late spring and summer. One possible explanation for this marked seasonality is that the claims may be less active or inactive at lower water temperatures, not exposing the siphons or feet upon which the bearded seals feed. Thus, although the clams are abundant, they are unavailable to the seals during colder periods of the year.

Internal Characteristics and Morphology

The skull of bearded seals is wide, comparatively short and more massive than in other ice-associated phocids of the North Pacific region (Fig. 7a, b, c). These seals commonly break through thin ice with their heads. There is no saggital crest. Teeth are comparatively large and, except for the upper canines, are weakly rooted. The dental formula is typical of phocids:

$$I\frac{1\text{-}2\text{-}3}{0\text{-}2\text{-}3} \, C \, \frac{1}{1} \, PC \, \frac{1\text{-}2\text{-}3\text{-}4\text{-}5}{1\text{-}2\text{-}3\text{-}4\text{-}5}$$

on each side for a complete complement of 34. Anomalies are not uncommon and usually involve lack of an incisor or presence of supernumerary postcanines.

The palate is more arched and therefore higher than in other phocids and the mandibles are deep. These features result in a comparatively large buccal cavity, an adaptation for a modified type of suction feeding (presumed). Mean condylobasal length for 63 adult Pacific bearded seals was 222.8 mm and zygomatic breadth was 130.3 mm (Manning, 1974). The rostrum is broad and rounded, supporting fleshy nasal pads with a large number of vibrissae (Fig. 2).

In a study of skull variation between the two recognized subspecies, *E. b. barbatus* and *E. b. nauticus*, Manning (1974) found that there was no variation between the sexes. Skulls from the Atlantic subspecies were significantly larger than in the Pacific form in the majority of measurements he analyzed. However, nasal breadth was significantly smaller in the Atlantic form. He concluded, based on his studies of skulls, that the recognition of two subspecies was justified. Conversely, Kosygin and Potelov (1971) concluded, based on analysis of 416 skulls, that division into two subspecies was not justified.

Fay (1967) reported that a bearded seal he examined had 15 thoracic vertebrae and pairs of ribs, typical of the order Pinnipedia.

The trachea of bearded seals is different from that of other North Pacific phocids in that most of the rings are incomplete. On the average there are 69 rings (64-81) of which only the first 6-12 are complete. In the remainder there is a broad elastic membrane between the ends of the flattened rings. Ends of the rings are most widely separated in the central portion of the trachea (Sokolov et al., 1968; Burns, unpublished). The function or functions of this type of trachea are unknown but may be related to production of the unique vocalizations in this animal. The hyoid arch is complete.

Lungs in the bearded seal are multilobular and generaly resemble those of ringed seals, having apical, cardiac, diaphragmatic and postcardiac lobes on the right lung and lacking the postcardiac on the left. These are different from the unlobulated lungs of ribbon seals (Sokolov et al., 1968).

In five large bearded seals, average length of the intestine was 2502 cm. The duodenum averaged 33.6 cm; small intestine, 2164.6 cm; and large intestine, 303.8 cm.

Average heart weight in adults was found to be about 1110 g.

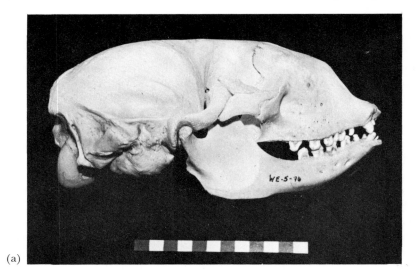

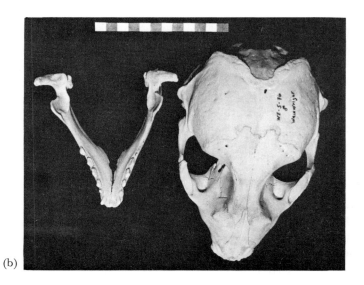

FIG. 7 Skull of a 15-month-old Pacific bearded seal: (a) lateral view, (b) dorsal view.

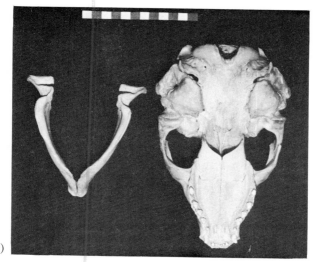

FIG. 7 (c) Ventral view.

Physiology

The physiology of bearded seals has essentially not been studied, probably because they are relatively difficult to capture alive, except for pups during the early stages of the short nursing period. Also, they grow to a large size. These seals have not been maintained in laboratories and only a very few physiologists have had access to fresh specimen material required for most physiological studies.

Fay et al. (1967) found that the modal number of chromosomes in bearded seals ($2N$) was 34.

In our work we found the haematocrit in 17 bearded seals to average 56.4%, ranging from 52.5% in a nursing pup to 60.3%. The muscle tissue is purple-coloured, indicating a high myoglobin content and the total blood volume (not measured) seems to be high.

All of our general impressions indicate that bearded seals exhibit the entire array of diving adaptations found, for instance, in harbour *(Phoca vitulina richardsi)*, spotted, ribbon and ringed seals. Reviews of these adaptations can be found in the numerous publications of L. Irving (1934-74), P. F. Scholander (1940-74) and R. W. Elsner (1964-79). Additional sources of information include Scheffer (1958), King (1964), Harrison and King (1980), Harrison et al. (1968) and Andersen (1969).

Disease

An exhaustive list of the helminth parasites in bearded seals is beyond the scope of this report. Interested readers are referred to the various writings of K. I. Skriabin, S. L. Delyamure (1955) and King (1964). Current studies of the helminth parasites and other pathologies in bearded seals in the Bering-Chukchi region are being conducted by American and Soviet investigators.

Fay *et al.* (1978) reported that the causes of natural mortality in these seals, other than predation by polar bears, are essentially unknown. The only major pathological findings in samples from the living population included helminthiasis of the liver and associated secondary bacterial invasion. This occured in five of 96 specimens examined. They reported other conditions including acute dermatomycosis, focal necrosis of the liver, trauma from unknown causes, biliary fibrosis, hepatitis and gastroduodenal ulcers.

The most commonly occurring helminth parasites in eight seals from the Bering-Chukchi region examined by Fay *et al.* (1978) included *Diphyllobothrium cordatum*, *D. lanceolatum*, *Pyramicocephalus phocarum* and *Corynosoma validum* which occurred in all; *Diphyllobothrium* sp. and *Phocanema dicipiens* which occurred in five; *Orthosplanchnus fraterculus* in four and *Contracaecum osculatum* in two. San Miguel sea-lion virus/VES virus was not found in one tested bearded seal and 11 seals were negative in tests for *Leptospira*.

A total of three bearded seal pups, one still-born and two new-born, with injuries resulting in death from internal haemorrhage was reported by Fay *et al.* (1978). Between 1962 and 1969, I found four dead pups, three still-born and one new-born, which had apparently died of similar causes. The incidence of death from trauma in term foetuses (during birth?) and shortly after birth seems high in relation to the limited opportunities we have had to detect this type of natural mortality.

Pesticides and heavy metals are present in tissues of bearded seals. Galster and Burns (1972) reported, DDT, DDD, DDE, dieldrin and PCBs in adipose tissues of polar bear, ringed seal, spotted seal, walrus, bearded seal and Steller sea lion *(Eumetopias jubata)* from the Bering-Chukchi region. Small concentrations of pesticides were present in nearly all samples and only PCBs occurred in what were thought to be high concentrations. The contaminant burden of the different species was greater than differences between areas. Of the pinnipeds, bearded seals had the highest concentrations of DDT residues ($0.330\ \mu g\ g^{-1}$) and sea lions the lowest (0.026 μg g^{-1}). Levels of dieldrin were low and varied from 0.5 to 0.1 the levels of DDT. Concentrations of PCBs were similar in all the pinnipeds and averaged $1.78 \pm .52 \mu$g g^{-1}.

Accumulations of mercury were present in samples of liver, muscle and fat of the Bering Sea pinnipeds. The average concentration of mercury in the examined tissues from bearded seals was 0.95 μg g^{-1}. It was four times more concentrated in the liver than in muscle or adipose tissue (Galster, 1971). Mercury and selenium were present in bearded seals from the Canadian Arctic (Smith and Armstrong, 1978).

Petrochemical contaminant levels in seals from the Bering-Chukchi region have, as yet, not been determined.

Acknowledgements

I wish to thank K. Frost and L. Lowry of the Alaska Department of Fish and Game for their contributions to this and other marine mammal studies in Alaska, and especially for their assumptions of tasks which provided time for me to write this chapter. The late Anthony Koezuna of King Island was attracted to bearded seals in a way which I was never able to understand. His knowledge, tutorial skills, interest and companionship kept us on the ice or in our kayaks whenever there was a chance to see or hear these seals. Dr F. H. Fay has assisted my endeavours in many ways, perhaps most importantly by our discussions and his suggestions over many years. L. Shults, University of Alaska, has been a pleasurable and helpful field companion and provided the photograph used in this chapter. Many members of the Alaskan Department of Fish and Game have contributed to these studies of marine mammals. Their contributions are appreciated. From 1962-75 this work was supported by funds from Federal Aid in Wildlife Restoration Projects. Funding since 1975 was mainly from the Bureau of Land Management, Outer Continental Shelf Environmental Assessment Program, Contract Number 03-5-022-53.

References

Allen, J. A. (1880). History of North American pinnipeds. US Geol. and Geogr. Surv., Washington. Misc. Publ. 12 xvi. 785 pp.

Andersen, H. T. (Ed.) (1969). "The Biology of Marine Mammals", Academic Press, New York and London.
Benjaminsen, T. (1973). Age determination and the growth and age distribution from cementum layers of bearded seals at Svalbard. *Fisk Dir. Skr. Ser. Hav Unders.* **16**, 159-170.
Burns, J. J. (1967). The Pacific bearded seal. Alaska Dept. Fish and Game, Juneau. 66 pp.
Burns, J. J. (1970). Remarks on the distribution and natural history of pagophilic pinnipeds in the Bering and Chukchi Seas. *J. Mammal.* **51**, 445-454.
Burns, J. J., Ray, G. C., Fay, F. H. and Shaughnessy, P. D. (1972). Adoption of a strange pup by the ice-inhabiting harbor seal, *Phoca vitulina largha. J. Mammal.* **53**, 594-598.
Burns, J. J. and Eley, T. J. (1978). The natural history and ecology of the bearded seal *(Erignathus barbatus)* and the ringed seal *(Phoca hispida)*. *Annu. Rep.* BLM/OCSEAP Contract 03-5-022-53. 61 pp.
Bychkov, V. A. (1971). A review of the conditions of the pinniped fauna of the USSR. Sci. Principles of the Conserv. of Nature, Moscow. 59-74. (Transl. Bureau Foreign Language Division, Canada, No. 0929.)
Chapskii, K. K. (1938). The bearded seal *(Erignathus barbatus*, Fabr.) of the Kara and Barents Seas. *Trans. Arctic Inst., Leningrad* **123**, 7-70. (Transl. Dept. Sec. State of Canada.)
Chapskii, K. K. 1966. Contemporary situation and the task in renewal of marine hunting industry resources. *In* "Marine Mammals", (Eds V. A. Arseniev, B. A. Zenkovich and K. K. Chapskii), Nauka. Moscow, USSR. (Orig. not seen).
Delyamure, S. L. (1955). "Helminthofauna of Marine Mammals: (Ecology and Phylogeny)", Izd. Akad. Nauk SSSR, Moscow. (Transl. Israel Prog. Sci. Transl.) 517 pp.
Dubrovskii, A. (1937). On the nuptial cry of the bearded seal. *Priroda* **4**. (Orig. not seen.)
Fay, F. H. (1967). The number of ribs and thoracic vertebrae in pinnipeds. *J. Mammal.* **48 (1)**, 144.
Fay, F. H. (1974). The role of ice in the ecology of marine mammals of the Bering Sea. *In* "Oceanography of the Bering Sea", (Eds D. W. Hood and E. J. Kelley), 383-389. Inst. Mar. Sci., Univ. of Alaska, Fairbanks.
Fay, F. H., Rausch, V. R. and Feltz, E. T. (1967). Cytogenetic comparison of some pinnipeds (Mammalia: Eutheria). *Can. J. Zool.* **45**, 773-778.
Fay, F. H. and Ray, C. (1968). Influence of climate on the distribution of walruses, *Odobenus rosmarus* (Linnaeus). I. Evidence from thermoregulatory behavior. *Zoologica* **53**, 1-32.
Fay, F. H., Dieterich, R. A. and Shults, L. M. (1978). Morbidity and mortality of marine mammals. *Annu. Rep.* BLM/OCSEAP Contract 03-5-022-56. 38 pp.
Fedoseev, G. A. and Bukhtiyarov, Yu. A. (1972). The diet of seals of the Okhotsk Sea. Fifth All-Union Conf. Studies of Marine Mammals (USSR), Part 1:110-112. (Transl. F. H. Fay.)

Feltz, E. T. and Fay, F. H. (1966). Thermal requirements *in vitro* of epidermal cells from seals. *Cryobiology* **3 (3),** 261-264.

Galster, W. R. (1971). Accumulation of mercury in Alaskan pinnipeds. *In* Proc. 22nd Alaska Sci. Conf., Fairbanks, 76. (Abstract.)

Galster, W. R. and Burns, J. J. (1972). Accumulation of pesticides in Alaskan marine mammals. Proc. 23rd Alaska Sci. Conf., Fairbanks. (Abstract)

Harrison, R. J. and King, J. E. (1980). "Marine Mammals", 2nd edn. Hutchinson, London. 192 pp.

Harrison, R. J., Hubbard, R. C., Peterson, R. S., Rice, C. E. and Shusterman, R. J. (Eds). (1968). "The Behavior and Physiology of Pinnipeds", Appleton-Century-Crofts, New York. 411 pp.

Heptner, V. G. (1976). Mammals of the Soviet Union. Vol. 2. Pinnipeds and toothed whales. Publishing House for Higher Schools, Moscow. 718pp. (in Russian.)

Johnson, M. L., Fiscus, C. H., Ostenson, B. T. and Barbour, M. L. (1966). Marine mammals. *In* "Environment of the Cape Thompson Region, Alaska", (Eds N. J. Wilimovsky and J. N. Wolfe), 897-924. US Atomic Energy Commission, Oak Ridge, TE.

Kenyon, K. W. (1962). Notes on the phocid seals at Little Diomede Island, Alaska. *J. Wildl. Manage.* **26 (4),** 380-387.

King, J. E. (1964). "Seals of the World", British Museum (Natural History), London.

Kosygin, G. M. (1966). Some data on the feeding of the bearded seals in the Bering Sea during the spring-summer months. *Izv. TINRO* **58,** 153-157. (Transl. U.S. Bureau Comm. Fish).

Kosygin, G. M. (1971). Feeding of the bearded seal *Erignathus barbatus nauticus* (Pallas) in the Bering Sea during the spring-summer period. *Izv. TINRO* **75,** 144-151. (in Russian.)

Kosygin, G. M. and Potelov, V. A. (1971). Age, sex and population variability of the craniological characters of the bearded seals. *Izv. TINRO* **80,** 266-288.

Lowry, L. F., Frost, K. J. and Burns, J. J. (1977). Trophic relationships among ice inhabiting phocid seals. *Annu. Rep.* BLM/OCSEAP Contract 03-5-022-53. 59 pp.

Lowry, L. F., Frost, K. J. and Burns, J. J. (1978). Trophic relationships among ice inhabiting phocid seals. *Annu. Rep.* BLM/OCSEAP Contract 03-5-022-53. 68 pp.

Manning, T. H. (1974). Variations in the skull of the bearded seal. *Biol. Pap., University of Alaska, Fairbanks.* **16,** 1-21.

McLaren, I. A. (1958). Some aspects of growth and reproduction of the bearded seal, *Erignathus barbatus* (Erxleben). *J. Fish. Res. Bd. Can.* **15,** 219-227.

Mohr, E. (1952) "Die Robben der Europaischen Gewasser", Monographien der Wildsaugetiere, Verlag Dr Paul Schops, Frankfurt am Main. (Partial transl., R. L. Rausch.)

Pikharev, G. A. (1940). Some data on the feeding of the Pacific bearded seal. *Izv. TINRO* **20**, 101-120. (In Russian.)

Plekhanov, P. (1933). Determination of the age of seals. *Soviet North* **4**, 111-114. (In Russian.)

Potelov, V. A. (1975). Reproduction of the bearded seal (*Erignathus barbatus* in the Barents Sea. *Rapp. P. -v. Réun. Cons. Int. Explor. Mer* **169**, 554.

Ray, C., Watkins, W. A. and Burns, J. J. (1969). The underwater song of *Erignathus* (bearded seal). *Zoologica* **54**, 79-83.

Scheffer, V. B. (1958). "Seals, Sea Lions and Walruses", Stanford University Press, Stanford, CA.

Sleptsov, M. M. (1943). On the biology of reproduction of Pinnipedia of the far east. *Zool. Zhur.* **22**, 109-128. (In Russian.)

Smith, T. G. and Memogana, J. (1977). Disorientation in ringed and bearded seals. *Can. Field-Nat.* **91**, 181-182.

Smith, T. G. and Armstrong, F. A. J. (1978). Mercury and selenium in ringed and bearded seal tissues from arctic Canada. *Arctic* **31**, 75-84.

Sokolov, A. S., Kosygin, G. M. and Shustov, A. P. (1968). Structure of the lungs and trachea of Bering Sea pinnipeds. *Izv. TINRO* **62**, 252-263. (In Russian.)

Stirling, I., Archibald, R. and DeMaster, D. (1977). Distribution and abundance of seals in the eastern Beaufort Sea. *J. Fish. Res. Bd. Can.* **34**, 976-988.

Tavrovskii, V. A. (1971). Mammals of the Yakutia. Izdalel'stvo Nauk., Moscow. (In Russian.)

Tikhomirov, E. A. (1961). Distribution and migration of seals in waters of the Far East. *In* "Transactions of the Conference on Ecology and Hunting of Marine Mammals", (Eds E. H. Pavlovskii and S. K. Kleinenberg), 199-200. Akad. Nauk SSR, Ikhtiol. Comm., Moscow. (Transl. US Fish Wildl. Serv.)

Tikhomirov, E. A. (1966). On the reproduction of seals belonging to the family Phocidae in the North Pacific. *Zool. Zhur.* **45**, 275-281. (Transl. Fish. Res. Bd. Can.)

7

Hooded Seal
Cystophora cristata Erxleben, 1777

Randall R. Reeves and John K. Ling

Genus and Species

The hooded seal has a somewhat confusing taxonomic history. There has never been doubt as to the monotypic nature of the genus *Cystophora* (Allen, 1880), but its familial affinities have been difficult to establish. Two modern synoptic reviews of the order Pinnipedia (Scheffer, 1958; King, 1964) recognized three subfamilies of the family Phocidae Brookes, 1828 (earless seals): the Phocinae Gill, 1866; the Monachinae Trouessart, 1897; and the Cystophorinae Gill, 1866. The Cystophorinae encompassed two genera: *Cystophora* Nilsson, 1820 and *Mirounga* Gray, 1827. Their linkage rested on three striking similarities between hooded seals and elephant seals, namely: (1) possession of an inflatable nasal sac, (2) reduction of the incisors to 2/1, and (3) similarity in the shape of the postcanine teeth.

King (1966) examined and reviewed skeletal features and some soft parts hitherto ignored, and established 17 good characters likening

Cystophora to other northern phocids and *Mirounga* to other southern phocids. She resolved to abandon the Cystophorinae altogether, thereby reassigning *Cystophora* to the Phocinae (northern phocids) and *Mirounga* to the Monachinae (southern phocids). King dismissed dental characters as "unreliable" and attributed the development of nasal sacs in both genera to convergent evolution. The findings of Burns and Fay (1970) supported King's position. They further proposed the monotypic tribe Cystophorini as one of three groupings within the Phocinae (along with Phocini and Erignathini). The distinguishing features of the Cystophorini are the incisor formula and the relatively broader anterior half of the interorbital septum.

External Characteristics and Morphology

Coloration

The young develop a creamy grey coat *in utero* which is shed before birth and replaced by a short slate-blue pelage (the "blueback" stage). Ventral coloration is lighter, becoming cream. The face is dark.

After the yearling coat is shed at the first annual moult, mottled pelage patterns begin to appear. These consist of dark brown, brownish-black, or black patches against a silvery grey background all over the body, more pronounced on the back and upper sides. The patterns are mottled and irregular. Upon entering the water hooded seals lose much of the mottled colour pattern and assume a more uniform dark grey hue.

Dimensions

Males (Fig. 1a) are larger than females (Fig. 1b) at maturity. On average males reach 260 cm in standard length compared with 200 cm for females. Males weigh 192-352 kg (Ognev, 1962); adult females generally weigh 145-300 kg (Shepeleva, 1973). The pelt and blubber together weigh 60-150 kg, and the mean blubber width is 3.4-4.5 cm.

Distribution and Migration

The normal distribution of hooded seals is limited to the arctic and subarctic North Atlantic (Fig. 2). Subadult males occasionally reach the western Beaufort Sea (Burns and Gavin, 1980), but hooded seals are not common west of Baffin Bay and Davis Strait. Vibe (1950) reported

FIG. 1 (a)Adult male hooded seal; (b) adult female hooded seal with pup. (Photographs by Fred Bruemmer).

on their seasonal appearance in good numbers near Cape York, and Kapel (1975) confirmed their regular inclusion in the seal catch by West Greenlanders at various stations. Hooded seals do not, as a rule, penetrate into Hudson Bay. Whelping areas in Davis Strait at about 64° N, off Labrador, and inside the Gulf of St Lawrence are occupied at least seasonally (Sergeant, 1974). Stray individual females have successfully pupped on the Maine coast as recently as 1974 (Richardson, 1975), and extralimital occurrences of juveniles are on record for much of the US eastern seaboard south to Florida (summarized by Sergeant, 1974).

The hooded seal is abundant along the south coast of Greenland (Kapel, 1975) and congregates in great numbers to moult in Denmark Strait. Scoresby Sound was regarded by Pedersen (1942) as the northern boundary of the species' range off East Greenland, but Rasmussen (1960) asserted that both old and young are commonly found in pack ice northeast of Greenland to about 77° N. The most important whelping site in the eastern sector is the pack ice near Jan Mayen, and outside the breeding and moulting seasons seals range between Bear Island and Svalbard. They are generally absent from the shallow Barents Sea, although stragglers have been reported as far east as the mouth of the Kara Sea and the Yenisey River (Ognev, 1962). Occasional strays are known from European coasts south to the Bay of Biscay (King, 1964).

The hooded seal's range and relative abundance apparently vary dramatically with changes in climate and ice cover. Rasmussen (1960) demonstrated by analysis of catch curves that the population of hooded seals breeding at Newfoundland declined significantly during the first part of this century. He interpreted the simultaneous increase in catches at the Jan Mayen breeding ground as evidence that the bulk of the stock had relocated from west to east. A marked reduction in availability of suitable whelping ice at Newfoundland was, according to Rasmussen, the chief impetus for this shift. Vibe (1967), using historical catch data for South and West Greenland, arrived at a similar conclusion—that from 1860 to 1910, the so-called "drift-ice pulsation stage", hooded seals abounded in Davis Strait; but that after 1910, as "East Greenland Ice" decreased in Davis Strait and the climate moderated, they began to disappear from there. Since 1965 there has been a marked increase in the catch at both Newfoundland and south-west Greenland, but this has been interpreted by Kapel (1975) as evidence of the effectiveness of protection afforded hooded seals at Denmark Strait since 1961, rather than as a response to changing climate or ice conditions.

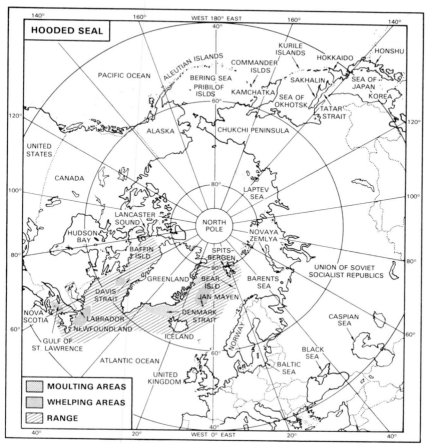

FIG. 2 Map showing distribution of the hooded seal.

Timing and routes of migration

The lives of hooded seals appear to be governed by a regular schedule of migration and dispersal that can be thought of as beginning at the summer moulting grounds on pack ice in Denmark Strait between 66° and 68° N (Rasmussen, 1960) and east of Greenland between 72° and 76° N (Sergeant, 1974, 1976). Judging by the absence of any other known major moulting area and the scant evidence from tag recoveries, virtually the entire population of adult hooded seals assembles here, reaching a peak of abundance between 15 June and 15 July, and remaining into August (Rasmussen, 1960).

After moulting, the seals disperse widely, presumably in search of food. Part of the population moves south and westward around Cape Farewell and then north along the coast of West Greenland, some reaching the Thule district by late August and September (Kapel, 1975). Others spread out north and east of Denmark Strait, becoming especially common in pack ice between Greenland and Spitsbergen during late summer and early fall (Rasmussen, 1957).

The winter whereabouts of hooded seals east of Greenland is poorly known, but according to Kapel (1975) a few are killed during that season in fjords near Angmagssalik. In the western sector many hooded seals appear to feed on the Grand Banks off Newfoundland during winter (Rasmussen, 1960); Sergeant (1976) reported that immature animals are netted in early winter off Labrador and northeast Newfoundland.

In February adults begin to concentrate near thick sea ice, where whelping and mating take place in the second half of March. Until very recently, only two major whelping patches were known—the largest in the "West Ice" near Jan Mayen and the other at the "Front" off Newfoundland. A small number of seals are known to pup in the Gulf of St Lawrence. In 1974 a large but long-forgotten whelping patch at 62-64° N, 56-60° W in Davis Strait was rediscovered (Sergeant, 1974).

After breeding, most adult seals work their way back towards Denmark Strait to moult, feeding heavily *en route*. This movement creates a strong influx of seals to the drift ice off South Greenland in May and June, presumably consisting of animals that have whelped and bred in Davis Strait or off Newfoundland (Kapel, 1975). Seals from Jan Mayen presumably proceed to feeding banks north of Iceland and east and south of Angmagssalik (Rasmussen, 1960). Small numbers are believed to move north into Davis Strait or Baffin Bay to moult; others moult well north of Denmark Strait (Rasmussen, 1960).

The migratory pattern described above applies principally to adults. There is evidence suggesting that young of the year generally avoid the large moulting patch(es) and that immatures occur only at the fringes of the whelping and breeding patches (Rasmussen, 1960). Sergeant (1976) indicated that young hooded seals in their first summer "have a greater tendency to wander outside the normal range than young harps."

Stock identity

The relationship and degree of mixing between the Jan Mayen, Davis Strait, and Newfoundland populations are still not fully understood. Although there once were thought to be separate populations, limited tagging studies carried out since the early 1950s suggest a single, well-integrated population that mixes regularly during the moult in Denmark Strait (Rasmussen, 1960; Rasmussen and Øritsland, 1964; Sergeant, 1974). Individuals tagged off Newfoundland or in the Gulf of St Lawrence have been recaptured off south-west, south, and south-east Greenland (Sergeant, 1974); several seals tagged at Denmark Strait have been recovered at West Greenland and Newfoundland (Anon., 1977). It should be noted, however, that no tagged Jan Mayen hooded seals have been recovered in or *en route* to Newfoundland or Davis Strait, and vice versa.

The coincidence in timing of births at all three major whelping grounds adds strength to the argument for a panmictic stock (Sergeant, 1974). Homing, however, in the sense of allegiance to a particular breeding area, may still ensure the genetic isolation of discrete stocks. Genetic interchange is, perhaps, most likely between the Davis Strait and Newfoundland "stocks". Sergeant (1974) believed the former to provide recruits for the heavily exploited herd at the "Front" off Newfoundland.

Electrophoretic studies of blood proteins and analysis of pancreas amylase have been inconclusive, although considerable intraspecific variation was detected (Naevdal, 1971). Cranial studies of variation in hooded seal populations are in progress (Sergeant, 1974).

Abundance and Life History

Hooded seals are considerably less abundant than their close neighbours on the North Atlantic sea ice—harp seals *(Phoca groenlandica)*—and this apparently has always been true. However, it has been extremely difficult, because of their preference for heavy ice far from shore, to develop population estimates for hooded seals. Sergeant (1974) guessed that about 50 000 seals were present at the Davis Strait whelping patch in March, 1974. When relocated in 1976 this patch was judged to be considerably smaller (Anon., 1976), but in 1977 an aerial count showed a minimum of 13 000 pups, implying a total Davis Strait breeding population of at least 42 000 (MacLaren Atlantic Limited, 1977).

The whelping population at the "Front" off Newfoundland is generally too scattered and inaccessible for aerial censusing, and most estimates for it have been derived statistically from catch curves. In

1977 there were thought to be about 77 000 females one year of age or older, implying the production of about 38 000 pups there that year (Anon., 1978). This level is believed to represent approximately a 50% increase since the early 1960s.

The Jan Mayen ("West Ice") breeding patch has not been censused. However, in summer 1959 Øritsland (1959) succeeded in censusing from the air a moulting patch in Denmark Strait consisting of about 230 000 seals. Øritsland (1960) gave 505 000 as a tentative figure for the entire North Atlantic population. Rasmussen (1960), citing Scheffer's (1958) world estimate of 300 000-500 000, expressed the belief that the actual number is well in excess of half a million.

Exploitation and conservation

Hooded and harp seals, because they occur in many of the same areas and breed within days or weeks of each other, have traditionally been hunted, and more recently managed, jointly. Rasmussen (1957, 1960) and Sergeant (1976) have provided particularly useful reviews. The seal fishery at Jan Mayen began in the eighteenth century, primarily as a "filling-up" exercise by European whalers. Although hooded seals were hunted intensively on moulting grounds in Denmark Strait as early as 1874, they did not become a significant part of the catch at Jan Mayen until the 1920s. Commercial sealing at Newfoundland became a major late-winter occupation in the late eighteenth century and has persisted since then. Norway, and to a much smaller degree Canada and the Soviet Union, have done most of the commercial hooded seal hunting in this century. In addition, Greenlanders have always hunted them as available—both for subsistence and sale of pelts (Vibe, 1950; Kapel, 1975). In recent years this native harvest has amounted to several thousand seals (mainly one year or older) per annum.

Commercial sealing effort has focused on new-born pups, and the main products before the 1930s were oil and leather. More recently fine furs have become fashionable, and the "blueback" coat of the pup is regarded as the most desirable hair seal pelt on the European market (Sergeant, 1976). The capture of the pup often requires the killing of the aggressive mother, and this has made the proportion of adults in the catch at pupping sites rather high: 28% at Newfoundland in 1895-1913; 33% in 1914-31; and 35% in 1932-61 (Mercer, 1976). On the moulting grounds in Denmark Strait the catch was believed to be fairly representative of the entire population, although age composition there has been known to vary considerably (Rasmussen, 1960). Commercial catches at Newfoundland by quinquennia since 1945 are listed in Table 1.

TABLE 1 Average annual reported Canadian and Norwegian hooded seal catches at Newfoundland and in the Gulf of St Lawrence, by quinquennia between 1946 and 1975.

Quinquennium	Canada	Norway	Total
1946-50	5 707	355	6 062
1951-55	2 373	3 598	5 972
1956-60	822	5 633	6 455
1961-65	1 471	3 572	5 043
1966-70	2 209	11 334	13 543
1971-75	1 351	10 594	11 945

(From: International Commission for Northwest Atlantic Fisheries, Statistical Bulletin 18, Dartmouth, Nova Scotia, 1970; and subsequent volumes of same.)

Attempts by sealing nations to regulate the take of hooded seals have been summarized by Sergeant (1976). In 1958 opening and closing dates for sealing at the Jan Mayen breeding grounds and the Denmark Strait moulting patch were set by Norway, in agreement with the Soviet Union (Anon., 1959), which had joined the Jan Mayen fishery in the 1950s. Both nations also agreed to refrain from summer hunting in the pack ice north of Jan Mayen (Rasmussen, 1960).

Norway's summer hunt at Denmark Strait was closed in 1961. According to Rasmussen (1960), the quality of pelts from moulting seals was low anyway, and this hunt was economical only so long as it was supplemented by a catch of Greenland sharks *(Somniosus microcephalus)*. The Norwegian fishery at Jan Mayen, aimed principally at bluebacks, persists, but since 1972 a quota of 30 000 seals has been in effect (Sergeant, 1976).

In the north-west Atlantic, commercial sealing has, since 1967, been conducted under terms set by the International Commission for Northwest Atlantic Fisheries (ICNAF), and more recently (since January 1, 1977) by unilateral Canadian action resulting from 200-mile limit legislation. Canada and Norway had already agreed in 1961 to a closing date for the Newfoundland fishery, and since 1968 an opening date has also been enforced. The Gulf of St Lawrence was closed to commercial sealing in 1972. An annual catch quota of 15 000 hooded seals was instituted in 1974 for the international (i.e. Norwegian-Canadian) hunt at the "Front" off Newfoundland, and it remained at this level through the 1979 sealing season. Also, current regulations restrict the catch of adult females at Newfoundland to 5% of the total catch (M. C. Mercer, in litt., April 12, 1979). There is no quota for the Davis Strait whelping patch; it is, in effect, protected for the moment. Hunting by Greenlanders and Canadian natives is unregulated.

Predation

Man is, without a doubt, the most serious predator, although polar bears *(Ursus maritimus)* probably do considerable damage at whelping grounds (MacLaren Atlantic Limited, 1977). It is also possible that Greenland sharks prey at least on young or disabled individuals.

Feeding

The hooded seal is a pelagic, deep-diving animal normally found in deep water or hauled out on heavy drift ice. It does not appear to compete with other pinnipeds for either food or space. It follows a rather taxing regimen of feast periods alternating with fasts, demonstrated by the condition of animals caught at different seasons. Rasmussen (1960) found mean blubber thickness in adults to vary from 4.8 cm at the beginning of the breeding season to 3.6 cm after the pups were weaned. Between early April and mid-June their condition improves, and Rasmussen found mean blubber thicknesses as high as 5.19 cm on animals newly arrived at Denmark Strait to moult. They become considerably thinner during the moult, mean blubber thickness declines to 3.05 cm by late June. After moulting, the seals feast so that by the following breeding season they have once again reached prime condition. Barren females are, predictably, in noticeably better condition at the end of the breeding season than females that have pupped, and immatures are generally fatter than adult males, for which the exigencies of mating are energetically costly (Shepeleva, 1973).

Because hooded seals fast while breeding and moulting (the periods during which they are sampled most thoroughly), their feeding behaviour and dietary preferences are not well known (King, 1964). Squid *(Gonatus fabricii)* and redfish *(Sebastes marinus)* may be important prey (Sergeant, 1976), and octopus, herring, capelin, halibut, cod and even shrimps, mussels, and starfish are reportedly consumed as well.

Population structure and mortality rates

The sex ratio at birth is close to 1:1, which probably does not change greatly with age under natural conditions (Øritsland, 1964; Øritsland and Benjaminsen, 1975). Such a situation may be expected in a monogamously breeding species.

In a seal population that has experienced a century of nearly continuous exploitation, it is difficult to determine natural mortality rates. Even estimates of total mortality have been elusive for hooded seals, as most samples are subject to age- or sex-related bias. Øritsland and Benjaminsen (1975), using catch curves and mean ages of fully recruited age-groups, calculated total annual mortality for adult females (older than 6 years) at Newfoundland as 16%; adult males (older than 10 years), as 23%. They explained the difference by referring to a sizable hidden (i.e. unreported) hunting mortality for adult males, which are most likely to be found near the edges of ice floes where they are often shot and lost as they slip into the water. Mean total annual mortality of adults at Newfoundland may be 22% (based on 1971-74 catch curves), from which a natural mortality rate of 11% can be derived (Anon., 1976).

Maximum age in both sexes is scarcely more than 35 years (Øritsland and Benjaminsen, 1975).

Parasites, diseases and injuries

The helminth parasites known to infest hooded seals were listed by Dailey (1975). There is no reason to suppose that such infestations are lethal for individuals whose overall health is good.

King (1964) reviewed what little is known about disease afflicting hooded seals, most of it based on experience with captives. Tuberculous lungs and a severe infection of the cranial cavity were noted. A bacterial infection of salmonellosis was implicated in septicaemia in one hooded seal (Ridgway et al., 1975).

Rasmussen (1960) suggested that ice accidents and storms may injure or kill wild hooded seals, especially pups.

Internal Anatomical Characteristics

Skull

In comparison with other phocids, the hooded seal has an extremely short cranium (Fig. 3), with a very long, wide snout (King, 1972). Most of the rather extensive modifications of the skull appear to be associated with the species' inflatable nasal appendage (see section on "Nasal Sac", p. 185).

Like other phocids with small crania and large snouts, *Cystophora* has its interorbital area toward the back of the skull. *Cystophora* and *Mirounga* are unique in that the dorsal junction of the maxilla and jugal projects laterally and forms a shoulder rather than a smooth outline. The orbit proper does not appear to begin until the start of the jugal. King suggested that this adaptation may enable the seal to see round its obtrusive nasal sac.

The fronto-nasal area is elevated in *Cystophora*, as it is in *Mirounga* and *Halichoerus*. The nasal openings are wider in *Cystophora* than in any other phocid. The premaxillae do not meet the nasals; they have a distinct dorsal ridge. The narial basin is "key-hole shaped" rather than oval as in other phocids (except *Halichoerus*). The nasals project beyond the front edge of the maxilla by about a third of their length, which creates significantly more "overhang" above the narial basin than exists in other phocids. The amount of maxilla between the preorbital process and the rear edge of the narial basin is much reduced in *Cystophora*, creating a creased or "pushed back" appearance of the maxilla.

The palate is considerably elongated posteriorly, more so than in any other phocid. The hamular process of the pterygoid is much closer to the anterior border of the tympanic bulla than to the front edge of the

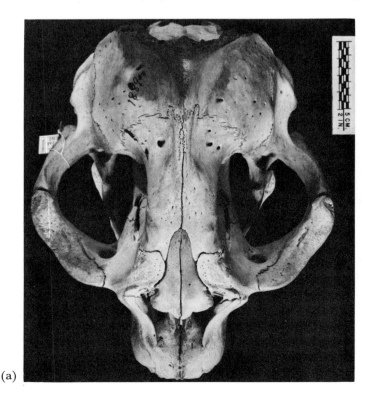

(a)

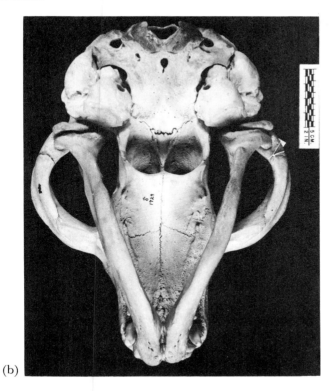

(b)

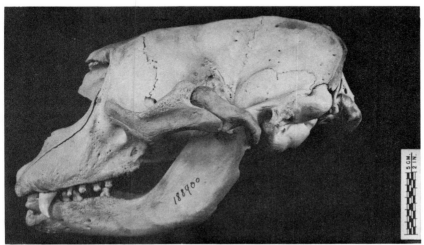

FIG. 3 Hooded seal skull: (a) dorsal view; (b) ventral view; (c) lateral view; (Photographs from Smithsonian Institution, Washington, D.C.)

zygomatic arch. In this character *Cystophora* resembles *Mirounga* and *Lobodon*.

In *Cystophora* the mastoid region is visible on the dorsal aspect of the skull, in common with all northern phocids (and *Monachus*), whereas ventrally part of the petrosal is visible projecting into the foramen lacerum posterius (King, 1966). The ear ossicles are similar to those of other northern phocids.

Skeleton

Except as noted, the following description is taken from King (1966), who carried out detailed comparison of, *inter alia*, skeletal characters in *Cystophora* and *Mirounga*.

The vertebrarterial foramen faces posteriorly. The atlantal facet of the axis vertebra tends to form one continuous surface with the two articulating surfaces. There is a well-formed spine along the entire length of the scapula. A supracondylar foramen is present on the humerus which also tends to develop a supinator ridge. The palmar side of the cuneiform has a distally projecting ledge. Metacarpals of digits 1 and 2 are of almost equal size (69:61 mm), and there is no reduction in size of the intermediate phalanx of digit 5. The heads of the metacarpals, moreover, have a longitudinal ridge which extends on to the palmar surface. The proximal surfaces of the proximal phalanges have a deep indentation on their palmar border and a deeply concave articular surface corresponding with the ridge on the metacarpal heads. Interphalangeal articulations are trochlear.

The ilium is everted, with a deep lateral excavation. In the femur the pit for the origin of the popliteus muscle is more pronounced; there is a distinct trochanteric fossa; and the greater trochanter is generally higher than the head. The post-tibial fossa tends to be more pronounced. Metatarsal 3 is about two-thirds the length of the first. *Cystophora* also has the ability to flex its terminal phalangeal joint of the pes to afford a better grip on ice. Claws are large on both fore and hind flippers.

Tracheal rings are complete in *Cystophora,* and the larynx is similar to those of other northern phocids (Schneider, 1963).

Teeth

The teeth are small and the tooth row short (King, 1972). The dental formula (Ognev, 1962) is

$$I\frac{2}{1} C \frac{1}{1} PC \frac{5}{5}$$

Genetics

The hooded seal has a chromosome number of $2n = 34$ (Árnason, 1972), and its karyotype is virtually identical to that of *Mirounga angustirostris* (Árnason, 1974).

Nasal sac

The hooded or bladdernose seal derives its name from the inflatable nasal sac, which is developed only in males. It reaches its fullest size in sexually mature bulls which use it for display during courtship fighting. Until recently, the nasal appendage of the hooded seal and the proboscis of the elephant seals were prime characters used by systematists to link these two genera in the one subfamily Cystophorinae. That there are differences in the structure of the nasal sacs in these two genera, however, has been realized for some time (King, 1966). For example, the hood of *Cystophora* is inflated by air, whereas the proboscis of *Mirounga* is enlarged by muscular action. Moreover, a fiery red bladder may be forced out of one nostril by air pressure within the hood acting upon the nasal septum.

Berland (1958, 1966) examined and described the hood of *Cystophora* in detail and wrote (1958): "In males having a well-developed hood, the mucous nasal septum which divides the nasal cavity into two equal halves is also very well-developed and is rather thick and elastic, the cartilaginous nasal septum remaining short. The extrusion of the bladder may be explained as follows: by closing one nostril, and by blowing out air through the nose, the mucous nasal septum is made to bulge into the opposite half of the nasal cavity and is extruded through the opposite nostril as a red bladder. The extrusion of this bladder appears to be confined to the breeding and mating season in the spring." Miller and Boness (1979) diagrammed male threat postures involving hood extrusion.

Behaviour

Diving and swimming capabilities

In his now classical studies on diving in mammals and birds, Scholander (1940) investigated the physiological aspects of diving in two species of seals, *Halichoerus grypus* and *Cystophora cristata*. Using animals only a few months old in the restrained state, Scholander recorded very similar cardiovascular responses in both species.

A month-old hooded seal descended to a depth of 75 m on its first recorded dive. Such diving ability is enhanced by the lungs becoming practically atelectatic without losing contact with the chest wall (Scholander, 1940). Fatal air embolism was induced in a forced submersion to 300 m in 3 min and resurfacing 9 min later. Scholander believed that the seal had not readied itself for the dive by expelling all of the air from its lungs.

Pronounced bradycardia was observed in *Cystophora*; a decrease in heart rate from 100 to 10 beats per min was not uncommon. Oxygen consumption drops to a third during a dive since the capacity for oxygen storage cannot be increased by more than about 5%. Oxygen consumption in resting seals, however, is high: 200-300 ml min^{-1} in a 29 kg animal. Stored oxygen is sufficient for a dive lasting only 5-6 min, but submergence for periods up to 18 min is made possible by the reduced consumption rate.

The precocious hooded seal pup is remarkably well equipped to swim and dive. Shepeleva (1973) reported on one removed surgically from its mother several days before natural birth. It was "capable of moving about like an adult within 30 minutes". During its first days of life it swam, dived, and crawled onto the ice.

Sound production

Schevill *et al.* (1963) recorded calls of a young male hooded seal in captivity, and Terhune and Ronald (1973) described the calls of wild adults and young in both air and water. The latter are believed to be the only recorded sounds of non-captive hooded seals.

Adult male calls in air coincide with the inflation and deflation of the proboscis (sic), but as they are of low intensity these calls are believed to be only incidental to the spectacular visual display of the inflated bladder. Underwater calls include series of rapid (110 s^{-1}) clicks and low-frequency pulsed sounds. Onomatopoetically, these calls, the *a*, *b*, and *c*, sound like "grung", "snort", and "buzz" respectively.

The calls of adult females and pups are pulsed at varying rates and amounts which have not been correlated with simultaneous behavioural observations. Thus the sounds of females and pups may simply be variations on the same kind of call.

Temperature regulation

The hooded seal is a cold-adapted species. Shepeleva (1973) made a detailed study of skin, body, and organ temperatures under various environmental conditions and was impressed by the ability of pups as well as adults to maintain a high body temperature (ranging between 35.4° and 37.6 °C). As an example, a seven-day-old pup lying in a strong wind with the air temperature −9.5 °C had a skin temperature of only 3.0 °C on the membrane of its hind flippers, but its core temperature remained at 36.4 °C. The temperature of the flipper membranes is governed by vascular reactions, whereas much of the rest of the body has the insulation of the thick blubber layer. Under windy or heavy frost conditions pups retract their heads and press their limbs tightly against the body to reduce body surface and therefore heat loss. A shivering reflex was induced in pups exposed to wind or immersed in cold water. The area around the nose consistently has the highest skin temperature, due to exhalation, and thus is probably the site of greatest heat loss.

Reproduction

Parturition

Birth takes place primarily during the second half of March (Sergeant, 1976), although pupping at Newfoundland has been observed as early as 11 March and as late as 5 April (Rasmussen, 1960) and at Jan Mayen starting on 10 March and continuing up to 10 April (Øritsland, 1964). Birth usually occurs at the centre of a large ice floe (Terhune and Ronald, 1973). Twinning has not been reported for this species. The female attends the pup on the ice and is herself courted by at least one (or as many as seven) adult males, which often remain in the water (Shepeleva, 1973; Sergeant, 1976). Such groupings are usually referred to as "families", although the appropriateness of this term is questionable since the pair bond is not known to be particularly stable. The female is aggressive in defending her pup, and such behaviour poses a problem for management by making it difficult to spare the mother while capturing the blueback. Haigh and Stewart (1979), while attempting to narcotize and tag hooded seals on the whelping ice, found the critical distance in approaching females to be about 5 m.

Size of the neonate

Length at birth is about 87-115 cm (mean 100 cm) (Rasmussen, 1960), weight about 23-30 kg (Shepeleva, 1973).

Lactation

Lactation lasts for 7-12 days, with 3-4 nursing episodes per day (Shepeleva, 1973). The pup consumes about 2 kg of milk daily. According to Shepeleva, the pup's intestine remains closed for the entire nursing period. A dense plug consisting of fur and faeces is formed near the anus at the time of the embryonal moult, ensuring retention and maximum utilization of the mother's milk.

Ridgway et al. (1975), using Sivertsen's (1941) data on hooded seal milk, showed it to be relatively dilute (49.8% water) compared to that of other polar pinnipeds. (They indicated 40.4% fat, 6.7% protein, and 0.86% ash.) Findings of Shepeleva (1973), however, suggest that Sivertsen's milk composition figures are misleading (no doubt due to a limited sample). Shepeleva regarded hooded seal milk as relatively rich: 45-65% fat. The nutrient transfer from fasting mother to suckling amounts to about 12 kg, which is reflected almost equally by the female's weight loss and the pup's weight gain (in the form of pelt and blubber).

Growth

Early growth in the hooded seal was monitored by Shepeleva (1973). He found that virtually all the nutrients acquired from mother's milk are converted to a thick blubber layer in the suckling. The "condition factor" (an index relating body circumference to body length) of neonates is comparable to that of yearlings and adults. It is highest in pups at weaning.

Males and females grow at a similar rate during the first year of life, reaching about 133 cm by one year of age (Rasmussen, 1960). At five years males average 190 cm, females 177 cm; at 10 years, 218 vs 193 cm. Rasmussen's largest male, a 19-year-old, measured 250 cm; his largest female, a 32-year-old, 212 cm. Extreme size variation exists within every age and sex class.

Sexual maturity

Sexual maturity in females (indicated by the first ovulation) occurs when the seals are from two to nine years old, but over 50% are mature at three years and give birth to their first pup at four years (Øritsland, 1964, 1975). More than 90% are mature at six years.

Reproductive rates based on the occurrence of corpora lutea of pregnancy and lactation varied in a sample taken at Jan Mayen and

Denmark Strait from 77.8% at three years to 100% at six years, with a mean value of 98.2% for seals aged six years and older. The average reproductive rate for all mature females was 94.7% (Øritsland, 1964). In a sample taken at Newfoundland, age at first ovulation varied from three years (57.9%) to nine years (100%); maximum pregnancy rates were 96.9% for age-groups three to ten and 98.8% for older seals (Øritsland, 1975). These rates are very high compared with similar data for species treated elsewhere in this volume.

Male hooded seals reach sexual maturity at four to six years, but do not achieve full social status until they are considerably older (Øritsland and Benjaminsen, 1975). There is little information on the male reproductive cycle.

Reproductive physiology

The anatomy of the female reproductive system is not appreciably different from that described in other species (Øritsland, 1964).

The ovaries grow rapidly in females two to four years old, but growth slows in older animals. Small follicles first appear in the second year, and the number of follicles is maximum in the ovaries of seals two years old. By four years of age the ovaries have the same size distribution of follicles as in older age groups. The mean diameter of the largest follicle increases with age up to five years, but does not change appreciably thereafter. The average number of corpora lutea and corpora albicantia show an annual increase of approximately 0.7 up to seven years of age. In older seals, however, the addition of new corpora lutea is offset by the resorption of old corpora albicantia.

Ovarian function alternates with a two-year cycle of activity. Ovulation and mating occur some 12 days post partum, and the corpus luteum of pregnancy increases in size both before and after implantation. Maximum size is reached just before parturition. At parturition, follicular activity is minimal in the ovary that supported the recent pregnancy. In the other ovary the Graafian follicle increases in size to a maximum just before ovulation. Field observations indicate that young females ovulate for the first time after the breeding season has ended.

Gestation

After fertilization the blastocyst remains free in the reproductive tract for a period of up to 16 weeks—probably less in primiparous females. Foetuses have not been found in June and July, when a number of unimplanted blastocysts have been recorded. The gestation period, then, measured from implantation to parturition, is not more than 240-250 days.

Breeding behaviour

As noted above, mating takes place about two weeks after parturition. According to Rasmussen (1960): "On the same ice floe beside a hood female a male is usually found which awaits the end of suckling when pairing can take place. Usually this hood male is a large, old animal which by virtue of its strength, size and appearance chases away the competing younger males which swim round the ice floe and constantly try to reach the female." Such challengers haul out at the edges of the floe (Øritsland and Benjaminsen, 1975); and boisterous displays, often involving snout expansion, and fights ensue (Ognev, 1962; Miller and Boness, 1979). Reports by sealers indicate that copulation is accomplished in the water (Øritsland, 1964). According to Miller and Boness, the shortness of the lactation period may allow some degree of sequential polygyny during the brief pupping season.

Captivity

Scholander (1940) used *Cystophora* in his experimental studies of diving in seals. The animals were about three months old when captured, and they quickly adapted to handling and artificial feeding on fish. In addition to these, a number of adult hooded seals have been kept successfully in zoos. The first one reported appears to have been a male maintained for about one and a half years at the Bremerhaven Zoo (Ehlers, 1965). Klös (1966) mentioned that at least five hooded seals had been held in European zoos since 1954.

After a period of fasting, captive hooded seals develop a good appetite for dead herring. For example, a 275 cm male weighing 375 kg was fed 30 kg of fish per day at 1000 and 1600 hours (Ehlers, 1965).

Blix *et al.* (1973) have made what appear to be the first scientific studies of dietary and health problems in captive *Cystophora*. They successfully reared three hooded seals captured at an age of about two to three months. The animals lived for more than a year on a diet consisting of herring flour, herring oil, water, vitamins, and minerals. There appears to be a relationship between vitamins B1 and B-complex deficiency and the inability to digest fish. Administration of these vitamins before feeding fish to the seals invariably eliminated any loss of condition. Magnesium sulphate also appears to be an important dietary component in preventing tetanies attributable to hypomagnesaemia.

Blood Values

Direct bleeding gave Scholander (1940) a blood volume of 9-10% of body weight in two hooded seals weighing 25 and 29 kg. He suggested 15% as being a more likely figure, but Elsner (1969) believed it to be nearer 12% of body weight. Oxygen capacity of the blood of *Cystophora* has been given as 24.30 ml per 100 ml (Scholander, 1940) and 36 vol % (Clausen and Ersland, 1969). The RBC count is 4.8×10^6 per mm^3; Hct and Hb, 63% and 26.4 g per 100 ml. Other values are given by Clausen and Ersland.

References

Allen, J. A. (1880). "History of North American Pinnipeds: A Monograph of the Walruses, Sea-Lions, Sea-Bears and Seals of North America", US Dept. Inter. Misc. Pupl. 12.
Anonymous (1959). Norwegian-Soviet sealing agreement, 1958. *Polar Rec.* **9**, 345-348.
Anonymous (1976). Report of Scientific Advisers to Panel A (Seals), Bergen, Norway, 9-10 December 1975. In "Redbook 1976", 197-198. International Commission for Northwest Atlantic Fisheries (ICNAF), Dartmouth, Nova Scotia.
Anonymous (1977). Report of Scientific Advisers to Panel A (Seals), Copenhagen, Denmark, 11-13 October 1976. In "Redbook 1977", 95-99. International Commission for Northwest Atlantic Fisheries (ICNAF), Dartmouth, Nova Scotia.
Anonymous (1978). Report of the *Ad Hoc* Working Group on Seals. In "Redbook 1978", 17-20. International Commission for Northwest Atlantic Fisheries (ICNAF), Dartmouth, Nova Scotia.
Árnason, U. (1972). The role of chromosomal rearrangement in mammalian speciation with special reference to Cetacea and Pinnipedia. *Hereditas* **70**, 113-118.
Árnason, Ú. (1974). Comparative chromosome studies in Pinnipedia. *Hereditas* **76**, 179-226.

Berland, B. (1958). The hood of the hooded seal, *Cystophora cristata* Erxl. *Nature (London)* **182**, 408-409.

Berland, B. (1966). The hood and its extrusible balloon in the hooded seal—*Cystophora cristata* Erxl. *Norsk Polarinst., Årbok (1965)*, 95-102.

Blix, A. S., Iversen, J. and Påsche, A. (1973). On the feeding and health of young hooded seals *(Cystophora cristata)* and harp seals *(Pagophilus groenlandicus* in captivity. *Norw. J. Zool.* **21**, 55-58.

Burns, J. J. and Fay, F. H. (1970). Comparative morphology of the skull of the ribbon seal, *Histriophoca fasciata*, with remarks on systematics of Phocidae. *J. Zool.* **161**, 363-394.

Burns, J. J. and Gavin, A. (1980). Recent records of hooded seals, *Cystophora cristata* Erxleben, from the western Beaufort Sea. *Arctic* **33**, 326-329.

Clausen, G. and Ersland, A. (1969). The respiratory properties of the blood of the bladdernose seal *(Cystophora cristata)*. *Resp. Physiol.* **7**, 1-6.

Dailey, M. D. (1975). The distribution and intraspecific variation of helminth parasites in pinnipeds. *Rapp. P. -v. Réun. Cons. Int. Explor. Mer* **169**, 338-352.

Ehlers, K. (1965). Records of the hooded seal, *Cystophora cristata* Erxl. and other animals at Bremerhaven Zoo. *Int. Zoo Yearb.* **5**, 148-149.

Elsner, R. (1969). Cardiovascular adjustments to diving. *In* "The Biology of Marine Mammals", (Ed. H. T. Andersen), 117-145. Academic Press, London and New York.

Haigh, J. C. and Stewart, R. E. A. (1979). Narcotics in hooded seals *(Cystophora cristata)*: a preliminary report. *Can. J. Zool.* **57**, 946-949.

Kapel, F. O. (1975). Recent research on seals and seal hunting in Greenland. *Rapp. P. -v. Réun. Cons. Int. Explor. Mer* **169**, 462-478.

King, J. (1964). "Seals of the World", British Museum (Natural History), London.

King, J. (1966). Relationships of the hooded and elephant seals (genera *Cystophora* and *Mirounga*). *J. Zool.* **148**, 385-398.

King, J. E. (1972). Observations on phocid skulls. *In* "Functional Anatomy of Marine Mammals", Vol. 1, (Ed. R. J. Harrison), 81-115. Academic Press, London and New York.

Klös, H.-G. (1966). A brief note on a hooded seal *Cystophora cristata* at West Berlin Zool. *Int. Zoo Yearb.* **6**, 263.

MacLaren Atlantic Limited (1977). Report on the Davis Strait aerial survey 77-1 for Imperial Oil Ltd, Aquitaine Co. of Canada Ltd and Canada Cities Services Ltd, Arctic Petroleum Operators Association, Project No. 134, December, 1977. MacLaren Atlantic Limited, Dartmouth-Fredericton, Moncton. 14 pp. (typescript).

Mercer, M. C. (1976). Mammals. *In* "Living Marine Resources of Newfoundland-Labrador: Status and Potential", (Ed. A. T. Pinhorn), 46-51. Fish. Res. Bd. Canada, Bulletin 194.

Miller, E. H. and Boness, D. J. (1979). Remarks on display functions of the snout of the grey seal, *Halichoerus grypus* (Fab.), with comparative notes. *Can. J. Zool.* **57**, 140-148.

Naevdal, G. (1971). Serological studies on marine mammals. *Rapp. P. -v. Réun. Cons. Int. Explor. Mer* **161**, 136-138.

Ognev, S. I. (1962). "Mammals of the USSR and Adjacent Countries, Vol. III: Carnivora (Fissipedia and Pinnipedia)", Israel Program for Scientific Translations, Jerusalem. (transl. from Russian; first published in Russian in 1935).

Øritsland, T. (1959). Klappmyss (Hood Seal). *Fauna, Oslo* **12**, 70-90.

Øritsland, T. (1960). Flyleting etter klappmyss på fangstfeltet i Danmarkstredet. *Fauna, Oslo* **13**, 153-162.

Øritsland, T. (1964). Klappmysshunnens forplantningsbiologi (The hooded seal female's reproductive biology). *Fisken og Havet 1964* **(1)**, 1-15. *(Fiskets Gang.* **50**, 5-19).

Øritsland, T. (1975). Sexual maturity and reproductive performance of female hooded seals at Newfoundland. Research Bulletin, International Commission for Northwest Atlantic Fisheries (ICNAF), No. 11, 37-41.

Øritsland, T. and Benjaminsen, T. (1975). Sex ratio, age composition and mortality of hooded seals at Newfoundland. Research Bulletin, International Commission for Northwest Atlantic Fisheries (ICNAF), No. 11, 135-143.

Pedersen, A. (1942). Säugetiere und Vögel. *Medd. Grønl.* **128 (2)**, 1-119.

Rasmussen, B. (1957). Exploitation and protection of the East Greenland seal herds. *Norsk Hvalfangst-tid.* **46 (2)**, 45-59.

Rasmussen, B. (1960). Om Klapmyssbestanden i det nordlige Atlanterhav (On the stock of hood seals in the northern Atlantic). *Fisk. og Havet* **1**, 23 pp. (Fisheries Research Bd. of Canada Transl. Ser., No. 387, 28 pp. typescript).

Rasmussen, B. and Øritsland, T. (1964). Norwegian tagging of harp seals and hooded seals in North Atlantic waters. Fiskeridirektoratets Skrifter Serie Havundersøkelser [Reports on Norwegian Fishery and Marine Investigations] **13 (7)**, 43-55.

Richardson, D. T. (1975). Hooded seal whelps at South Brooksville, Maine. *J. Mammal.* **56**, 698-699.

Ridgway, S. H., Geraci, J. R. and Medway, W. (1975). Diseases of pinnipeds. *Rapp. P. -v. Réun. Cons. Int. Explor. Mer* **169**, 327-337.

Scheffer, V. B. (1958). "Seals, Sea Lions and Walruses. A review of the Pinnipedia", Stanford University Press, Stanford, CA.

Schevill, W. E., Watkins, W. A. and Ray, C. (1963). Underwater sounds of pinnipeds. *Science (N.Y.)* **141**, 50-53.

Schneider, R. (1963). Der Kehlkopf der Klappmütze. *Anat. Anz.* **112**, 54-68.

Scholander, P. F. (1940). Experimental investigations on the respiratory function in diving mammals and birds. *Hvalråd. Skrift.* **22**, 1-131.

Sergeant, D. E. (1974). A rediscovered whelping population of hooded seals *Cystophora cristata* Erxleben and its possible relationship to other populations. *Polarforschung* **44 (1)**, 1-7.

Sergeant, D. E. (1976). History and present status of populations of harp and hooded seals. *Biol. Conserv.* **10**, 95-118.

Shepeleva, V. K. (1973). Adaptation of seals to life in the arctic. *In* "Morphology and Ecology of Marine Mammals", (Eds K. K. Chapskii and V. E. Sokolov), 1-58. Wiley, New York. (Transl. from Russian by H. Mills, for Israel Program for Scientific Translations, Jerusalem.)

Sivertsen, E. (1941). On the biology of the harp seal *Phoca groenlandica* Erxl. Investigations carried out in the White Sea 1925-1937. *Hvalrådets Skrifter* **26**, 1-166 + 11 plates.

Terhune, J. M. and Ronald, K. (1973). Some hooded seal *(Cystophora cristata)* sounds in March. *Can. J. Zool.* **51**, 319-321.

Vibe, C. (1950). The marine mammals and the marine fauna in the Thule district (Northwest Greenland) with observations on ice conditions in 1939-41. *Medd. Grønl.* **150 (6)**, 1-116.

Vibe, C. (1967). Arctic animals in relation to climatic fluctuations. *Medd. Grønl.* **170 (5)**, 1-227.

8
Monk Seals
Monachus Fleming, 1822

Karl W. Kenyon

Genus and Species

Three species of monk seal, widely separated geographically comprise this genus: the Mediterranean monk seal, *Monachus monachus* (Hermann, 1779); the Caribbean (or West Indian) monk seal, *M. tropicalis* (Gray, 1850); and the Hawaiian monk seal, *M. schauinslandi* Matschie, 1905. They are all contained in the Family Phocidae. *Monachus* contains the most primitive of all seals. The fossil record indicates that the ancestors of the three species originated in the North Atlantic at a very early date and that the Hawaiian monk seal became separated from the Atlantic-Caribbean monachines as early as 15 million years ago (Repenning and Ray, 1977). Perhaps because of their primitiveness, the monk seals seem far more sensitive than other phocids to the intrusion of man into their environment. In recent years, and concomitant with the rapid spread of human activity to even the

FIG. 1 (a) An Hawaiian monk seal as it emerged from a dive at Tern Island, French Frigate Shoals, Leeward Hawaiian Islands. (b) Hawaiian monk seal female with new-born pup.

most remote and isolated areas, all three species have shown alarming population declines. One species, the Caribbean monk seal, probably became extinct during the 1950s (Kenyon, 1977).

Because the monk seals (Fig. 1) evolved on remote oceanic islands (Caribbean and Hawaiian monk seals) or uninhabited waterless (without fresh water) islands and along mainland coasts (Fig. 2) backed by extensive deserts (the Mediterranean monk seal) they were never exposed to land predators and thus did not develop the ability to flee; they are genetically tame. Thus, the relict genus *Monachus* falls easy prey to human intruders who, if they wish, may walk up to a resting monk seal and easily club it to death.

Monk seals suffer disturbance by human presence despite genetic tameness. Pregnant females and mothers with nursing young appear greatly upset when approached. That such disturbance causes increased mortality among pups has been observed in Hawaiian monk seals.

The Mediterranean Monk Seal

The Mediterranean sea, the north-west coast of Africa, and its offshore islands, have been inhabited by sea-going man for at least 5000 years. Probably the monk seal was able to survive because many of the islets are waterless and remote from human settlements. Also, the rocky cliffs backed by the Sahara desert along the African coast offered seclusion. Limestone cliffs having sea-caves along Mediterranean coasts offered remote and hidden pupping areas. With the modern increase in motorized vessels, the widespread use of SCUBA gear and, in recent years, a growing intensity in fishing activity, the most remote refuges of the monk seal are being invaded by man. A high rate of abortion, perhaps related to disturbance of pregnant females, has been observed. Seals entangled in fishing gear may be drowned or killed by fishermen. Together these factors appear to be steadily pushing the monk seal towards extinction. Unless reserves can be established and human invasion of the monk seals' last retreats prevented it is questionable whether the species will survive the twentieth century.

In their paper "The Recent Status of *Monachus monachus*, the Mediterranean Monk Seal", D. Sergeant, K. Ronald, J. Boulva and F. Berkes (1978), have reviewed and summarized all available data on this seal both from published and unpublished sources. The following information is mostly extracted from that paper and Simon's (1966).

Morphology of Mediterranean Monk Seals

Coloration

At birth the pup is black but the pelage may be marked with white spots or splotches. The natal pelage is moulted at the age of four to six weeks. The new juvenile pelage is silvery grey dorsally and light ventrally. Adult pelage colour is variable but usually dorsally dark and light below. Some individuals appear mostly dark while others are pale silvery white with variable dark blotching. Some animals exhibit a large white belly patch. Variable colour patterns appear characteristic of the species, as they are similar among members of widely separated colonies.

Dimensions

Little data are available on body size. The maximum nose-tail length is recorded as 278 cm, (8 ft, 10 in). Two adult males measured 238 and 262 cm, (7 ft, 10 in, and 8 ft, 7 in). Two adult females measured 235 and 278 cm, (7 ft, 8 in, and 9 ft, 1 in). The maximum recorded weight is 400 kg (880 lb). The more usual adult weight would appear to be about 250-300 kg (550-660 lb). A new-born pup (navel not healed) measured 90 cm (35.5 in) in nose-tail length and weighed 26 kg (57 lb). Six new-born young averaged 84 cm (33 in) in length, and three pups averaged 20 kg (44 lb) in weight.

Reproduction of Mediterranean Monk Seals

Mating was observed to take place underwater on 1 August 1976 at Deserta Grande Island. A large male visited a group of five seals and found one receptive female. This observation leads to the conclusion that the species is polygynous (Sergeant *et al.*, 1978). The period of gestation is presumed to be about 11 months and the period of lactation about six weeks. Seven pups were observed among 61 older animals at the Cap Blanc Colony in 1974 and 1975. This observation suggested a production rate there of 11.5%.

The season of birth is prolonged. Births are recorded within the seven-month period from May to November with a peak occurring in September-October.

Pupping, in recent years at least, occurs mostly in caves, some of them having underwater entrances. It is suggested that the use of caves may have been caused by human intrusion into the monk seal's habitat and that the preferred pupping habitat is on beaches under cliffs. The sole remaining colony using beaches is at Cap Blanc. The success of this colony and the population declines noted in areas where the seals resort to sea-caves to bear their young support this hypothesis. Also, six aborted foetuses were found over a three-year period in a grotto having an underwater entrance. It has recently been observed that in areas along the Turkish coast where few caves occur that might be suitable for pupping, only old seals can be found.

Depth of diving

Circumstantial evidence gleaned from fishermen indicates that Mediterranean monk seals usually take fish from and/or damage nets set in water 30 m (*c*. 100 ft) or less in depth. One adult male, however, was reportedly taken on tackle set at 75 m (*c*. 250 ft). Seals were recorded following fishing vessels from five to 20 nautical miles off the coast of Spanish Sahara. Here depths on the Continental Slope were 615 m (over 2000 ft) and in-shore 38-45 m (125-150 ft). Thus, maximum diving capability is unknown but it appears that monk seals prefer to feed in water less than 100 ft deep when possible.

Feeding habits

Information from fishermen and from stomach contents indicate that monk seals take a variety of organisms. The following are recorded: barbouni *(Mullus surmuletus)*, synagrida *(Dentex)*, gopa *(Bops boups)*, kephalos *(Mugil cephalus)*, chicharres (a carangid, *Trachurus* sp.), pargo (a sparid), ray (Rajidae) and octopus.

Captivity

This seal is said to tame easily in captivity, young animals become tame and accept food more quickly than adults. Fasting from five days to two or three weeks after capture is usual. Preferred food of captives appears to be eels but sardines *(Sardina* and *Sardinella)* are also eaten. One seal consumed 10% of its body weight of *Sardina* in one day. The monk seal, however, does not appear to adapt well to captivity and seldom survives for much more than a year. Death in one case followed a miscarriage. At death the seal showed "symptoms of pneumonia".

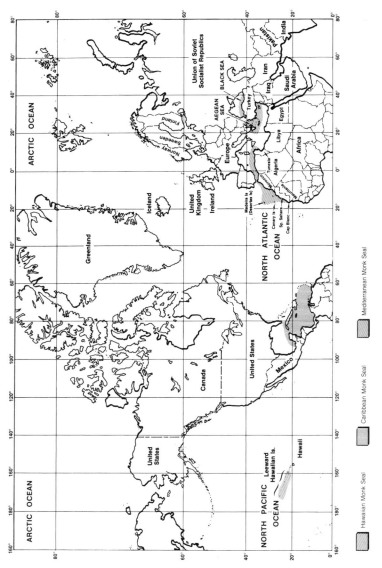

FIG. 2 Distribution map of Monk seals. The Caribbean population is apparently extinct.

Distribution and Abundance

"The original range included the southern and western coasts of the Black Sea, the coasts and islands of the Mediterranean Sea, the coast of north-western Africa south-westward to Cap Blanc, Mauritania, and the Madeira and Canary Islands." (Rice, 1977). Although this seal is distributed throughout much of its primaeval range (Fig. 2) its numbers are greatly reduced.

Today three centres of population are found in the Mediterranean. The most important is in the Aegean Sea where seals occur in the Dodecanese Islands of Greece and along the adjacent coasts of Turkey, among the Cyclades, Northern Sporades, in the sea of Marmara, westward toward Crete, the Pelopponesus and Ionian sea and along the western part of the south coast of Turkey. A second area, along the Mediterranean coast of Morocco and the Algerian coast of Tunisia, supports a lesser population. A third minor concentration is found in the eastern Mediterranean including the coasts of south-central Turkey, Cyprus and Lebanon. In the Atlantic a colony survives at the Dsertas Islands and another at Rio de Oro, about 30 miles north of Cap Blanc, Mauritania, formerly the Spanish Sahara (Maigret *et al.*, 1976). There are reports of individual seals seen at widely scattered locations.

During the 1971-76 period a number of observers visited areas where monk seals could still be found. Their combined estimates totalled 625 seals. To arrive at this figure Sergeant *et al.* (1978) used an estimate of 150 seals for the Aegean (p. 206). A population of 310 seals for the Greek islands of the Aegean was found by Marchessaux and Duguy (1977). In making a general statement, most observers agree that the present population probably is somewhere between 500 and 1000 seals, and that the population is declining.

The Caribbean Monk Seal
(Probably extinct)

Distribution and Abundance

"In historic times its range included shores and islands of the western Caribbean Sea, the Greater Antilles, the northern Lesser Antilles, the Bahamas, the Yucatan Peninsula, and the Florida Keys. In prehistoric times it ranged north to South Carolina." (Rice, 1977).

History

Columbus discovered the Caribbean monk seal on his second voyage to the new world in 1494. It was, however, 350 years before a specimen reached a museum and the scientific description of the species was made by Gray (1850). Being genetically tame the seals were easily taken and they were relentlessly killed for their skins and oil. Rice (1973) has presented an exhaustive historical review by geographical areas where and when the seals were taken and then not found again. The seal had become so scarce by 1887 that J. A. Allen called it an "almost mythical species".

The last authentic report that Rice (1973) was able to find was from C. Bernard Lewis, Director of the Institute of Jamaica. He said that until 1952 there was a small colony of monk seals at Seranilla Bank (in the middle of the Caribbean about half way between Jamaica and Honduras). "Seal" sightings have been reported since that time in scattered locations in the Gulf of Mexico and Caribbean but none have been authenticated as *Monachus*. Several escaped California sea lions *(Zalophus californianus)*, however, have been found in the Gulf of Mexico (Gunter, 1968). These errant captive sea lions may have been mistaken by some observers for monk seals.

Natural history

Very little is known of the biology of the Caribbean monk seal. From the information that is recorded it appears that it was in many respects similar to the other species of *Monachus*.

The peak of the pupping season was probably December. In this respect it differs from the other two species. From 1 to 4 December 1886, H. L. Ward (1887) killed five pregnant females all having near-term foetuses and saw a mother nursing a very young pup. He also noted that the females (like the other two species) had two functional pairs of mammae.

Aerial survey

In the hope of discovering a surviving colony of Caribbean monk seals the US Fish and Wildlife Service conducted an aerial survey from 19 to 25 March 1973. All areas where the seal had been reported, or suspected to be, within the past 100 years were carefully examined during the 3960 statute mile (6377 km) survey. In preparation for this survey correspondence was carried on with a number of individuals having experience within the former habitat of the monk seal. None of them had seen seals in areas they had visited but advised looking at other areas where they suspected the seals might still exist.

During the survey we found: "At every island group visited, either fishing vessels or shrimp trawlers at anchor, or fishermen and their shacks on shore or the remains of abandoned fishermen's camps were observed. However, there was no indication of the existence of monk seals. At Seranilla Bank, where this seal was last authentically reported and the hope of finding seals was greatest, three fishing boats (about 18 m long) were observed at anchor with a total of 13 canoes tied to their sterns. On shore at Beacon Cay the remains of a fisherman's camp was found. Beacon Cay is the largest island on Seranilla Bank and offers the most suitable beaches for monk seals."

"My conclusion from the 1973 survey is that the Caribbean monk seal has been extinct since the early 1950s. The fact that I saw no monk seals was not as important as the fact of the ubiquitous human presence." (Kenyon, 1977).

The Hawaiian Monk Seal

The Northwestern (Leeward) Hawaiian Islands are tiny, waterless oceanic bits of coral sand and rock. They are separated by an often harsh marine environment from the main or high Hawaiian Islands. Prehistoric Polynesian migrants never permanently colonized them. Evidence exists, however, that early seafarers did stop briefly at least on Kure, where they left the Polynesian rat *(Rattus exulans)*, and at Necker, where they left stone idols and built walls. It was not until the early nineteenth century, however, that the monk seal's habitat was invaded by western man.

In 1824 the sealing brig "Aiona" was thought to have taken the last monk seal. This most ancient and primitive of all seals had existed as "a living fossil" in isolation for 15 000 000 years (Repenning and Ray, 1977) free of human contact. It had evolved without the need to flee from land predators and fell an easy victim to the sealers' clubs. But some seals, probably those at sea, escaped the kill and another sealing vessel, the bark "Gambia", reportedly took 1500 skins in 1859.

The species must have been near extinction for in 1911 two biologists, Dill and Bryan (1912), saw no seals during a six-week visit to Laysan Island. A contemporary, Max Schlemmer, who lived on Laysan for 15 years killed seven seals during that period. One of these was the specimen upon which Matschie (1905) based the scientific description of the species.

The remnant population remained essentially undisturbed until in the early 1940s World War II brought human activity to the Leewards. Thus began a new era in the history of the monk seal. A Naval Station was built at Midway Atoll. While monk seal populations thrived at other atolls, the colony at Midway had dwindled to about 70 animals by 1957 and by 1968 a regularly breeding colony there had ceased to exist (Kenyon, 1972).

With an expanding fishery now developing in the waters near the Leeward islands the future prospects for the monk seal are unknown.

External Characteristics and Morphology

Body size

At birth the pup weighs 35-40 pounds (16-18 kg) and measures about 40 in (100 cm) in length. At weaning, after a 35- to 40-day nursing period, the pup is enormously fat; the birth weight has quadrupled to about 140 lb (64 kg) but body length shows little change. After weaning (Fig. 3), while the young seal is learning to fend for itself, it steadily loses weight. Juveniles, estimated to be about a year old, weigh about 100 lbs (45 kg). An adult male of about average size weighed 380 lb (172.4 kg) and measured slightly more than 7 ft (214.2 cm) in standard (nose-tail) length. A large adult (probably pregnant) female measured 7 f 8 in (233.7 cm) in length and her weight was estimated at about 600 lb (272.2 kg). Pregnant females, shortly before parturition, are usually exceedingly fat.

The mother monk seal nurses her pup from four abdominal nipples, which are extruded only when the pup is nursing (Kenyon and Rice, 1959). The bearded seal *(Erignathus barbatus)* (Mohr, 1952) and the monk seal are apparently the only phocids having four functional teats.

Organ weights

An adult male monk seal weighed 380 lb. The weights of certain organs and their percent of total body weight are: liver and gall bladder, 11.5 lb (3%); intestines (and contents), 19.5 lb, (5.1%); kidneys, 2.5 lb, (0.06%); heart, 1.6 lb, (0.042%).

FIG. 3   Newly weaned Hawaiian monk seal in grey pelage.

Pelage colour and moult

The birth coat is woolly and black. About the end of the six-week nursing period the black pelage has fallen out and been replaced by juvenile pelage. This is silvery grey on the back and sides and creamy white on the belly, chest and throat. Under exposure to sunlight and sea water the pelage gradually changes to dull brownish dorsally and yellowish ventrally. During successive moults a similar colour sequence occurs except that in the adult male the fresh new pelage is decidedly brownish, rather than silvery grey. Only during the first post-natal moult do the pelage hairs fall out individually. During successive annual moults the epidermis peels off (Fig. 4) with the old hair to reveal the bright new pelage that has grown under it. Only the elephant seals share with the monk seals this almost unique epidermal moult (Kenyon and Rice, 1959).

FIG. 4 Moulted pelage from Hawaiian monk seal showing moulted epidermis with hair and roots below. (× 5)

Dentition

The gums of the monk seal pup are smooth at birth, the tiny milk teeth having been resorbed before birth. In one individual, at age 27 days all permanent lower teeth had erupted and by age 46 days all teeth were approximately 3-5 mm above the gums (Kenyon and Rice, 1959). The adult dentition (Fig. 5) is

$$1\frac{2}{2} C \frac{1}{1} PC \frac{5}{5} \times 2 = 32$$

Age

The age of monk seals may be ascertained by counting growth layers in the cementum on sectioned roots of the canine teeth. Using this method an adult female and male were found to be aged 11 and 20 years respectively (Kenyon and Fiscus, 1963). More recently, the ages of seals that died on Laysan Island in 1978 (see Diseases, p. 213) were studied. The oldest was a male, length 210 cm (6 ft 11 in), aged 30 years (R. L. DeLong, personal communication).

Distribution and Abundance

Distribution

The Hawaiian monk seal breeds regularly on five atolls and islands in the Leeward (Northwestern) Hawaiian Islands (Fig. 2). Until the late 1950s it regularly bore young at a sixth location, Midway Atoll. Since that time lone females have come ashore sporadically and born a few pups. In recent years (since about 1976) one or two pups have been born on tiny, rocky Necker Island. This seal does not undertake any seasonal migration. Individual animals, however, are infrequently seen in the main (high) islands of Hawaii (Kenyon and Rice, 1959) and at Johnston Atoll. The important breeding areas are mostly under the protection of the Hawaiian Islands National Wildlife Refuge, administered by the US Fish and Wildlife Service. These are (from east to west): French Frigate Shoals, Laysan Island, Lisianski Island, Pearl and Hermes Reef, and Kure Atoll (the only breeding area not in the Refuge.) Kure is under the jurisdiction of the State of Hawaii and is used as a US Coast Guard Loran Station.

Abundance

The first comprehensive counts of monk seals were made using aircraft, boats, and on foot, in 1957 when a total of 1031 seals were recorded (Kenyon and Rice, 1959). Rice (1960) continued the field counts in the 1957-58 period for a total count of 1206 seals. Since that time counts of seals were conducted at irregular intervals. All of these indicated that a population decline was in progress.

The most recent comprehensive counts of the Hawaiian monk seal population are

Date	Seals counted	Reference
18 Mar.-9 Apr. 1976	565	DeLong et al. 1976
8-25 Apr. 1977	625	DeLong and Brownell, 1977
12-26 Jul. 1978	502	Fiscus et al. 1978

The greatest declines have occurred at the western extremity of the breeding range. At Kure Atoll a count of 142 seals was obtained in the spring of 1958 (Rice, 1960). Among a number of counts there in 1977

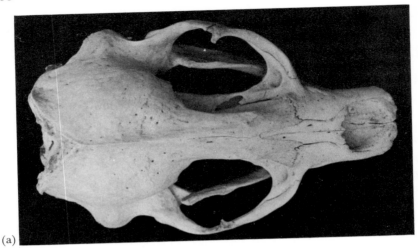

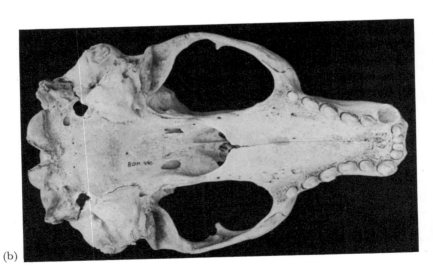

FIG. 5 Skull of Hawaiian monk seal male aged 20 years from Midway Atoll (KWK 79-9-7). (a) Dorsal view; (b) ventral view.

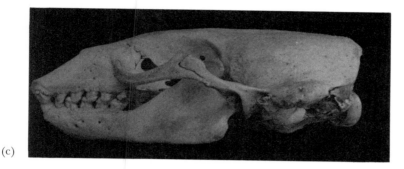

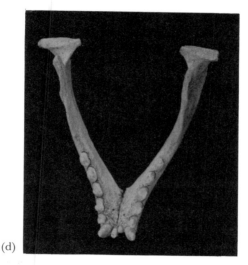

FIG. 5 (c) Lateral view; (d) lower jaw, dorsal view.

the highest was 52 seals on 6 May (Ruehle and Johnson, 1977). At Midway Atoll, Rice (1960) counted 68 seals. By 1968 only an occasional seal was sighted there and since then sporadic sightings and an occasional birth have been reported. No mass dying of "sick" seals has been reported from either Kure or Midway by the personnel stationed there. It thus appears that human disturbance of pregnant females and females with young has caused reduced survival of pups and is responsible for the population declines (see Reproduction, p. 210).

The next breeding atoll eastward is Pearl and Hermes Reef. Here Rice (1960) counted 338 seals in 1958. In July 1978 only 26 seals were found (Fiscus *et al.*, 1978). The reason (or reasons?) for the decline is unknown.

Population declines were also noted by the same observers at Lisianski Island (from 281 seals in 1958 to 85 in 1978) and at Laysan Island (from 326 seals in 1958 to 113 in 1978).

The two eastern-most breeding grounds, on the other hand, have shown population increases. Until recent years only a few seals were seen sporadically at rocky Necker Island. During the late 1970s several pups were born there and a total of 30 seals was counted in July 1978 (Fiscus *et al.*, 1978).

At French Frigate Shoals Rice (1960) recorded only 43 seals in 1958. In mid-July 1978 the count was 196 seals (Fiscus *et al.*, 1978).

The causes of observed population declines are only partially understood. Mortality from predation by sharks, decreased pup survival resulting from human disturbance, and the "sickness" observed at Laysan Island are all factors. Studies in progress and future studies will, hopefully aid in the preservation of this species which was officially designated as endangered in the US *Federal Register* dated 23 November 1976.

In conclusion, it appears that the Hawaiian monk seal population has declined from about 1350 seals in 1958 to less than 700 individuals in 1978.

Reproduction and Population Dynamics

During spring and summer months adult males constantly cruise along favourite basking beaches searching for receptive females. The species is polygynous. Because adult males outnumber adult females by about three to one (Kenyon, 1979; and B. and P. Johnson, personal communication) each basking female is frequently disturbed. Mating takes place in the water, and in one observed incident a female was severely damaged by the teeth of several males that attempted copulation (R. L DeLong, personal communication).

Pupping usually occurs on coral sand beaches backed by *Scaevola* shrubs where the mother takes her single pup for shelter during the night. Successful pupping also takes place on beaches where sheltering vegetation does not exist and in recent years pups have been born on the rocky beaches at Necker Island. It appears, however, that when human activity forces pregnant females to desert traditional pupping beaches, as at Kure Atoll (Kenyon, 1972), and bear their pups on shifting sand spits, pup survival is greatly reduced.

The season of pupping is prolonged, lasting for about eight months from late December to mid-August with a peak of births in the March-May period (Kenyon, 1966). The gestation period is not precisely known but is presumed to be about one year. The sex ratio at birth is approximately equal.

During the 1964 breeding season Wirtz (1968) observed the breeding activity of 44 tagged females and found that 15 (34%) bred in both seasons, 14 (32%) only in 1964 and 15 (34%) only in 1965. The birth rate was estimated by Rice (1960) to be about 16.3%. More extensive data on frequency of pupping and birth rate is now being assembled by B. and P. Johnson (personal communication).

The mother monk seal does not leave her pup in order to feed herself during the approximately five- to six-week nursing period. She subsists on the energy stored in her blubber (Kenyon and Rice, 1959).

During the nursing period the mother monk seal is sensitive to human disturbance. Available evidence indicates that such disturbance has serious consequences. At Kure Atoll, after it was occupied in 1960 by a US Coast Guard Station, the mortality rate among pups was at least 19% and probably 27% (Wirtz, 1968) and at Midway Atoll, also occupied by man, 39% of the pups born there died before weaning (Rice, 1964). By contrast, at uninhabited Laysan Island, where observers exercise maximum care not to disturb seals, mortality was much lower. During the 1977, 1978, and 1979 pupping seasons, 103 pups were born there and only nine (9%) died before weaning (Johnson and Johnson, 1979).

Causes of Mortality

Predation

Monk seals bearing large scars on their bodies are frequently seen on basking and pupping beaches, and severly bitten seals have also been found dead. Some wounds are apparently inflicted by sharp coral while the seals are foraging. Others, show a pattern of gashes typical of shark dentition, are certainly inflicted by sharks.

Among 235 seals examined on beaches in 1969 a total of 29 (12%) bore scars believed to have been inflicted by sharks (Kenyon, 1973).

Tiger sharks *(Galeocerdo cuvier)*, grey reef sharks *(Carcharinus amblyrhynchos)* and reef white-tip sharks *(Triaenodon obesus)* are common near the monk seals' pupping and basking beaches. Monk seal remains were recovered from the stomachs of three tiger sharks at French Frigate Shoals by Taylor and Naftel (1978). Also, a tiger shark was observed by Balazs and Whittow (1979) while it consumed a monk seal. They did not, however, see the shark kill the seal. There can be little doubt that predation by sharks is an important cause of mortality in monk seals.

FIG. 6 (a) Hawaiian monk seal wrapping itself in a shark fishing line, French Frigate Shoals, Leeward Hawaiian Islands. (b) Adult male Hawaiian monk seal showing scar probably made by fishing line since it resembles those known to be caused by fishing line in the Northern fur seal.

Fishing gear

Entanglement in fishing gear may cause seal mortality. A monk seal was observed to investigate a shark fishing line set near Tern Island, French Frigate Shoals. The line was secured on land and extended to a buoy about 60 m from shore. On several occasions the seal wrapped the line around its body (Fig. 6(a)). Seals are attracted to scraps of discarded fish nets on basking beaches and entangle themselves in the netting and attached lines. In several observed instances of line entanglement, the seal eventually freed itself unharmed. Monk seals have been photographed (Fig. 6 (b)) on basking beaches bearing scars typical of those known to be inflicted on Northern fur seals by fish nets and attached lines (Fiscus and Kozloff, 1972). No information is available on the extent of possible monk seal mortality at sea caused by entanglement in fishing gear. There is one undocumented incident of a Hawaiian monk seal that drowned in a fish trap (G. Naftel, personal communication). In future fishing operations care should be taken to design gear to be used within the habitat of the Hawaiian monk seal that will minimize risk of entanglement that might drown or injure seals.

Diseases

On Laysan Island between December 1977 and July 1978, "sick" monk seals came ashore and remained there, showing various degrees of debilitation for days or weeks until they died. A total of 22 seals died (Johnson and Johnson, *in* Fiscus *et al.*, 1978). Most of the seals exhibited variable degrees of lung and heart congestion. It was suspected that ciguatoxin (a poison originating in a dinoflagellate and concentrated in the flesh of fish) known to occur in a variety of fishes in the Leeward Hawaiian Islands, might have affected the seals. A study team investigated the mortality in May, 1978 and found an additional seven dead seals on other islands (DeLong, 1978). Analysis of preserved liver samples from two seals were positive for ciguatoxin. An experiment was conducted to discover if ciguatoxin would intoxicate a similar seal, the northern elephant seal (a close and not endangered relative of the monk seal). Eels from the Leeward Hawaiian Islands, known to contain ciguatoxin, were fed to two elephant seals. Both died after showing symptoms similar to those observed in dying monk seals. Further studies of the histopathology in the two species of seals involved in this study are in progress (DeLong and Gilmartin, 1979).

On 1 February 1977 a monk seal came ashore and died on Tern Island, French Frigate Shoals, Leeward Hawaiian Islands. It was buried by US Coast Guard personnel and exhumed on 15 March 1977. This adult male measured 211.5 cm , (6 f, 11.25 in), nose-tail length. Most of the teeth were missing, including the lower canines. The upper canines were both broken, only the roots remained in the skull. Also, osteolysis had occurred around some of the postcanines, indicating that abscesses had formed and the teeth were lost some time before death. Three vertebrae in mid-back showed considerable osteolysis, indicating an infection in this area. We concluded that the observed pathological conditions and old age caused the death of this seal (Kenyon and Rauzon, 1977).

Behaviour

Vocalizations

From birth the pup utters a bleating sound—"mwaa-mwaa-mwaa". When disturbed it utters an explosive "aaah" or "gaah". Adult sounds fall into two categories. (1) "Bubbling" sounds seem to originate deep in the throat. These sounds may be uttered either with the mouth open, as during threat display, or with the mouth closed, as when mildly alarmed by an intruder. The sounds are audible up to a distance of about 50 ft. The sounds resemble the bubbling of water being poured rapidly from a jug and also have been likened to a series of rapid belches. (2) "Bellowing" sounds are uttered with open mouth, as when a mother is defending her pup. Bellows are also uttered when seals are interacting, as when a non-receptive female is approached by a male or when one drives another from the beach. The mother seal displays affection for her young by uttering low throaty moaning or growling sounds (Kenyon and Rice, 1959).

Defensive behaviour

When a mother monk seal is approached by another seal or a human intruder she at first snorts or bellows. If the intruder continues to approach, the mother charges, mouth open and bellowing or bawling (Fig. 7). If the intruder does not retreat, she may inflict wounds with her teeth.

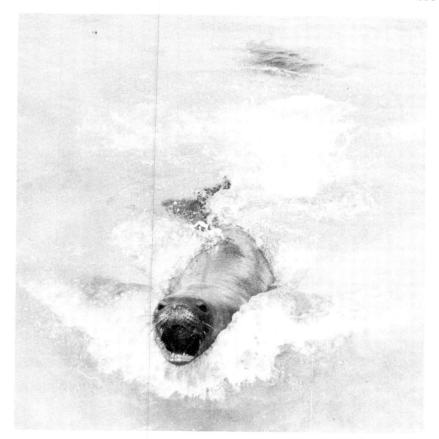

FIG. 7 Mother monk seal charges the photographer. Typical defensive behaviour with open-mouth threat while bellowing. Pup is dimly visible underwater behind mother at the top of the photograph.

If a large dominant male approaches a subdominant seal (female or younger male) resting on a beach, the resting animal snorts and may vocalize. As the intruder nears, the resting seal rolls on its side, usually perpendicular to and ventral side toward the intruder (Fig. 8). Simultaneously the free front flipper is raised perpendicular to the body. This is submissive posture. Open-mouth sparring accompanied by snorting, bubbling sounds and bawling may follow. The same submissive posture is taken toward human intruders.

FIG. 8 Adult male monk seal, foreground, approaches an adult female which demonstrates typical submissive-defensive posture. Note the usual dark colour of the male. Green seal turtles (behind female) frequent the same beaches as monk seals. Lisianski Island, Leeward Hawaiian Islands.

Social behaviour

In general, monk seals are solitary both in the water and on shore. When loose groups form on beaches it appears that they gather more because of favourable local environmental conditions than from any sense of gregariousness. Except for mothers with pups, resting seals avoid bodily contact. They are not positively thigmotactic as are, for example elephant seals *(Mirounga)* and California sea lions *(Zalophus)*.

Diving and swimming capabilities

Much is yet to be learned about aquatic habits and behaviour of monk seals. A seal tagged on Laysan Island travelled a linear distance of 1013 km at sea and was recorded at Johnston Atoll by Schreiber and Kridler (1969). This observation and others of monk seals on oceanic islets in the Leeward Hawaiian Chain, such as Nihoa and Necker, where little or no pupping occurs, indicate that the seals are capable of finding food and travelling long distances in the open ocean. Prolonged periods of diving activity were first observed near Tern Island, French Frigate Shoals. Recognizable individual seals continued diving (presumably for food) for a number of consecutive hours. One individual made 25 dives in 4.4. h, averaging about 5.7 dives per hour. The dives lasted from about five to 14 minutes (mean 9.5 m). A minute (more or less) was spent at the surface between dives (Kenyon and Rauzon, 1977). More extensive observations, indicating that some dives may last for 20 min or more, were made by Mark Rauzon. While SCUBA diving near monk seal breeding islands, Taylor and Naftel (1978) saw monk seals exhale in coral caverns well beneath the surface. The exhaled breath formed a large bubble of breathable air in the roof of the cave. They conclude that the cave with its stored air supply may be used by the seals as a refuge from sharks when escape to a beach might be impossible.

The pup is able to swim weakly at birth. It gains strength during practice swimming sessions several times daily in the close company of its mother during the nursing period.

Feeding habits

Quite frequently, monk seals disgorge their stomach contents while resting on shore. Information on feeding habits is mainly based on these food remains and on the hard parts of food organisms in faecal deposits. These reveal that during the periods when monk seals remain near the breeding and resting atolls they feed on fishes and invertebrates both within atoll lagoons and in deeper water outside the lagoons. Species commonly eaten include: the spiney lobster *(Panulirius marginatus)*, Conger eel *(Ariasoma bowersi)*, moray eels *(Gymnothorax* sp. and *Echidna* sp.), flatfish *(Bothus mancus* and *Bothus pantherinus)*, scorpenids *(Scarpaenopsis gibbosa?)*, larval fish *(surgeon fish?)*, and octopus *(Cephalopod)*. Monk seals are known to travel long distances in the open ocean, so presumably they are able to prey on pelagic species, but nothing is known of their pelagic food habits.

Thermoregulatory behaviour

It has been supposed that monk seals must have some special physical adaptation to life in a tropical climate. It was reasoned that this must be so because their blubber thickness is comparable to that of seals living in frigid climates. The deep body temperature, 35.5 °C in the case of one resting seal (Kenyon and Rice, 1959) is lower than that recorded for some other pinnipeds. In his extensive studies of thermoregulation in the Hawaiian monk seal, Whittow (1978) found that seals kept cool on hot windless days by resting on damp sand at the water's edge and by making wallows into cool sand layers. They did not remain on the dry, hot, upper beach levels except during cool windy or cloudy weather. He concluded: "For the most part, the seals lay in postures that exposed their ventral pale-colored hair coat to the atmosphere. The temperature of this surface was significantly lower than that of the darker dorsal coat. The seals were extremely inactive while ashore; their respiratory pattern included long periods of breath-holding, and the heart rate during breath-holding was low. These features were considered to be compatible with a low level of metabolic heat production and to be adaptive to heat exposure." (p. 47).

Captivity

At least nine Hawaiian monk seals have been held in captivity at the San Diego Zoo and the Waikiki Aquarium since 1951. Most of these lived only a few months to a little more than a year. An exception was a male seal that was brought to the Waikiki Aquarium as a juvenile. This animal adapted well to captivity, learned to perform certain entertaining behaviours and remained in good health until shortly before his death at the age of 13 years (Taylor, 1979). Another seal taken as an adult refused food and was liberated. The causes of death in captivity included gastric ulcers, acute pneumonia, vegetative endocarditis and peritonitis (Kenyon and Rice, 1959).

References

Allen, J. A. (1887). The West Indian seal (*Monachus tropicalis*, Gray). *Bull. Am. Mus. Nat. Hist.* **2 (1)**, 1-34.

Balazs, G. H. and Whittow, (1979). First record of a tiger shark observed feeding on a Hawaiian monk seal. *Elepaio* **39 (9)**, 107-109.

DeLong, R. L. (1978). Investigations of Hawaiian monk seal mortality at Laysan, Lisianski, French Frigate Shoals, and Necker Island, May 1978. Natl. Oceanic Atmos. Admin., Natl. Mar. Fish. Serv., Northwest and Alaska Fish. Center, Seattle, Wash., Interim Rep., 5 July 1978.

DeLong, R. L., Fiscus, C. H. and Kenyon, K. W. (1976). Survey of monk seal *(Monachus schauinslandi)* populations of the Northwestern (Leeward) Hawaiian Islands. Natl. Mar. Fish. Serv., Northwest Fish. Centr. Prog. Rept.

DeLong, R. L. and Brownell, R. L. (1977). Hawaiian monk seal *(Monachus schauinslandi)* habitat and population survey in the Northwestern (Leeward) Hawaiian Islands, April 1977. Natl. Mar. Fish. Serv., Northwest and Alaska Fish. Ctr. Proc. Rept.

DeLong, R. L. and Gilmartin, W. G. (1979). Ciguatoxin Feeding Experiments with a Model Phocid. *In* "Monk Seal Workshop Report", U.S. Mar. Mamm. Commission, 30 Aug. 1979. Processed report of Nov. 1979.

Dill, H. R. and Bryan, W. A. (1912). Report of an expedition to Laysan Island in 1911. *Bull U.S. Bur. Biol. Surv.* **42,** 9.

Fiscus, C. H. and Kozloff, P. (1972). Fur seals and fish netting. *In* "Fur Seal Investigations, 1971", App. E., 122-132. Natl. Mar. Fish. Serv., Northwest Fish. Center, Mar. Mamm. Biol. Lab., Seattle, Wash.

Fiscus, C. H., Johnson, A. M. and Kenyon, K. W. (1978). Hawaiian monk seal *(Monachus schauinslandi)* survey of the Northwestern (Leeward) Hawaiian Islands, July 1978. Natl. Oceanic Atmos. Admin., Natl. Mar. Fish. Serv. Northwest and Alaska Fish. Center, Seattle, Wash., Proc. Rept., Aug. 1978.

Gray, J. E. (1850). Catalogue of the specimens of mammalia in the collection of the British Museum. Pt. 2, "Seals", British Museum (Natural History), London.

Gunter, G. (1968). The status of seals in the Gulf of Mexico, with a record of feral otariid seals off the United States Gulf coast. *Gulf Res. Rept.* 2(3): 301-308.

Johnson, B. W. and Johnson, P. A. (1979). Preliminary results of the 1979 Monk Seal Studies on Laysan Island. *In* "Monk Seal Workshop Report", U.S. Mar. Mamm. Commission, 30 Aug. 1979. Proc. Rept. of Nov. 1979.

Kenyon, K. W. (1966). Marine Wildlife Observations, Leeward Hawaiian Islands, 8-27 September 1966. Unpub. Rept. in U.S. Fish and Wildlife Serv. files.

Kenyon, K. W. (1972). Man versus the monk seal. *J. Mammal* **53 (4),** 687-696.

Kenyon, K. W. (1973). Hawaiian monk seal *(Monachus schauinslandi).* IUCN, Survival Service Commission. IUCN Pub. New Series Supplementary Paper **39,** 88-97.

Kenyon, K. W. (1977). Caribbean monk seal extinct. *J. Mammal.* **58 (1),** 97-98.

Kenyon, K. W. (1979). Hawaiian monk seal observations at Kure Atoll, 10-22 May 1979. Unpub. Rept. in U.S. Fish and Wildlife Serv. files.

Kenyon, K. W. and Rice, D. W. (1959). Life history of the Hawaiian monk seal. *Pacific Sci.* **13,** 215-252.

Kenyon, K. W. and Fiscus, C. H. (1963). Age determination in the Hawaiian monk seal. *J. Mammal.* **44,** 280-281.

Kenyon, K. W. and Rauzon, M. J. (1977). Hawaiian monk seal studies, French Frigate Shoals, Leeward Hawaiian Islands National Wildlife Refuge, 15 Feb. to 5 Apr. 1977. Proc. Rept. Div. Coop. Res., U.S. Fish and Wildl. Serv., 20 May 1979.

Maigret, J., Trotignon, J. and Dugay, R. (1976). Le phoque moine *Monachus monachus* (Hermann 1779), sur les côtes meridionales du Sahara. *Mammalia (Paris)* **40,** 413-422. (The monk seal on the southern coasts of the Sahara. Translation seen only.)

Marchessaux, D. and Duguy, R. (1977). The monk seal, *Monachus monachus* (Hermann, 1779), in Greece. *Mammalia (Paris)* **41 (4)**, 419-439. (Translation seen only.)

Matschie, G. F. P. (1905). Eine Robbe von Laysan. *S. B. Ges. Naturf. Fr. Berl.* 254-262.

Mohr, E. (1952). Die Robben der europäischen Gewässer. Band 12. Paul Schöps, Monographien der Wildsäugetiere, Frankfurt am Main.

Repenning, C. A. and Ray, C. E. (1977). The origin of the Hawaiian monk seal. *Proc. Biol. Soc. Wash.* **89 (58)**, 667-688.

Rice, D. W. (1960). Population dynamics of the Hawaiian monk seal. *J. Mammal* **41**, 376-385.

Rice, D. W. (1964). The Hawaiian monk seal. *Nat. Hist.* **73**, 48-55.

Rice, D. W. (1973). Caribbean monk seal *(Monachus tropicalis)*. IUCN, Survival Service Commission. IUCN Pub. New Series Supplementary Paper, **39**, 98-112.

Rice, D. W. (1977). A list of the marine mammals of the world (Third Edition). NOAA Tech. Rept. NMFS SSRF - 711. U.S. Gov. Printing Off., Wash. D.C. 20402. iii + 15.

Ruehle, J. and Johnson, A. M. (1977). Observations of monk seals and other wildlife, Kure Atoll, 10 Feb. to 13 May 1977. Unpub. Rept. U.S. Fish and Wildlife files.

Schreiber, R. W. and Kridler, E. (1969). Occurrence of an Hawaiian monk seal *(Monachus schauinslandi)* on Johnston Atoll, *Pacific Ocean. J. Mammal* **50**, 841-842.

Sergeant, D., Ronald, K., Boulva, J. and Berkes, F. (1978). The recent status of *Monachus monachus*, the Mediterranean monk seal. *Biol. Conserv.* **14**, 259-287.

Simon, N. (1966). Mediterranean monk seal. Red data book, vol. 1, International Union for Conservation of Nature and Natural Resources Surv. Serv. Comm. 1110 Morges, Switzerland.

Taylor, L. R. (1979). Farewell to Friday. Kilo i'a, 8:3. (A publication of the Friends of the Waikiki Aquarium.)

Taylor, L. R. and Naftel, G. (1977). How to avoid shark attack (if you happen to be a Hawaiian monk seal). *Oceans* **10 (6)**, 21-23.

Taylor, L. R. and Naftel, G. (1978). Preliminary investigations of shark predation on the Hawaiian monk seal at Pearl and Hermes Reef and French Frigate Shoals. Final rept., U.S. Mar. Mamm. Commission contr. No. 7AC011, NTIS PB285-626.

Ward, H. L. (1887). Notes on the life history of *Monachus tropicalis*, the West Indian seal. *Am. Nat.* **21**, 257-264.

Whittow, G. C. (1978). Thermoregulatory behavior of the Hawaiian monk seal *(Monachus schauinslandi)*. *Pac. Sci.* **32 (1)**, 47-60.

Wirtz, W. O., II. (1968). Reproduction, growth, and development and juvenile mortality in the Hawaiian monk seal. *J. Mammal* **49**, 229-238.

9

Crabeater Seal

Lobodon carcinophagus
(Hombron and Jacquinot, 1842)

Gerald L. Kooyman

Genus and Species

The first allusion to the crabeater seal is an illustration of it in the published works of Bellingshausen's 1819-21 expedition (Bertram, 1940). The seal was not named in this work, and the type specimen is believed to have been collected between the South Sandwich and South Orkney Islands by members of J. Dumont d'Urville's French Antarctic expedition of 1837-40. According to Scheffer (1958) the actual description by J. B. Hombron and H. Jacquinot was published in or after 1842. The binomial given the species was *Phoca carcinophaga*. In 1844, Gray established the genus as *Lobodon* (lobed-tooth). In 1853 H. Jacquinot and J. Pucheran published a description of the crabeater seal with the binomial *Lobodon carcinophaga*. Later the species name was masculinized to *carcinophagus* (crabeater) (Berg, 1898). It has also been called *Ogmorhinus carcinophagus* Trouessart, 1897.

External characteristics and morphology

Bertram (1940) measured the length of 62 females and 41 male seals. He found the nose-to-tail length of most mature females ranged between 216 and 241 cm. The longest one was 262 cm. The length of most males ranged from 203 to 241 cm. The longest male was 257 cm. Few weights are available but the adults probably range in weight from 200 to 300 kg. Compared to most seals they are lithe (Fig. 1) although not nearly so much as the leopard seal.

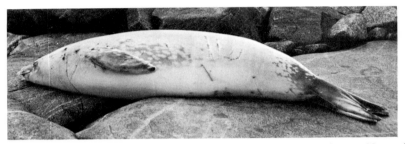

FIG. 1 A young crabeater seal hauled out on a rocky shore. Note the numerous scars along the flanks and neck and the blonde pelage.

Moult occurs in January and February. After the moult the animals are mainly dark brown dorsally and grade to blonde ventrally. The colour on the back and sides is not uniform but contains large brown patches interrupted with lighter brown, particularly on the sides. The flippers are darkest of all. As the fur ages, the animal gradually lightens until it is almost a uniform blonde (Fig. 1), hence the name "the white Antarctic seal" (Wilson, 1907). Frequently there are deep scars in the pelt. The size and spacing of many of these scars suggest that they result from wounds inflicted by leopard seals (Siniff & Bengston, 1977).

Distribution

All Antarctic seal censuses show that crabeater seals confine themselves almost exclusively to the pack ice. However, in the summer months when the ice is breaking up next to the continent some seals pass

FIG. 2 Distribution map of the crabeater seal. This species may be found anywhere south of the Antarctic convergence, but it is especially associated with pack ice. The arrows indicate single observations north of the convergence.

through these areas. Every summer some seals, especially young animals, come south into the Ross Sea. Lindsey (1937) observed them in the Bay of Whales (79° S) and they are seen every year in McMurdo Sound (77° 50′ S). A number of observations in this area from 1962-1969 have been collated by Stirling and Kooyman (1971).

Occasionally seals become trapped in these regions as water north of them freezes. In 1967 in McMurdo Sound two young crabeater seals were seen beside a Weddell seal hole in early November. The nearest open water was about 60 km to the north (Stirling and Kooyman, 1971). It is likely that these seals were caught in the fall or winter freeze and spent the winter in the area. Less fortunate are some of those that wander inland. A number of carcasses have been found inland, particularly in the Dry Valleys near McMurdo Sound. They are called dry valleys because no snow settles there during the winter. One seal carcass was found on the Ferrar Glacier at 1100 m (Wilson, 1907). This is perhaps an altitude record for a seal. The inland record is held by a live male pup which was picked up by a helicopter crew on 12 December 1966 on Crevasse Valley Glacier; 76° 50′ S, 145° 30′ W. This location was 113 km from open water and at an altitude of 920 m. A few days earlier another seal that was identified as a crabeater seal was seen on the Balchem Glacier at 76° 25′ S, 145° W. This location is 88 km from the coast and at an altitude of 920 m (Stirling and Rudolph, 1968).

Northern range limits have been summarized by Scheffer (1958). Crabeater seals are occasionally seen in New Zealand, Tasmania, Southern Australia and South America. Three or four animals have been found near the mouth of the Rio de la Plata, South America, at a latitude of approximately 35° S. (Vaz Ferreira, 1956; Ximénez et al., 1972.) The most northern sighting is from South Africa (Courtenay-Latimer, 1961).

Abundance

Since it is likely that the crabeater seal is the most abundant pinniped and that someday commercial harvesting on a large scale may be considered, some elaboration on population estimates is worthy of discussion.

Although it is believed to be an extremely abundant seal, there are few estimates of actual population size. Those estimates should be treated with caution since they are all based on samples which are very small relative to the total population and little is known about various behaviour patterns of the seals which would influence these counts. Furthermore, there has been little in the way of calibration of censusing techniques in the Antarctic. The *caveat* explicit in the statement of R. Carrick (1964) should be noted: "It is necessary to establish the accuracy of the sample before proceeding to extrapolate."

A summary of all censuses appears in Table 1. Scheffer (1958) estimates the world population to be two to five million. The method of his estimate is not clear. Eklund and Atwood (1962) censused seals from an icebreaker while traversing the Ross Sea and then later a portion of the Indian Ocean sector of Antarctic waters. Most of the seals counted were within 457 m of the ship, but occasionally counts

TABLE 1 Population density and total number estimates by various authors

Author	Date of publication	Locality	Area surveyed (km^2)	Number counted	Density (seals km^{-2})	Estimated world population
Scheffer	1958	—	—	—	—	2 to 5 × 10^6
Eklund and Atwood	1962	Ross Sea and Indian Ocean	—	—	0.47	5 to 8 × 10^6
Øritsland	1970	Scotia Arc	—[a]	904	0.12	
Ray	1970	Ross Sea	1 293[a]	275	1.74	
Siniff et al.	1970	Weddell Sea	921[a]	1 139	0.56[b]	
Erickson et al.	1971	Weddell Sea	851[a]	3 477	1.00[b]	50 to 75 × 10^{6c}

[a] Area surveyed by helicopter was: Øritsland-100%, Ray-100%. Siniff-34%, Erickson, 47%.
[b] Mode count.
[c] Since reduced to 15 × 10^6 (Gilbert and Erickson, 1977).

were attempted out to 2744 m from the ship. The observers were on the flying deck of the ship, as well as in the crow's nest. Counts were made from 0600 to 2100 hours. During the census the ship traversed 960 km of pack ice in the Ross Sea and 232 km in the Indian Ocean sector. From this data, density estimates were made and then extrapolated to the total antarctic population, which was estimated to be five to eight million seals.

Other investigators who have censused crabeater seals are: Øritsland (1970) in the Scotia arc area where he flew 1650 km, counted 904 seals and arrived at an overall density of 0.12 seals per km^2. These seals were mostly crabeaters which several investigators have observed to comprise over 80-90% of the seals of the pack ice (Eklund and Atwood, 1962; Øritsland, 1970; Erickson et al., 1971). In the Ross Sea, Ray (1970) made an aerial census from a helicopter in which he surveyed 1293 km^2 and counted 275 seals. An interesting deviation from other studies is that south of Coulman Island (73° S, 169° W) few crabeater seals were seen and they were not the dominant species. Even north of Coulman Island densities were low compared to elsewhere.

In a somewhat similar manner to Eklund and Atwood's study the most ambitious census of all was begun in 1968 in the Weddell Sea (Siniff et al., 1970) and continued through 1969 and 1970 (Erickson et al., 1971). This is of particular interest because the concentration of seals here may be the greatest in all the Antarctic waters (Laws, 1964). The shipboard census was conducted from the same ship and flying bridge as Eklund and Atwood's study. The census strip was 200 m on either side of the ship. Counts were made at various time intervals over the 24-hour period. Total seals counted were: 1139 in 1968 and 3477 in 1969. The average density for the two years was 0.78 crabeater seals per km^2.

The estimate in the Weddell sea is thought to be low by the authors because of an apparent preference by the seals to haul out at particular hours of the day. They were not always able to make counts at the optimum time. In fact, out of 34 days of counting only eight were done at optimum hours in 1968. In 1969 there were 14 out of 24 days in which counts were made at optimum hours (Erickson et al., 1971). They noted that 97% of the seals observed were crabeaters. The density estimates of 1.00 crabeater seals per km^2 were about twice the average of 1968. Since these figures represent an average over various hours of the day the 1969 value was recalculated according to the optimum haul out hours and the adjusted density was 6.74 seals per km^2. Utilizing satellite ice photographs and considering the different densities observed under particular ice conditions the authors estimated the total Weddell Sea crabeater seal population to be 8.2 million in 1968 and 10.6 million in 1969. From these figures they speculate further that the total Antarctic population may be 50-75 million (Table 1), but after censusing a broader area of Antarctic waters these census takers have reduced their estimates to 30 million (Erickson and Hofman, 1974). More recently there has been a further reduction in the estimates of the total population to about 15 million (Gilbert and Erickson, 1977.)

Anatomy and Physiology

Although many crabeater seals have been collected for various reasons, there are surprisingly few reports on their anatomy and physiology. Indeed, it seems somewhat of a paradox that so little biology of any kind is known about perhaps the most abundant large mammal in the world.

General features of the skull can be noted from Figs 3 (a)-3 (c). The dentition is extremely interesting (Fig. 4). The postcanine teeth of which there are five in both the upper and lower jaws, are complex. The number of cusps, which are very pronounced, are usually one anterior and three posterior to the antial point. The incisors are small. There are two pairs of upper and one pair of lower incisors. The canines are small. The lower canines are smaller than the uppers.

Cardiovascular system

The hepatic sinus, the large venous reservoir just posterior to the diaphragm and contiguous with the liver, is small compared to the Weddell seal, *Leptonychotes weddelli*. The caval sphincter is well developed but not as large as that of the Weddell seal.

Several aspects of the heart and aorta have been noted by Drabek (1975). The heart weight of three adults weighed from 964-1020 g. This was an average of 0.56% of total body weight. The general shape of the heart was similar to other Antarctic seals. The ventricles are unusually broad and flat compared with those of terrestrial carnivores. The internal dimensions and wall thickness of the aortic bulb are the smallest of all the Antarctic seals.

Some properties of the blood of five young seals have been reported by Tyler (1960). Red blood cell counts averaged 4.41×10^6 mm^{-3}. The

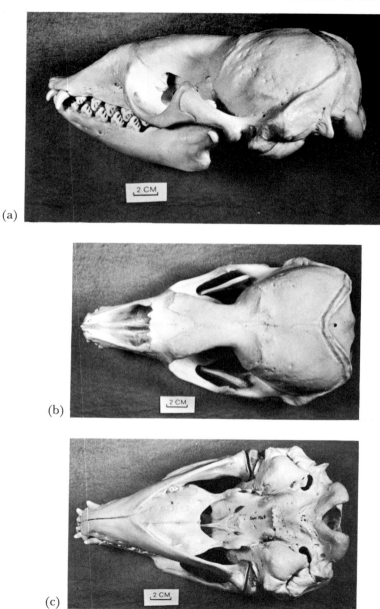

FIG. 3 The skull of a crabeater seal: (a) lateral view; (b) dorsal view; (c) ventral view.

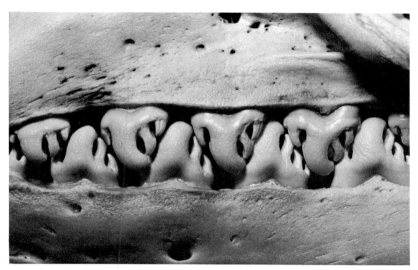

FIG. 4 A lateral view of a crabeater seal skull showing a close-up of the molars and premolars. Since the gums of the animal would fill the spaces above and below the apices of the cheek teeth the occlusion is perfect and anything passing out of the mouth must filter through the spaces between tubercles of the teeth.

RBC diameter was about 9 µm. The blood haemoglobin concentration was 18.2 g per 100 ml. The mean corpuscular haemoglobin concentration was 41.3%.

Respiratory system

There are ten sternal, four intercostal and two floating ribs. The tracheal length is nearly 16% of the nose-to-tail body length. In two adults there were 47 and 54 tracheal rings. The lateral width of the trachea was 4 cm at the larynx. This broadened to 5.5 cm at 10 cm caudal to the larynx, and 3.5 cm at the bifurcation. The dorsal-ventral diameter was 2 cm (Kooyman, unpublished data). Unlike the other three species of Antarctic seal, the tracheal cartilage is "U" shaped and rigid enough to maintain a lumen in the trachea when it is completely relaxed.

Also, unlike the other three species of Antarctic seal the lungs are lobed. The left lung is divided into two lobes and the right lung is divided into two lobes plus a small medial or accessory lobe.

Digestive system

The length of the small intestine of two young animals of total body length of 195 cm and 185 cm was 15.9 and 15.7 m. The large intestine was 0.76 and 0.72 m, respectively (Krylov, 1972).

Behaviour

Locomotion

The ability with which crabeater seals can move over snow and ice has impressed all who have observed its sprints. Wilson (1907) comments on its agility when harassed by dogs, and Lindsey (1937) gives a vivid account of the speed it can attain. He describes a chase in which a young man had to sprint at top speed in order to keep up with the seal.

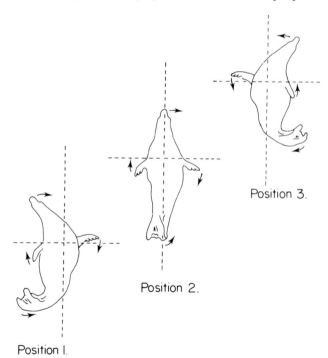

FIG. 5 The hind and foreflipper thrusting type of locomotion of the crabeater seal drawn from cine-film. Arrows indicate the direction of movement of the head and flippers (from O'Gorman, 1963).

O'Gorman (1963) estimated the speed of one animal in a short sprint at 10 km h^{-1} and he cites reports of speeds as great as 25 km h^{-1}. According to his survey of reports on other seals no other comes close to these speeds. When the seal is sprinting the head is held high and swings from side to side in about an 80° arc. The forelimbs are thrust alternately against the snow, the pelvis is moved from side to side while thrusting against the snow and the hind flippers are held together and off the snow. (Fig. 5).

Food habits

As described earlier the cheek teeth of crabeater seals are complex, perhaps the most complex of any carnivore. Spaces between the tubercules intrude deeply into the tooth. The main cusp of the upper and lower teeth fit between each other and the occlusion is perfect. When the mouth is closed the only gap is between each tubercle (Fig. 4). The main constituent of the stomach contents is euphausiids (Wilson, 1907). Presumably the teeth function as sieves to strain the invertebrates out of the water.

The only observations of feeding behaviour known to me are those of G. S. Wilson (personal communication). While over-wintering at the Argentine station of Almirante-Brown, Wilson noted that at night some invertebrates were attracted to a light over the pier. He observed crabeater seals catching some of these invertebrates. The seals seemed to be catching the prey one by one.

Social behaviour

During an extensive survey of seals in the Weddell Sea in the summers of 1968-71, the size of groups were noted (Erikson *et al.*, 1971). Most crabeater seals observed, 563 animals, were alone. There were 204 pairs and 109 trios counted. Larger groups were less common but some up to 1000 occur (Siniff *et al.*, 1979). Over a six-year-period at McMurdo Sound only two large groups have been reported. This was a group of 15 in January, 1963, and 10 in January, 1966 (Stirling and Kooyman, 1971). These observations are of animals resting on ice and they may be simply a collection of animals selecting the most convenient resting place and do not necessarily represent any kind of social unit. However, when the group of 10 animals was disturbed it broke up into groups of three. The trios stayed close together and one of these groups contained a pup. The two adults kept the pup between them.

Few observations of interactions between these seals have been made. Frequently, as described above, when animals lying on the ice together are disturbed they move together often rubbing or jumping into each other. A trio, consisting of an adult female, pup and adult male, has been noted several times (King, 1957; O'Gorman, 1963; Øritsland, 1970; Corner, 1972; Siniff and Reichle, 1976). Corner was able to account for one such trio over a two-week period in October, 1964. The animals were found on fast ice 15-25 km from open water. Unfortunately, the group was occasionally disturbed by the observers and during one of these episodes on the sixth day the trio broke up when the pup became separated from the adults. The adults remained together another eight days. During this period there seemed to be much aggression between the pair. During these bouts of fighting the male seemed to sustain cuts around the mouth and flippers. Similar aggression was noted by O'Gorman (1963). More detailed observations have been conducted recently (Siniff et al., 1979). They have confirmed other reports that a trio is common. The only other species of seal known to form a trio consisting of a male, female and pup is the hooded seal, *Cystophora cristata*.

Sound production

Out of water sounds of crabeater seals seem to be few and simple. When disturbed there is often much hissing and blowing through the nose, but none of the musical chirps or trills that Ross and Weddell seals emit. To my knowledge no one has recorded any sounds made underwater by crabeater seals.

Reproduction

Copulation probably occurs from October through December. Gestation is about nine months and there is a pregnancy rate of about 80% of the adult females (Bertram, 1940). Pupping probably occurs from September through November. Since ships rarely traverse the Antarctic pack ice as early as this in spring only a few new-born pups had been seen until cruises specifically organized to find and study crabeater seals occurred (Siniff et al., 1979).

While drifting across the Weddell Sea during Shackleton's Endurance expedition, Worsely saw eight or nine young animals, two of which may have been less than one month old (Bertram, 1940). The first new-born pup ever collected was obtained by Racovitza (1900) in September. It had thick, long fur the same colour as the adults. Its length was 114 cm. The motor skills of the animal seemed precocial. The mother and pup separated after two or three days. This was apparently interpreted as normal and the belief has been fostered by others. Bertram (1940) expressed doubts about such a short lactation and more recently it has been suggested, based on growth rings of the canines, that lactation lasts about five weeks (Laws, 1958).

The growth rate is rapid and nearly complete after two years. However, there is a small increment for several more years (Laws, 1958). The length of animals one to five years old is about 198-208 cm. The average length of animals five to eight years old is 221 cm, and 226 cm for those 12 years and older. Sexual maturity probably occurs in the second or third year, but according to Lindsey (1937) and Bertram (1940) some animals may breed at 12-14 months.

The oldest animal obtained in Laws' (1958) survey of 792 animals was 19 years. The oldest in Øritsland's sample of 218 animals was 29 + years.

Diseases

One of the more interesting reports of mortality of a marine mammal was based on observations in King Gustav Channel, at the western base of Antarctic peninsula (Laws and Taylor, 1957). In August, 1955, about 3000 crabeater seals were seen to be trapped on sea ice 5-25 km from open water. Access to sea water was available due to some large pools and cracks. By mid-October most of the animals were dead; the average mortality rate of the various groups was 85%, and 97% in the largest group of 1000 animals. Some facts about these animals were: (1) the dead animals still had a blubber thickness of 4 cm and there was no obvious cause of death; (2) there were numerous abortions; and (3) most of the animals were less than nine years old, or younger than other populations of crabeaters censused. The cause of death was surmised to be due to disease, possibly a virus. However, this is only an assumption for lack of a better explanation. A local Weddell seal population was not affected, nor were men and dogs who ate the meat from some of these animals.

Another report of what might be termed a "synthetic" disease is that of Sladen et al. (1966) of the presence of small amounts of DDT residues in a crabeater seal.

References

Berg, C. (1898). *Lobodon carcinophagus* (H.J.) Gr. en el Rio de la Plata. *Comun. Mus. Nac. B. Aires* 15 (from Scheffer, 1958).

Bertram, G. C. L. (1940). "Biology of the Weddell and Crabeater Seals", British Graham Land Exped. 1934-37 Vol. 1, 1-79. British Museum (Natural History), London.

Carrick, R. (1964). Southern seals as subjects for ecological research. *In* "Biologie Antarctique", (Eds R. Carrick, M. Holdgate and J. Prévost), 421-432. Hermann, Paris.

Corner, R. W. M. (1972). Observations on a small crabeater seal breeding group. *Br. Ant. Sur. Bull.* **30**, 104-106.

Courtenay-Latimer, M. (1961). Two rare seal records for South Africa. *Ann. Cape. Prov. Mus.* **1**, 102.

Drabek, C. M. (1975). Some anatomical aspects of the cardiovascular system of antarctic seals and their possible functional significance in diving. *J. Morphol.* **145**, 85-106.

Eklund, C. R. and Atwood, E. L. (1962). A population study of Antarctic seals. *J. Mammal.* **43**, 229-238.

Erickson, A. W. and Hofman, R. J. (1974). Antarctic Seals. Antarctic Map Folio Ser. (Ed. V. C. Bushnell), Folio 18, Antarctic Mammals, Amer. Geogr. Ser., New York.

Erickson, A. W., Siniff, D. B., Cline, D. R. and Hofman, R. J. (1971). Distributional ecology of Antarctic seals. *In* "Symposium of Antarctic Ice and Water Masses, Proceedings", (Ed. Sir George Deacon), 55-76. Scientific Community Antarctic Research, Brussels.

Gilbert, J. R. and Erickson, A. W. (1977). Distribution and abundance of seals in the pack ice of the Pacific sector of the southern ocean. *In* "Adaptations within Antarctic Ecosystems", (Ed. G. A. Llano), 703-748. Smithsonian Institution, Washington, D.C.

Gray, J. E. (1844). The seals of the southern hemisphere. (p. 1-8, pls. 1-10, 14-17), *In* "The Zoology of the Voyage of H.M.S. 'Erebus' and 'Terror' ", (Eds. J. Richardson and J. E. Gray), 2 Vols, London, (from Scheffer, 1958.)

King, J. E. (1957). On a pup of the crabeater seal, *Lobodon carcinophagus*. *Ann. Mag. Nat. Hist.* **10 (12)**, 619-624.

Krylov, V. I. (1972). *Sov. Ant. Exped., Info. Bull.* **8 (5)**, 279-283.

Laws, R. M. (1958). Growth rates and ages of crabeater seals, *Lobodon carcinophagus* (Jacquinot and Pucheran). *Proc. Zool. Soc. Lond.* **130**, 275-288.

Laws, R. M. (1964). Comparative biology of Antarctic seals. *In* "Biologie Antarctique", (Eds R. Carrick, M. Holdgate and J. Prévost), 445-454. Hermann, Paris.

Laws, R. M. and Taylor, R. J. F. (1957). A mass dying of crabeater seals, *Lobodon carcinophagus* (Gray). *Proc. Zool. Soc., Lond.* **129**, 315-323.

Lindsey, A. A. (1937). The Weddell seal in the Bay of Whales, Antarctica. *J. Mammal.* **18**, 127-144.

O'Gorman, F. (1963). Observations on terrestrial locomotion in Antarctic seals. *Proc. Zool. Soc. Lond.* **141**, 837-850.

Øritsland, T. (1970). Sealing and seal research in the south-west Atlantic pack ice, Sept.-Oct. 1964. In "Antarctic Biology", (Ed. M. W. Holdgate), Vol. 1, 367-376. Academic Press, London and New York.

Ray, C. (1970). Population ecology of Antarctic seals. In "Antarctic Ecology", (Ed. M. W. Holdgate), Vol. 1, 398-414. Academic Press, London and New York.

Racovitza, E. G. (1900). La vie des animaux et des plantes dans l'Antarctique. *Bull. Soc. Roy. Belge de Geographie, Brussels,* **24**, 177-230.

Scheffer, V. B. (1958). "Seals, Sea Lions and Walruses, A Review of the Pinnipedia", Stanford University Press, Stanford, CA.

Siniff, D. B. and Bengston, J. L. (1977). Observations and hypotheses concerning the interactions among crabeater seals, leopard seals, and killer whales. *J. Mammal.* **58**, 414-416.

Siniff, D. B., Cline, D. R. and Erickson, A. W. (1970). Population densities of seals in the Weddell Sea, Antarctica. In "Antarctic Ecology", (Ed. M. W. Holdgate), Vol. 1, 377-394. Academic Press, London and New York.

Siniff, D. B. and Reichle, R. A. (1976). Biota of Antarctic pack ice: R/V Hero cruise, 75-76. *Antarctic J.* **11**, 61.

Siniff, D. B., Stirling, I., Bengston, J. L. and Reichle, R. A. (1979). Social and reproductive behavior of crabeater seals *(Lobodon carcinophagus)* during the austral spring. *Canad. J. Zool.* **57**, 2243-2255.

Sladen, W. S. L., Menzie, D. M. and Reichel, W. L. (1966). DDT residues in Adelie penguins and a crabeater seal from Antarctica. *Nature (London)* **210**, 670-673.

Stirling, I. and Kooyman, G. L., (1971). The crabeater seal *(Lobodon carcinophagus)* in McMurdo Sound, Antarctica, and the origin of mummified seals. *J. Mammal.* **52**, 175-180.

Stirling, I. and Rudolph, E. D. (1968). Inland record of a live crabeater seal in Antarctica. *J. Mammal.* **49**, 161-162.

Trouessart, E. -L. (1897-1905). Catalogus mammalium tam viventium quam fossilium. Tomas I (1897), Fascic. 2, 219-452. R. Friedländer und Sohn, Berlin, (from Scheffer, 1958.)

Tyler, J. C. (1960). Erythrocytes and hemoglobin in the crabeater seal. *J. Mammal.* **41**, 527-528.

Vaz Ferreira, R. (1956). Caracteristics generales de las islas Uruguagas habitadas por lobos marinos. *Montevides, Servicio Oceanografico y de Pesca, Trabajos Sobre Islas de Lobos y Lobos Marinos, No.* 1.

Wilson, E. A. (1907). *Lobodon carcinophagus.* The white Antarctic or crabeating seal. In "Mammalia (Whales and Seals)", Natl. Ant. Exped, 1901-04. Natural History, Zoology 2: ix-xii, 1-41. British Museum (Natural History), London.

Ximenez, A., Langgerth, A. and Praderi, R. (1972). Lista sistematica de los maniferos del Uruguay. *Anales del Museo Nacional de Historia Natural de Montevideo,* **7**, 1-49.

10
Ross Seal
Ommatophoca rossi Gray, 1844

G. Carleton Ray

Genus and Species

The Ross seal is the last discovered and least known of the pinnipeds. Many authors have remarked on this species, but it is not my intention to review all literature, as much that has been printed only repeats the observations of others and is not based upon personal observation; inevitably, conjecture has crept in. Therefore, I shall summarize the thrust of studies to date and include some personal observations published in brief previously (Ray, 1970) or not yet published.

Common names

Hofman *et al.* (1973) give "big-eyed seal" and "singing seal" as "common" names, in addition to "Ross seal". The former is not particularly appropriate as the eyes do not appear extraordinarily large externally. The latter is inappropriate as this species appears to be less

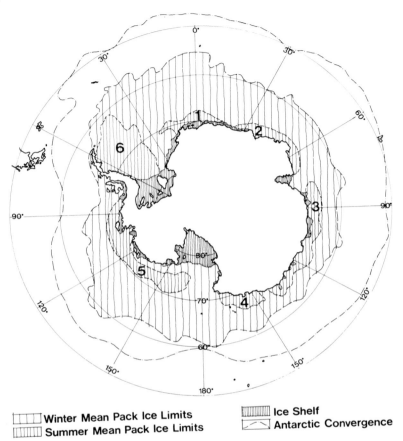

FIG. 1 Pack-ice habitats of the Ross seal, adapted from Erickson and Hofman (1974). Ross seals have been sighted throughout both summer and winter ice, strays occur northward to sub-Antarctic islands even to 53° S on the coast of Australia (*Marine Mammal News*, 1978). Individual sightings are given by Erickson and Hofman (1974), up to the date of their compilation.

Summer ice generally occurs in the areas marked 1-6 and most sightings occur at that time. Densities given by Erickson and Hofman's (1974) chart are approximately: Area 3 = 0.003-0.29 km^{-2}; Area 4 = 0.003-0.15 km^{-2}; Area 5 = 0.003-0.026^{-2}; Area 6 = 0.16-0.29 (outer boundary of area only). Other densities are: W. Ross Sea = 0.012-0.116 km^{-2} (Ray, 1970); Indian Ocean side of Antarctica = 0.088 (Eklund and Atwood, 1962). The highest densities yet recorded are: Area 1 = 2.9 km^{-2} (Wilson, 1975). Variations in abundance are probably a function of food, or some other factor, rather than ice distribution.

vocal than several other species (*Erignathus, Phoca groenlandica, Leptonychotes*, for example). In any case, these names are rarely used. "Ross seal" is the commonly used vernacular.

Scientific name and classification

The Ross seal was unknown until 1840 when Captain James Clark Ross's British Antarctic Expedition of 1839—43 discovered it at 68° S and 176° E (Fig. 1). Captain Ross secured two specimens and J. E. Gray (1844) named the species after him. *"Ommatophoca"* refers to the huge orbits, after the Greek "omma" for "eye".

Scheffer (1958) grouped the Southern Ocean phocids crabeater, Weddell, leopard, and Ross seals in the Tribe Lobodontini of the subfamily Monachinae. King (1966), contrary to Scheffer, included the elephant seal, *Mirounga*, in the Monachinae, but excluded the hooded seal, *Cystophora*, which had previously been thought to be closely related to *Mirounga*. C. E. Ray (1976) concurred with King, noting the highly diagnostic nature of the temporal bones at the generic level (see Internal Characteristics). Therefore, the Monachinae are tropical to south temperate to south polar in distribution, representing a different line of evolution from the northern phocids. This subfamily probably evolved from the Upper Miocene *Monotherium* of temperate seas; the tribes Monachini (monk seals) and Lobodontini then appear to have evolved within temperate-tropical and south polar seas respectively.

External Characteristics

Shape

The Ross seal is gracefully formed (Fig. 2). The head is large and the neck thick, but the body is more slender than that of the Weddell seal which the Ross seal superficially resembles.

Several observers have remarked that the head may be drawn back into the circular folds or rolls of heavily blubbered skin on the neck, but this is a characteristic of phocids generally. For example, Wilson (1907) remarked on the "pug-like expression of countenance . . . It was in each case excessively fat and its head was withdrawn into the circular folds or rolls of heavily blubbered skin on the neck to such an extent as to almost disappear from sight." However, several other seals, especially when young, exhibit exactly the same sort of behaviour. In the Ross seal's case, this habit is exaggerated but not to the point of rendering it "shapeless" (King, 1964a).

Size

This is a medium-sized seal of medium weight. It is, however, the smallest of Southern Ocean phocids. Wilson (1907) concluded, apparently on the basis of stretched skins, that the living seal could reach 331.5 cm. However, carcasses he measured reached but 236 cm, with a maximum girth of 150 cm.

Both Scheffer (1958) and King (1964a) cited size data, but without specific reference. Scheffer gave maxima of 223 cm for males and 229 cm for females. King stated that males are larger and may reach 300 cm. Hofman et al. (1973) and Hofman (1975) provided data on 11 males and 4 females as follows:

	Standard length	Estimated weight
Males	199 cm (168-208)	173 kg (129-216)
Females	213 cm (196-236)	186 kg (159-204)

I have measured three adult living females lying undisturbed on the ice at about 230 cm from nose tip to rear flipper tip, corresponding to about 190-200 cm standard length.) Maximum measurements indicate a slightly larger size for females, but age/length data would be required for confirmation, as specimens are presently very few. A problem is that authors do not always state how their measurements are taken, whether standard or curvilinear length or whether taken with the seal on its back or belly. Nevertheless, I would think that the Ross seal rarely, if ever, exceeds 250 cm in standard length, fully stretched out on its back.

Coloration and moult

Ray (1970) gave the colour and pattern as follows: "Dorsum dark with little spotting, sharply turning silvery on venter; streaked on side of head and throat and spotted or streaked on sides and flanks." The pattern is complex and varies among individuals, especially about the head and neck (Fig. 2). No other seal is so streaked. In some

FIG. 2 (a) Adult female Ross seal, showing size relative to adult male human. (b) Adult Ross seal in "typical" habitat near Cape Adare.

individuals, the streaks on the throat may approach a chestnut or chocolate colour. Hofman *et al.* (1973) noted the mask-like appearance caused by the merging of the streaked pattern about the eyes; such a pattern is, however, variable.

Colour in all seals is influenced by time of year and state of moult. In unmoulted summer Ross seals, the ground colours are tan to brownish. Polkey and Bonner (1966) gave the colour of a skin as "sandy brown"; their specimen was probably an unmoulted animal. The moult evidently occurs principally in January (Wilson, 1907; King, 1964a; Ray, pers. obs.) but may extend into February (Condy, 1977). I observed flaking of 2-3 mm diameter pieces of epidermis from the back, lower sides, rear flippers, and belly of an adult in January. Wilson (1975) has observed a similar shedding of skin and hair. Wilson (1907) also noted that there is occasional scarring on the neck and shoulders, which he thought to be a result of intraspecific fighting. Condy (1977) noted 3.2% of Ross seals were scarred, possibly as a result of attacks by *Orcinus orca* and/or *Hydrurga leptonyx* (P. R. Condy, pers. comm.).

Head

The snout is extraordinarily short, the mouth also short, and the head wide. The eye slits are no longer than in many other seals, so do not reveal the true eye size, which is proportionately larger than in any other seal; the eye may reach 7 cm in diameter (King, 1968) or 0.07-0.08% of the total body weight (Bryden and Erickson, 1976).

Polkey and Bonner (1966) stated that there are 5 principal mystacial vibrissae on each side, the longest being 42 mm. However, their **figure** shows 15 vibrissae to a side. Ling (1972) gave a vibrissae count of 15-17 to a side; vibrissae are 10-40 mm long, round in cross-section, and smooth in outline. The Ross seal, thus, has few vibrissae for a phocid and the vibrissae are the shortest of any phocid.

Body hairs

Similarly, the body hairs are the shortest of any phocid. Polkey and Bonner (1966) observed that the hairs are in units of a single primary hair with 1-2 secondary hairs, all emerging from one piliary canal, but originating from separate follicles. The primary is 120 μm in diameter and the secondaries about 50 μm. The skin is thicker and the hairs longer ventrally than dorsally and the primary hair densities range from 3.6 mm^{-2} on the top of the head to 1.9 mm^{-2} ventrally between the fore flippers.

Fore flippers

As for other lobodontines, the fore flippers of the Ross seal are elongated and have an oblique posterior border. King (1964b, 1968) placed the Ross seal as the most specialized seal in fore flipper modifications, including reduced claws and greatly elongated terminal phalanges. She suggested evolution toward an otarid-like mode of swimming. However, Hofman *et al.* (1973) found that both *Hydrurga* and *Lobodon* have longer fore flipper lengths proportionately to body length than the Ross seal. Bryden and Felts (1974) found that the bones and muscles of the forelimb are relatively small, suggesting that the fore flipper is not used extensively in water, and never out of water. Pierard and Bisaillon (1978) concluded that flipper modifications indicate the Ross seal to be a strong swimmer. The functional significance of fore flipper modifications of the Ross seal must remain somewhat conjectural until behavioural observations are extended (see Locomotion).

Rear flippers

Hofman *et al.* (1973) showed that the hind flipper may be almost 20% as long as the standard body length. For a young animal of 166 cm, I obtained a figure of 22%. This is proportionately the longest rear flipper length of any phocid and substantiates the argument that the lobodontines have accentuated a phocid-like mode of swimming, as opposed to having adopted an otarid-like method.

Internal Characteristics

As anatomical data are relatively easily acquired, this is a subject about which most is known for the Ross seal. I consider here only some major features which indicate the Ross seal's extreme specialization.

Skull and skeleton

Thompson (1911) gave detailed osteological notes on the type specimen at the British Museum (Natural History) which, unfortunately, is not a

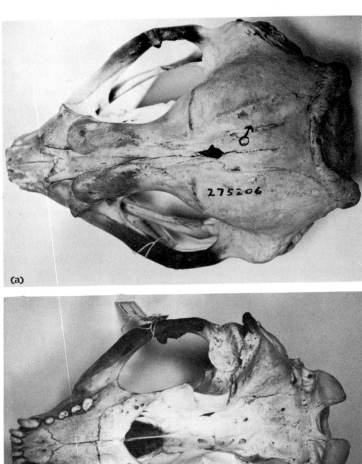

FIG. 3 Skull of adult, male Ross seal from USNM (Smithsonian). (a) dorsal view, showing inflated squamosal region; (b) ventral view, showing thickened and elongated auditory process of the bulla.

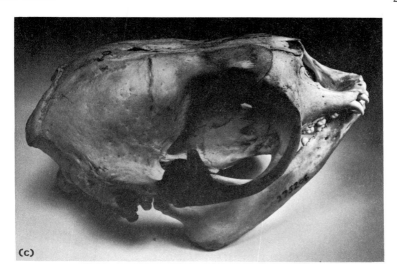

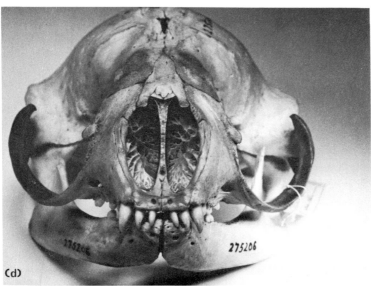

FIG. 3 (c) lateral view, showing orbit; (d) frontal view, showing maxilloturbinal bones.

complete skeleton. Bruce (1914) presented excellent photographs of skulls of lobodontine seals. Only a few good skulls are in modern collections (Fig. 3a-d). Pierard and Bisaillon (1978) have presented the most complete osteology of the Ross seal to date. They pointed out the close resemblance to *Leptonychotes* and emphasized King's (1965) earlier observation of the extreme development of the epiphyses of the terminal fore flipper phalanges, greater than for any other seal. Otherwise, the skeleton is evidently not extraordinary.

The skull can be mistaken for that of no other mammal. The eye socket is enormous. The maxilloturbinal bones are very densely packed suggesting an excellent olfactory sense or efficient warming of inhaled air or both. In fact, Pierard and Bisaillon (1978) state that "the voluminous and compact turbinates almost completely fill the nasal cavities, preventing water invasion of the air passages . . . air passage through such a cavity must be very scanty." However, my observations of living animals showed that Ross seals have little trouble breathing and exhale well through the nose. C. E. Ray (1976) pointed out other specializations of the skull: ". . . auditory process of the bulla is greatly thickened, long and prominent, there is a mastoid-like swelling of the squamosal regional dorsal to the external acoustic meatus and the mastoid region, and the petrosal is much reduced, especially the apex (absolutely smaller than in any other phocid)."

Teeth

Wilson (1907) spoke of the "degenerate" nature of the teeth. The incisors and canines are sharp and recurved, though small, and the cheek teeth are very reduced (see Figs 3 (b)-3 (d) and 4 (a) and (b)). The tooth formula is variable, being:

$$\frac{2}{2} : \frac{1}{1} : \frac{5\text{-}6}{4\text{-}6}$$

The teeth are presumed to be adapted for holding cephalopods, but it is impossible to say whether the cutting or rending of soft-bodied food is possible.

King (1964b) also notes that the cheek teeth are weak and variable in number, even to the point of frequently being loose in the living animal. I have had the opportunity to examine living Ross seals, even to the point of tactily examining their mouths and, whereas the cheek teeth barely penetrated the gums, I noted no loose teeth.

Palate, tongue, pharynx, hyoid and trachea

King (1964a,c, 1968, 1969) drew our attention to the several modifications of this region. First, the Ross seal has a very short, hard palate and, consequently, very long soft palate (Fig. 4). The latter comprises 60% of the total palatine length and extends far posteriorly to the level of the occipital condyles, rather than only to the glenoid cavity as in *Phoca vitulina*, for example. Consequently, the base of the tongue and the epiglottis are farther back than in other seals. Second, the swallowing muscles of the tongue and pharynx are well developed. Third, the mouth can be opened widely and large, powerful muscles shut the jaw forcefully. Bryden and Felts (1974) agreed with King that the hyoid bones, tongue, and pharyngeal muscles are well developed, but stated that the muscles of mastication are not so specialized and, therefore, the jaw may not be so forcefully closed. Nevertheless, both agreed that these anatomical features are probably adaptations for swallowing prey of large size.

The trachea also is very large. Bryden and Erickson (1976) stated that it is 0.46-0.48% of the total body weight. The single 165 cm Ross seal which I have dissected had a trachea 32 cm long with tracheal rings modified into ventral bars. The trachea was widest at its midpoint: i.e. at both the laryngeal and bronchial termini it was 5 cm in width, but at its midpoint it was almost 9 cm wide. This is the only seal with such a centrally expanded trachea. King (1964a, c, 1968) believed that the modification of the tracheal rings is also an adaptation for the swallowing of large morsels. King (1964c) stated: "In *Hydrurga* [the tracheal rings] have been reduced to ventral bars so that the trachea can be compressed quite flat by the expansion of the esophagus when penguins are being swallowed." However, it is now well known that *Hydrurga* does not swallow penguins whole, nor do other species of seals with flattened tracheal cartilages (*Erignathus, Leptonychotes*) swallow larger-sized chunks of food than other pinnipeds which do not have such tracheal modifications. Bryden and Felts (1974) suggested a role of sound production for the trachea. I have noted that the dorsal tracheal membrane is very distensible and agree with the latter conclusion (see Sounds).

Other anatomical characteristics

The Ross seal exhibits several other extreme modifications. King (1964a) described a cetacean-like kidney structure not shared by other

FIG. 4. Yearling Ross seal during sound production. Note inflated soft palate and subjacent tongue. (b) Ross seal, close-up of interior of mouth. Note small teeth and short, hard palate. The soft palate is not inflated fully in this case.

pinnipeds. Ling (1972) pointed out that the Ross seal and *Mirounga* are the only mammals with microcrine glands in association with vibrissae follicles, and that only the Ross seal has a common secretory canal for both holocrine and microcrine glands emptying into the tactile hair canal. This, Ling conjectured, may be to enhance tactile capability for sensory perception. Bryden and Erickson (1976) stated that the esophagus is the largest, proportionately, of all seals, and also may be more muscular and distensible. They hypothesized that the esophagus modification may be for sound production, which appears unlikely as there is little to be gained by the swallowing of air by a diving mammal. Also, the small and large intestines are short and heavy, relative to other phocids.

Finally, there is some anatomical indication of the diving capability of the Ross seal. The blood is 10% of total body weight, not so great as *Mirounga*, but greater than for most mammals (Bryden and Erickson, 1976). The caval sphincter, which prevents heart overloading during diving, is best developed in *Mirounga* and the Ross seal (Harrison and Tomlinson, 1964).

Distribution and Numbers

Numerous surveys of Southern Ocean phocids have yielded few sightings of Ross seals relative to other species. Captain Ross returned two specimens from his expedition. The next hundred years yielded 16 skeletons, 29 skulls, and a few skins (King, 1968). Bertram (1940) stated that less than 50 had been seen prior to his writing. Hofman *et al.* (1973) state that "less than 200 Ross seal sightings had accumulated prior to 1972". Since then, I would think that not more than 500 have been sighted by scientists: Gilbert and Erikson (1977), Hall-Martin (1974), Wilson (1975) and a few others. All of these assessments have been made by ship or aerial sighting techniques.

Øritsland (1970a) stated that Ross seals are "circumpolar in the pack." Scheffer (1958) and a few others thought that open pack was the preferred habitat. Most now believe that the Ross seal prefers heavy ice. For example, Hofman (1975) stated: "the Ross seal seems to prefer heavy, consolidated pack ice that is difficult to penetrate even with icebreakers." Gilbert and Erikson (1977) stated: the Ross seal is "most abundant in more dense ice concentrations dominated by larger floe sizes" and it prefers "the interior or the pack ice regions which are seldom visited by icebreakers". Condy (1977), on the other hand, noted variability in ice conditions in which Ross seals have been seen and stated that there must be other reasons for concentration than ice density: "numbers of Ross seals vary according to other factors not yet understood." He summarized data for the King Haakon VII Sea to show that Ross seals prefer smaller, smooth-surfaced floes there.

Ray (1970), Hall-Martin (1974), Wilson (1975), and Condy (1977) agree that concentrations may occur. I had too few data to substantiate my intuitive claim for localized abundance, but my observations and underwater acoustic recordings indicated a greater abundance of Ross seals near Cape Adare than farther to the south in the Ross Sea. The best data for local centers of abundance come from the King Haakon VII Sea; combined data of Hall-Martin (1974), Wilson (1975), and Condy (1977) arrived at 18.1% Ross seals of the total seals observed. This contrasts with a low of none seen by Siniff et al. (1970) to the more typical 1.9% and 2.4% (Gilbert and Erickson, 1977) and 0.1 to almost 5% (Erickson et al., 1974) in other sectors. Therefore, there seem to be some areas where Ross seals concentrate.

Figure 1 gives a habitat interpretation of Ross seal distribution, i.e. according to summer and winter ice distributions. The figure summarizes concentrations for different areas and it is clear that densities vary considerably from region to region. Ice distribution is obviously a major determinant of Ross seal distribution. However, it is but a part of the picture, food being one other important component. Whereas we know something of ice distribution, we know almost nothing of the Ross seal's food distribution. Furthermore, we have no idea whether the Ross seal enters a pelagic phase, for example, after the pattern of *Phoca groenlandica* or *Phoca fasciata*. There has apparently been only one attempt to locate Southern Ocean seal concentrations in the water, namely my attempt to do so by recording their voices (Ray, 1970).

Attempts to calculate population abundance at this time are premature. Scheffer (1958) arrived at a population estimate of 20-50 000, and King (1964a) at 20 000, but these were clearly guesses. Gilbert and Erickson (1977) calculated 37 462 Ross seals for the Amundsen and Bellingshausen Seas, on the basis of 137 sightings; their calculation of a total Southern Ocean population of 220 000 would appear difficult to verify. Hofman et al. (1973) gave a possible range of 104 000-650 000, based on the percent composition of the total seal population. It is my belief that we simply do not have enough sighting and behavioural data to make any reasonable estimate at present.

Life History

Breeding biology

The Ross seal appears to live a predominantly solitary life (Wilson, 1907; Scheffer, 1958; King, 1964a; Ray, 1970; Øritsland, 1970a; Gilbert and Erickson, 1977; Hall-Martin, 1974; Wilson, 1975; etc.). I have seen a loosely clustered group of three adult females and one young of the year female in January where the animals were hauled out within 100 m of one another, and about a kilometre from open water (see Sociability).

Øritsland (1970a) stated that the Ross seal breeds in November, moults while fasting in January, matures at 3-4 years for males and 2-7 years for females, and lives 12 years. However, reference to Øritsland (1970b) shows that all mature females of three years of age or greater were pregnant, indicating an age of maturity of 3 years for females. Further, the oldest seal he captured was 12 years old, indicating that his longevity figure is inconclusive. Tikhomirov (1975) stated that sexual maturity is reached at 3-5 years for females and that 9.8% of females are non-breeders. The latter condition he attributes as possibly due to heavy nematode infestation.

Tikhomirov (1975) examined Ross seals during November. He saw 8 females with suckling pups 109-138 cm long and 40-75 kg in weight. Wilson (1975) observed a female in January with a pup half her length. I observed a female near a yearling of about three-quarters her length early the same month. Tikhomirov estimated that birth takes place in early November, at least in the region of the Balleny Islands where he worked. He further estimated that copulation occurs after weaning in early to mid December and that implantation occurs 2.5-3 months later, by early March.

Food

Scheffer (1958) stated that Ross seals eat "soft-bodied cephalopods and fishes". Øritsland (1977) made the first systematic study of food preferences and found that the diet consisted of 64% cephalopods, 14% other invertebrates, including some krill, and 22% fish. He estimates the daily food requirement as about 12 kg a day for an adult seal.

Laws (1964) pointed out that the Ross seal's niche as a squid eater might minimize competition with other lobodontines for food, but this does not consider competition with other vertebrates. Rudmose-Brown (1913) estimated on the basis of size of squid beaks that Ross seals can devour squid nearly 2 m long. King (1964a, 1968) suggested that Ross seals can consume squids of up to 76 cm and 6.8 kg. It is obvious that the mouth is of large capacity and that both the long soft palate and the construction of the esophagus and trachea (see Internal Characteristics) may facilitate the ingestion of large prey. However, there is little indication whether the Ross seal swallows whole prey or rends it to pieces before swallowing.

Behaviour

The now familiar analogy between marine mammals and icebergs, of which so little can be seen, is particularly pertinent here. *Ommatophoca* is lodged in one of the most remote of habitats for man to enter. Our knowledge of the Ross seal's behaviour is fragmentary.

Reaction to intruders and threat

A most extraordinary feature of this seal's behaviour on the ice is its reaction to a human intruder. It almost totally ignores a human until it is approached to within 5-10 m or less. Upon close approach, it "raises its head, inflates the trachea and soft palate with air, and with open mouth, makes trilling and chugging sounds" (Ray, 1970). Figures 2 and 4 illustrate this distinctive behaviour. Although this behaviour appears to be a threat, I have placed my hand in a Ross seal's mouth during this display and was not bitten.

Sociability

Hall-Martin (1974) sighted only 3%, and Wilson (1975) 9%, of Ross seals in pairs. This would indicate a solitary habit, but little is known of reproductive behaviour. Scheffer (1958) called the Ross seal among the least gregarious of seals, along with *Hydrurga* and *Erignathus*. However, sighting seals alone on the ice may prove little. The nature of underwater sound propagation allows communication over long distances, which could provide for sociability or territoriality over large areas. *Erignathus* is highly vocal and evidently uses sound in territoriality (Ray *et al*., 1969), whereas the harp seal is also highly vocal and is very sociable (Møhl *et al*., 1975). As for the Ross seal, my experience is that when few or none could be seen on the ice during January, the hydrophone frequently revealed several in the water; Ray (1970) stated that: "the numbers of seals calling underwater always exceeded numbers seen locally on ice." This, of course, may merely indicate that this species is very aquatic in habits.

Hauling-out on ice

Thermoregulatory considerations and the need to grow new hair and tissue cause seals to haul out more during the moulting period than at other times of year (*cf*. Ray and Smith, 1968, for *Leptonychotes* and Feltz and Fay, 1967, for *Odobenus*). Wilson (1907) stated that the Ross seal "prefers to starve for a week or two than enter the water whilst the moult proceeds." Therefore, more seals would be expected to haul out during January's molting period than at other times of year. Condy (1977) noted "no particularly strong diurnal haul-out pattern." However, there was indication of a "main period of haul-out from approximately 08 h 00 to about 16 h 00."

Sounds

The sounds produced by the Ross seal have been noted by several early explorers. Rudmose-Brown (1913) remarked on the "curious, loud cry, which is quite characteristic and unmistakable". Fortunately, this seal produces similar sounds both in the water and in the air, allowing positive identification of underwater sounds (Fig. 5a,b). Ray (1970) has briefly described these sounds. Ray and Watkins (MS) are preparing a more detailed description.

The sounds and associated behavior were wonderfully described by E. G. Racovita (evidently unpublished, and quoted overleaf from Wilson (1907).

Racovitza	*GCR Translation*
Ce phoque possède aussi une aptitude curieuse que le distingue de ses congénères. Sa voix est beaucoup plus compliquée et les sons qu'il émet plus variés que ceux des autres phoques. Il peut gonfler son larynx et en outre son énorme voile de palais, de façon à constituer deux caisses de resonnance, deux poches contenant grande provision d'air. Cela lui permet d'executer des trilles et arpèges aussi sonores que bizarres. Lorsqu'on l'invite, il commence par gonfle son larynx en rabattant la tête en arrière. Il produit alors, la gueule ouverte, et son voile de palais distendu apparaissant comme une grosse boule rouge, un roucoulement semblable à celui d'une tourterelle entrouée. Puis il ferme la gueule et émet un gloussement de poule efrayée. Il expulse finalement avec violence, par les narines, sa provision d'air, et cela produit un renifflement comparable à celui que fait un cheval qui s'ebroue.	This seal also possesses a curious habit which distinguishes it from its relatives. Its voice is much more complicated and the sounds which it emits more varied than those of other seals. It can inflate both its larynx and its enormous soft palate so as to constitute two resonance chambers, two sacs containing a great provision of air. That permits it to execute trills and arpeggios as sonorous as they are bizarre. When one irritates it, it begins to inflate its larynx while bringing back its head posteriorly. It then produces, its mouth open and the soft palate distended to appear like a great red ball, a gurgling like that of a husky turtle dove. Then it closes its mouth and emits a clucking like an alarmed chicken. It finally violently expels its air supply through its nostrils, and that produces a snuffling comparable to that made by a horse snorting from fear.

Perhaps the Ross seal's sounds are not quite the barnyard Racovitza proclaims, but this is a remarkable display. Furthermore, the seal is so fearless that one is able to palpate its throat during the display and to photograph almost at will (Figs 2 and 4). King (1964a) states that the Ross seal uses its palate "rather like a bagpipe". The Ross seal produces these sounds as it greatly expands the trachea and soft palate, but, unlike a bagpipe, without losing any air from its body, except at the display's end, as Racovitza described. How, or even where, sound is actually produced, I could not state, but during sound

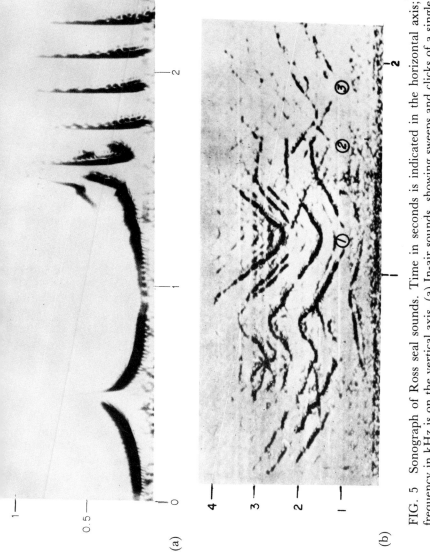

FIG. 5 Sonograph of Ross seal sounds. Time in seconds is indicated in the horizontal axis; frequency in kHz is on the vertical axis. (a) In-air sounds, showing sweeps and clicks of a single seal on ice. (b) Underwater sounds, showing sound production of three seals (indicated by encircled numbers).

production I could feel vibrations along the length of the seal's throat. Thus, wherever sound is produced, the trachea and palate appear to act as resonating chambers. The fact that no air is lost during sound production allows sounds to be made repeatedly underwater, although there is probably a limitation on the depth at which sounds of surface-like quality can be made because of compression.

Locomotion

My observations of Ross seals on ice indicate that locomotion is a slow hunching forward in which the fore flippers are not employed. As in *Leptonychotes*, the Ross seal crawls on its belly and chest. If pressed, it may use a rear flipper "swim" as does *Lobodon*, but more slowly and clumsily. Contrarily, Rudmose-Brown (1913) stated that the Ross seal seal is agile on ice and in the water, more so than *Leptonychotes*. He supposed that this agility is sufficient to provide escape from *O. orca*. O'Gorman (1963) disagreed, stating that the Ross seal is sluggish, possibly due to its "excessive fatness". He suggested that this species is ill-adapted for terrestrial locomotion and "does not use the sinuous method of progression" of *Lobodon*.

Ling (1964) stated that the Ross seal is a strong, manoeuvrable swimmer, but the basis for this statement is not clear. My experience is that the Ross seal dives as soon as it leaves the ice. There is no reason to believe that it swims differently than other phocids, namely by side-to-side sweeps of the rear flippers, with occasional use of the fore flippers for slow forward progression and for aid in manoeuvering.

Conclusions

There is little doubt that the Ross seal represents an extreme of adaptation among pinnipeds. It is also the least known, least observed, and last-discovered of all. This species is now only a little better known in its behaviour and ecology than during the Heroic Age of Antarctic Exploration three-quarters of a century ago.

Attempts to census seals on ice are of little value prior to understanding their ecology and behaviour and it may be safer and wiser not to make population estimates than to hazard guesses based on a paucity of data, or in the virtual absence of data on ecology and behaviour which make sighting data meaningful. Data on the underwater portions of seals' lives are especially needed. What is seen on the surface may be misleading, especially regarding social behaviour.

The Ross seal is protected from exploitation by the Convention for the Conservation of Antarctic Seals which has been in force since mid-1978. The protection of little-known species is mandatory, but this must not obviate need for study of habitat and niche, physiology and behaviour, without which we cannot predict the effect on seals of man's alteration of whole marine ecosystems.

Acknowledgements

I wish to thank the Office of Polar Programs, National Science Foundation, for the opportunity to observe living Ross seals. To the Office of Naval Research (Oceanic Biology) goes thanks for support of this writing. P. R. Condy, C. E. Ray, W. A. Watkins and D. Wartzok offered many suggestions on the manuscript.

References

Bertram, G. C. L. (1940). The biology of the Weddell and crabeater seals with a study of the comparative behaviour of the Pinnipedia. *Br. Mus. (Nat. Hist.) Sci. Repts* Brit. Graham Land Exped. 1934-37, 1, 1-139.

Bruce, W. S. (1914). On the skulls of Antarctic seals: Scottish Nat. Res. Exped. *Trans. Roy. Soc. Edin.* XLIX (Part II (4)), 345-6. 5 plates.

Bryden, M. M. and Erickson, A. W. (1976). Body size and composition of crabeater seals *(Lobodon carcinophagus)*, with observations on tissue and organ size in Ross seals *(Ommatophoca rossi). J. Zool. (London)* **179**, 235-247.

Bryden, M. M. and Felts, W. J. L. (1974). Quantitative anatomical observations on the skeletal and muscular systems of four species of Antarctic seals. *J. Anat.* **118 (3)**, 589-600.

Condy, P. R. (1977). Ross seal, *Ommatophoca rossi* (Gray, 1850), with notes from the results of surveys conducted in the King Haakon VII Sea, Jan./Feb. 1974-76. Symposium on Endangered Wildlife in Southern Africa, Pretoria, 1976. Proceedings, 1-16.

Eklund, C. R. and Atwood, E. L. (1964). A population study of Antarctic seals. *J. Mammal.* **43**, 229-238.

Erickson, A. W. and Hofman, R. J. (1974). Antarctic Seals. *In* "Antarctic Mammals", Folio 18, Antarctic Map Folio Series, Amer. Geog. Soc., New York, Ross Sea p. 8 and Seals, Sheet 4.

Erickson, A. W., Denney, R. N., Brueggeman, J. J., Sinha, A. A., Bryden, M. M. and Otis, J. (1974). Seal and bird populations of Adelie, Clarie and Banzare Coasts. *Ant. Jour.* **9 (6),** 292-296.

Feltz, E. T. and Fay, F. H. (1967). Thermal requirements *in vitro* of epidermal cells from seals. *Cryobiology* **3 (3),** 261-264.

Gilbert, J. R. and Erickson, A. W. (1977). Distribution and abundance of seals in the pack ice of the Pacific section of the Southern Ocean. *In* "Adaptations Within Antarctic Ecosystems: Proceedings of the Third SCAR Symposium on Antarctic Biology", (Ed. G. A. Llano), 703-740. Smithsonian Institution, Washington, D.C.

Gray, J. E. (1844, 1875). The seals of the southern hemisphere. *In* "The Zoology of the Voyage of H.M.S. 'Erebus' & 'Terror', under the Command of Captain Sir James Clark Ross, R.N., F.R.S., during the Years 1839 to 1843. "Part I. Mammalia. Longman, Brown, Green, and Longmans, London, 2 vols: pp. 1-8, pls 1-10 & 14-17 (1844) and pp. 9-12 (1875).

Hall-Martin, A. J. (1974). Observations on population density and species composition of seals in the King Haakon VII Sea, Antarctica. *So. Afr. J. Ant. Res.* No. 4, 34-39.

Harrison, R. J. and Tomlinson, J. D. W. (1964). Observations on diving seals and certain other mammals. *Symp. Zool. Soc. Lond.* **13,** 59.

Hofman, R. J. (1975). Distribution patterns and population structure of Antarctic seals. Ph. D. Thesis, University of Minnesota.

Hofman, R., Erickson, A., and Siniff, D. (1973). The Ross seal *(Ommatophoca rossi). In* "Seals". IUCN Publ. New Series, Supp. Paper No. 39, 129-139. International Union for the Conservation of Nature and Natural Resources, Morges, Switzerland.

King, J. E. (1964a). "Seals of the World", British Museum (Natural History), London.

King, J. E. (1964b). A note on the increasing specialization of the seal fore flipper. *J. Anat., Lond.* **98 (8),** 476-7.

King, J. E. (1964c). Swallowing modifications in the Ross seal. *J. Anat., Lond.* **99 (1),** 206-7.

King, J. E. (1965). Giant epiphyses in a Ross seal. *Nature (London)* **205,** 515-516.

King, J. E. (1966). Relationships of the Hooded and Elephant seals (Genera *Cystophora* and *Mirounga*). *J. Zool. (London)* **148 (4),** 385-398.

King, J. E. (1968). The Ross and other Antarctic seals. *Austr. Nat. Hist.* March, **16 (1),** 29-32.

King, J. E. (1969). Some aspects of the anatomy of the Ross seal, *Ommatophoca rossi* (Pinnipedia: Phocidae). Sci. Rep. Br. Antarct. Surv., No. 63, 1-54.

Laws, R. M. (1964). Comparative biology of Antarctic seals. *In* "Biologie Antarctique", (Ed. R. Carrick), 445-454. Hermann, Paris.

Ling, J. K. (1972). Vibrissa follicles of the Ross seal. *Br. Ant. Sur. Bull.* **27,** 19-24.

Møhl, B., Terhune, J. M. and Ronald, K. (1975). Underwater calls of the harp seal, *Pagophilus groenlandicus*. *Rapp. P.-v. Reun. Cons. int. Explor. Mer.* **169**, 533-543.

O'Gorman, F. (1963). Observations on terrestrial locomotion in Antarctic seals. *Proc. Zool. Soc. Lond.* **141 (4)**, 837-850.

Øritsland, T. (1970a). Biology and population dynamics of Antarctic seals. *In* "Antarctic Ecology", Vol. 1 (Ed. M. W. Holdgate), 361-366. Academic Press, London and New York.

Øritsland, T. (1970b). Sealing and seal research in the Southwest Atlantic pack ice, Sept.-Oct. 1964. *In* "Antarctic Ecology", Vol. 1 (Ed. M. W. Holdgate), 367-376. Academic Press, London and New York.

Øritsland, T. (1977). Food consumption of seals in the Antarctic pack ice. *In* "Adaptation Within Antarctic Ecosystems. Proc. Third SCAR Symposium on Antarctic Ecology", (Ed. G. A. Llano), 749-768. Smithsonian Institution, Washington, D.C.

Pierard, J. and Bisaillon, A. (1978). Osteology of the Ross seal *Ommatophoca rossi* Gray, 1844. Biology of the Antarctic Seas IX, Antarctic Research Series, Vol. 31, 1-24.

Polkey, W. and Bonner, W. N. (1966). The pelage of the Ross seal. *Br. Ant. Sur. Bull.* **8**, 93-96.

Ray, C. E. (1976). *Phoca wymani* and other tertiary seals (Mammalia: Phocidae) described from the eastern seaboard of North America. Smithsonian Contributions to Paleobiology No. 28, 1-36.

Ray, G. C. (1970). Population ecology of Antarctic seals. *In* "Antarctic Ecology", Vol. 1 (Ed. M. W. Holdgate), 398-414. Academic Press, London and New York.

Ray, G. C. and Smith, M. S. R. (1968). Thermoregulation of the pup and adult Weddell seal, *Leptonychotes weddelli* (Lesson), in Antarctica. *Zoologica* **53, (1)**, 33-46.

Ray, G. C. and Watkins, W. A. Sounds of *Ommatophoca Rossi*, the Ross seal. In preparation.

Ray, G. C., Watkins, W. A. and Burns, J. J. (1969). The underwater song of *Erignathus* (bearded seal). *Zoologica* **54 (2)**, 79-83.

Rudmose-Brown, R. N. (1913). The seals of the Weddell Sea: notes on their habits and distribution. Edinburgh, Scottish National Antarctic Expedition, Sci. Res. Voyage "Scotia", 1902-04, Vol. 4 (Zool.), pt. 13, 181-198.

Scheffer, V. B. (1958). "Seals, Sea Lions and Walruses", Stanford University Press, Stanford, CA, and Oxford University Press, London.

Siniff, D. B., Cline, D. R. and Erickson, A. W. (1970). Population densities of seals in the Weddell Sea, Antartica, in 1968. *In* "Antarctic Ecology", Vol. 1 (Ed. M. W. Holdgate), 377-397. Academic Press, London and New York.

Thompson, R. B. (1911). Scottish National Antarctic Expedition: Osteology of Antarctic Seals. *Trans. Roy. Soc. Edinburgh*, Vol. 47, Part I (8), 187-201.

Tikhomirov, E. A. (1975). Biology of the ice forms of seals in the Pacific section of the Antarctic. *Rapp. P. -v. Réun. Cons. Int. Explor. Mer* **169**, 409-412.

Wilson, E. A. (1907). Mammalia (whales and seals). *In* "National Antarctic Expedition 1901-04, Natural History", Vol. 2, Zoology. British Museum (Natural History), London.

Wilson, V. J. (1975). A second survey of seals in the King Haakon VII Sea, Antarctica. *S. Afr. J. Antarct. Res.* **2105**, 31-36.

11
Leopard Seal
Hydrurga leptonyx Blainville, 1820

Gerald L. Kooyman

Genus and Species

Allen (1905) discusses the history of the various names proposed for the leopard seal. The first generic name was *Stenorhinque* which was given by Cuvier in 1824. After several spelling modifications, it was discovered that the name was occupied. It had been used previously for a genus of crustacean and for two different genera of insects. In 1875 Peters suggested the generic name of *Ogmorhinus*. However, in about 1899 T. S. Palmer revealed that an appropriate generic application of *Hydrurga*, meaning "water worker" (King, 1964), had been applied to this animal by Gistel in 1848.

The species description was published by Blainville in 1820 in a paper entitled "Sea lion from the Falkland Islands." The species name given was *leptonyx*. According to King (1964) the meaning of *leptos* is small or slender, and of *onux* is claw.

Allen (1905) suggests that the type of skull may be in the Museum of the College of Surgeons, London. However, Scheffer (1958) believes the skull is in Paris in the *Musée Nationale d'Histoire Naturelle*. There is also some confusion as to the collection locality of the type specimen. From various text material Allen (1905) concludes it was collected in the Falkland Islands rather than "New" (South) Georgia Island.

External Characteristics and Morphology

More than any other species of phocid it appears built for speed. The animal is unusually slender in appearance. The fore flippers are comparatively long, and the head seems over-sized for the body. Indeed, the huge head with its large gape has suggested to many the appearance of a reptile (Fig. 1). There are only a few short vibrissae and the eyes are rather laterally positioned.

The colour in the mid-line of the back is dark grey to almost black. Along the sides it becomes lighter, fading through shades of grey to almost blue on the flanks. There is a diffuse but obvious demarcation along the sides where the dark dorsal coloration ends and the ventral lighter shades begin. This line is at the level of the eyes on the face. The lower flanks and belly are nearly silver.

According to Brown (1957) yearling animals range from 160-230 cm in length. At four years they are up to 380 cm in nose-to-tail-tip length (Table 1). The author does not report how he estimates age, or the sample size. One wonders about the 380 cm value because later in the report he says the largest animal measured was a pregnant female 333 cm long. The largest of five adult males was 315 cm.

The longest of four animals obtained by Wilson (1907) was a 333 cm female. Its maximum girth was 192 cm. Wilson reports that an animal collected on the Ross expedition of 1839-42 was 365 cm. The collections in the British Museum, most of which were obtained by Bertram and Hamilton, have been summarized (Hamilton, 1939). The largest of 18 males was 305 cm, and of 20 females there was one 358 cm in length. Lengths in relation to year class up to 4 years are given in Table 1.

No reports of precisely determined weights of adults are available. King (1964) says leopard seals weigh about 275 kg. Ray (1970) estimates the adult weight at 450 kg. Hamilton (1939) reports that the heaviest specimen collected by Ross was 385 kg. Two specimens described in the Scotia Reports, a 305 cm specimen weighing 275 kg

(a)

(b)

FIG. 1 (a) Leopard seal on ice. (b) A very large female leopard seal ashore on South Georgia. The gentoo penguin standing near the seal's head is 76 cm tall and gives some idea of scale. This animal was estimated to weigh about 500 kg. An elephant seal is in the background.

TABLE 1 Body lengths of different year classes of seals. Direct line measurements from the tip of the nose to the tail tip (From Brown, 1957; Hamilton, 1940).

Year	Length of males (cm)		Length of females (cm)	
	Brown	Hamilton	Brown	Hamilton
1	160-230	228	100-230	232
2	215-260	254	215-260	279
3	250-280	281	250-290	304
4	280-320	289	290-380	329

and one 309 cm in length, weighed 285 kg. These animals were weighed piecemeal and probably not enough was allowed for the blood loss. In 1977, I saw intermittently several female leopard seals beached on South Georgia. They were the largest leopard seals I have ever seen and must have weighed close to 500 kg. Unlike others I have seen they were not lean and trim, but quite fat (Fig. 1).

Distribution

Leopard seals are to be found throughout the antarctic pack ice and south to the edge of the continent. They occur year-round on the sub-antarctic islands of South Georgia (Matthews, 1929) and Heard Island (Brown, 1957). In fact, leopard seals may congregate in greater numbers around Heard Island (Fig. 2) than anywhere else (Brown, 1957). The highest counts made on the island were from July through September, 1952 when 1004, 859 and 936 seals, respectively, were counted. On one occasion 85 seals were counted in Atlas Cove, which is 2.4 km wide at its mouth and is 3.2 km in length. The animals are most abundant in July. Males and females are found in equal numbers. They are most scarce in November and December at which time the yearling population is 20% of the total compared to 4% in August. An interesting aspect noted from the counts is that more seals are ashore at night than in the daytime in June through September. This same sort of pattern has been noted in September counts of Weddell seals at McMurdo Sound, Antarctica (Kooyman, 1975).

FIG. 2 Distribution map of the leopard seal. This animal occurs everywhere south of the Antarctic convergence. There have been numerous sightings in New Zealand and Southern Australia. Elsewhere, the arrows indicate single observations north of the Antarctic convergence.

On some of the other subantarctic islands they are seasonal visitors. They are usually present from May to November on Kerguelen Island. They are common on Macquarie Island from July to November (Gwynn, 1953), and they are frequently seen in spring and early summer at the Falkland Islands (Hamilton, 1939).

The occurrence of leopard seals on the subantarctic islands mainly in winter suggests that this is due at least in part to the expanding pack-ice region where most of the population occurs. As the summer comes and the pack-ice zone contracts the animals move south with it. Little else is known about the movement of leopard seals in antarctic waters.

The leopard seal is apparently the greatest wanderer of the antarctic seals. It has been seen in the Cape of Good Hope and Cape Horn regions; one was reported at Tristan da Cunha, and several along the coast of New Zealand and Australia (Scheffer, 1958). The most northern record is an animal collected on Lord Howe Island, 31° 31' S (Hamilton, 1939).

Abundance

Reports of total population estimates are few and based on very few data. Scheffer (1958) suggested that there were 100 000-300 000 leopard seals. Eklund and Atwood (1962) estimated 152 500 seals based on counts in the Ross Sea and Indian Ocean. However, they doubt whether these two areas were representative of all antarctic waters. Two-thirds of the antarctic coast have been censused over the past few years and the total population for this region, which extends from the Weddell Sea to the Bellingshausen Sea, and to the Banzare Coast, is estimated to be between 100 000 and 200 000 animals (Erickson and Hofman, 1974). More recently the total pelagic population of Antarctica has been estimated to be 222 000 (Gilbert and Erickson, 1977).

Internal Characteristics and Morphology

Skeletal

Barrett-Hamilton (1901) and Hamilton (1939) have both described in detail the dimensions of the skull (Fig. 3). These are based on a series of specimens in the collection of the British Museum. Two features distinguish them from other seals: (1) the great length of the skull; and (2) the unusual teeth (Fig. 4). The longest skull is that of a female which is 43.1 cm. The longest male skull is 41.6 cm.

The teeth are very complex, second only to those of the crabeater seal. The molars have three very prominent tubercles with narrow clefts between. The canines are exceptionally long. The occlusion is nearly perfect between the incisors and canines. It is not quite as good in the molars.

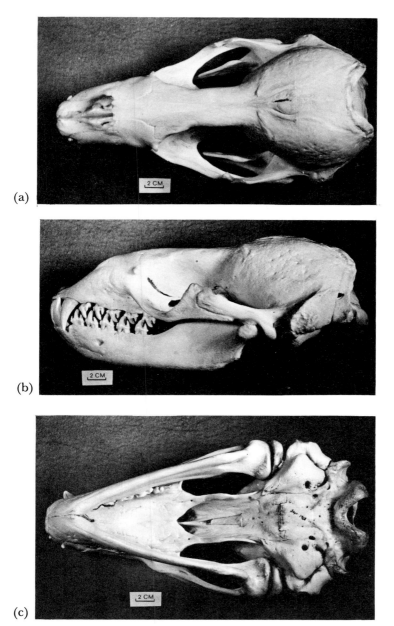

FIG. 3 (a) Dorsal, (b) lateral and (c) ventral views of a young leopard seal skull. Note the exceptional elongation of the snout and size of the canines.

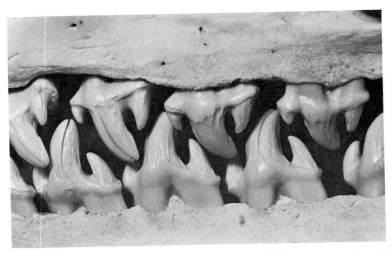

FIG. 4 A close-up view of the molars and premolars of a leopard seal. The occlusion is nearly as perfect as in the crabeater seal, but there are not as many tubercles on the teeth.

Viscera

There is little information available on the soft anatomy. Some special features of the cardiopulmonary system are all that have been noted. Compared to terrestrial mammals the ventricles of the heart are flattened in the dorso-ventral aspect and the apex of the ventricles is more rounded. Also, a well-developed aortic bulb is present (Drabek, 1975).

The trachea is unusual in its degree of flexibility. As in the Weddell and Ross seals this organ, when completely relaxed, will collapse to such an extent that the lumen in some parts of it is completely occluded. This feature was probably first described in the leopard seal by Murphy (1913). He suggested that it was important when the seal swallowed a large bolus of food. Since then it has also been speculated that such pliability of the trachea may be important as a means of pressure compensation when the animal dives to depths. The respiratory system must be compressed, otherwise structural damage to the tissue will occur. The degree of rigidity of the trachea and other parts of the respiratory system will influence the depth to which a seal can dive (Kooyman and Andersen, 1969). Such flexibility must also influence the nature of the seal's vocalizations.

Behaviour

Feeding habits

The food habits of leopard seals are perhaps the most catholic of any seal. The stomach and gut contents of 32 animals collected by various investigators before 1940 are summarized by Hamilton (1939). Most were penguins, but seal, squid and crustaceans were also in several animals. Gwynn (1953) reports that of 18 leopard seal stomachs examined at Macquarie Island, five contained cephalopods and one fish. Of 23 seals stomachs examined at Heard Island, 12 contained penguins (Fig. 5), three fish and two seal meat (Brown, 1957). Leopard seals routinely take fur seals at South Georgia (S. McCann, personal communication). Perhaps the most unusual animal eaten by any seal was the full-grown platypus found in the stomach of a male leopard seal captured in 1859 near Sydney, Australia (Troughton, 1951).

Because of the rather bizarre and visible nature of some of its eating habits, they are probably discussed more than those of any other seal, especially the habit of some individuals of catching penguins. There are at least two popular articles (Kooyman, 1965; Müller-Schwarze and Müller-Schwarze, 1971) and one scientific report (Penney and Lowry, 1967) on this feeding pattern. These three reports are based on observations made at Cape Crozier, Ross Island, Antarctica, perhaps the only place where this activity can be viewed consistently.

FIG. 5 A leopard seal in the unusual circumstance of beaching itself while pursuing an Adelie penguin. Note the impression of an over-sized head.

Penney and Lowry estimate that at this rookery of 300 000 Adelie penguins, seals capture 15 000 adult birds in a 15-week period. The number of seals involved is uncertain, but neither Penney and Lowry nor Kooyman observed more than four seals at any one time. However, it would seem extraordinary if so few seals were involved. Possibly there are no more than four active at any given time but there are seals coming and going from off-shore floes. Otherwise, the Penney and Lowry estimate of 15 000 birds in 15 weeks reduces to these four seals eating 36 penguins per day each, or from 80 to 160 kg of meat consumed per day, assuming that the birds weigh 4.5 kg each and that 50-100% of the bird is eaten.

The adult penguins are caught after underwater pursuits or as they fall back into the water after missing a leap onto the ice foot. In contrast, the fledgling chicks which initially are quite awkward in water are often taken at the surface. The seal simply swims over and grabs them.

After the bird is caught, most of the skin is peeled away by pinching it between the incisors and whipping the bird back and forth until the skin tears away. This procedure takes up to 15 min to complete. One particularly successful seal was able to catch seven penguins, of which it ate six in 70 min (Kooyman, 1965).

In order to give perhaps a better perspective of these recent accounts of seal activity at Cape Crozier some other facts should be mentioned. All the seals seen at Cape Crozier in which circumstances permitted the identification of the sex have been males. Also, along the continental shores at Wilkes station Penney saw several dozen leopard seals over a two-year period and these were all males. This apparent preponderance of males in high latitudes along the continent may reflect an example of a few animals seeking a food source unusually south of their normal range. In the lower latitudes of the antarctic peninsula region near Palmer Station there are many leopard seals and many penguin rookeries. Yet, sightings of seals attempting to catch penguins are rare, even when the chicks are fledging. Examination of leopard seal faeces found on ice floes show instead that the diet here consists largely of krill (Hofman, *et al.,* 1977; Kooyman, personal observation).

Sound production

The underwater vocalizations of leopard seals are usually of low or medium frequency and long duration, rather haunting, sonorous drones. At Cape Crozier they are repeated incessantly as the animal moves back and forth near the ice foot providing weird but beautiful background music to the events taking place. The two most common calls are long steady drones of about 300-3500 Hz (Poulter, 1968; Ray, 1970). The lowest-frequency call is so powerful that not only can it be heard through the air water interface, but if the seal is beneath the ice it can be felt.

Captivity

Leopard seals have been successfully kept in captivity in at least two zoos. The Taronga Park Zoo in Sydney, Australia had two for several years in the 1960s, and the Marineland of New Zealand at Napier, North Island collected one off a beach near the aquarium in 1973. In the last year or two (1978-79) leopard seals have been transported to the northern hemisphere and kept in at least two different places (Ocean Park Limited, Hong Kong, and Sea World of San Diego).

Reproduction

Based on growth rate of the os penis and histology of the testes, sexual activity begins in the third year and by the fourth year the males are sexually mature (Hamilton, 1939). Studies of ovaries by Hamilton indicate that the females begin ovulating in the second year and they commonly become pregnant in the third year.

Copulation between a captive pair was observed in January and November, 1966 at the Taronga Park Zoo, Sydney, Australia (Marlow, 1967). Mating, although never observed in the wild, is thought to occur from January through March because of the variable size of foetuses collected during the year (Hamilton, 1939). Foetal growth rate has been noted by Hamilton (1939) who was citing data collected by Bertram (1940), Gwynn (1953), and Brown (1957). These results are summarized in Table 2. According to Hamilton (1939) and Brown (1957) birth usually occurs in November. The circumstantial evidence indicates that gestation is about nine months. The only record of a new-born and its weight is that reported by Brown (1952) in which the pup weighed 29.5 kg and was 157 cm long. However, it was a still birth in mid-November and the mother had been held captive since 12 September. During that time she did not feed and drank only sea water.

TABLE 2 Foetal body lengths, weights and dates collected. (B) stands for data collected by Bertram and cited by Hamilton (1939). (Br) and (G) equal data collected by Brown (1957) and Gwynn (1953).

Date	Length (cm)	Weight (kg cm^{-1})	Collector
9 Mar	7.8		B
17 Mar	11.8		B
22 Mar	4.2		B
22 Mar	7.7		B
28 Mar	11.0		B
28 Mar	11.2		B
30 Mar	25.0		B
3 Apr	15.7		B
16 Apr	15		B
16 Apr	20		Br
30 Apr	36		B
14 May	41	0.94	Br
2 June	41		Br
12 June	56		Br
27 June	74	4.5	Br
16 July	77	5.5	Br
19 Aug	109	13.4	Br
19 Aug	109	13.4	G
29 Aug	111	12.7	G
13 Sept	109	16.4	G
24 Sept	113	15.4	Br
15 Nov	157	29.5	Br (Newborn)

Gwynn (1953) observed one pup each time on 7 January 1949 and 1951. Both still had small patches of lanugo. He reckoned that neither could have been more than two months old. One of these pups he believes must have been independent for some time because the gut contained numerous tapeworms. Wilson (1907) reports collecting a female in mid-January that was in full milk. In short, the usual duration of nursing of the pups is unknown. More information will be forthcoming on this subject in the near future due to recent efforts to obtain more information on leopard seal breeding habits in the pack ice near the antarctic peninsula. On these expeditions mothers and pups have been found (Siniff, personal communication).

References

Allen, J. A. (1905). The Mammalia of southern Patagonia. Repts. Princeton U. Exped. Patagonia, 1896-99, **3 (1)**, 86-88.
Barrett-Hamilton, G. E. H. (1901). Seals. In "Résultats du Voyage du S.Y. Belgica 1896-99", Rapp. Sci. Zool., Buschman, Antwerp.
Bertram, G. C. L. (1940). The biology of the Weddell and crabeater seals. *Sci. Rept. British Graham Land Expedition*, 1934-37; 1-139, British Museum, London.
Brown, K. G. (1952). Observations on the newly born leopard seal. *Nature (London)* **170**, 982-983.
Brown, K. G. (1957). The leopard seal at Heard Island, 1951-54. Austral. Nat'l. Antarctic Res. Exped. Interior Repts., **16**, 34 pp.
Drabek, C. M. (1975). Some anatomical aspects of the cardiovascular system of antarctic seals and their possible functional significance in diving. *J. Morphol.* **145**, 85-106.
Eklund, C. R. and Atwood, E. L. (1962). A population study of antarctic seals. *J. Mammal.* **43**, 229-238.
Erickson, A. W. and Hofman, R. J. (1974). Antarctic Seals. Antarctic Map Folio Ser. (Ed. V. C. Bushnell), Folio 18, Antarctic Mammals, Amer. Geographical Soc., New York.
Gilbert, J. R. and Erickson, A. W. (1977). Distribution and abundance of seals in the pack ice of the Pacific sector of the Southern Ocean. In "Adaptations within Antarctic Ecosystems", (Ed. G. A. Llano), 703-748, Smithsonian Institution, Washington, D.C.
Gwynn, A. M. (1975). The status of the leopard seal at Heard Island and Macquarie Island, 1948-50. Aus. Nat'l. Antarctic Res. Exped. Interim Rept. 3.
Hamilton, J. E. (1939). The leopard seal *Hydrurga leptonyx* (De Blainville). *Discovery Rept* **18**, 239-264.
Hofman, R. S., Reichle, R. A., Siniff, D. B. and Müller-Schwarze, D. (1977). The leopard seal *(Hydrurga leptonyx)* at Palmer Station, Antarctica. In "Adaptations within Antarctic Ecosystems", Proc. of third SCAR Symp. on antarctic biology, (Ed. G. A. Llano), pp. 769-82, Smithsonian Institution, Wash., D.C.
King, J. E. (1964). "Seals of the World", British Museum (Natural History) London.
Kooyman, G. L. (1965). Leopard seals of Cape Crozier. *Animals* **6**, 59-63.
Kooyman, G. L. and Andersen, H. T. (1969). Deep diving. In "The Biology of Marine Mammals", (Ed. H. T. Andersen), 65-94. Academic Press, New York and London.
Kooyman, G. L. (1975). A comparison between day and night diving in the Weddell seal. *J. Mammal.* **56**, 563-574.
Marlow, B. J. (1967). Mating behavior in the leopard seal, *Hydrurga leptonyx* (Mammalia: Phocidae), in captivity. *Austr. J. Zool.* **15**, 1-5.

Matthews, L. H. (1929). The natural history of the elephant seal. *Discovery Repts.* **1,** 234-55.

Müller-Schwarze, D. and Müller-Schwarze, C. (1971). Seeleoparden des Ross-Meeres sprangen uns an. *Das Tier 11, Jahrgang,* 38-43.

Murphy, R. C. (1913). The trachea of *Ogmorhinus* with notes on other soft parts. *Bull. Amer. Mus. Nat. Hist.* **32,** 505-506.

Penney, R. L. and Lowry, G. (1967). Leopard seal predation on Adelie penguins. *Ecology* **48,** 878-882.

Poulter, T. C. (1968). Underwater vocalization and behavior of pinnipeds. *In* "The Behavior and Physiology of Pinnipeds", (Ed. R. J. Harrison, R. C. Hubbard, R. S. Peterson, C. E. Rice and R. J. Schusterman), 69-84. Appleton-Century-Crofts, New York.

Ray, C. (1970). Population ecology of antarctic seals. *In* "Antarctic Ecology", Vol. 1, (Ed. M. W. Holdgate), 398-414. Academic Press, London and New York.

Scheffer, V. B. (1958). "Seals, Sea Lions and Walruses, A Review of the Pinnipedia", Stanford University Press, Stanford, CA.

Troughton, E. (1951). "Furred Animals of Australia", Angus and Robertson, Sydney. (As cited by Scheffer, 1958.)

Wilson, E. A. (1907). Mammalia (seals and whales). *In* "National Antarctic Expedition, 1901-04", Natural History, Zool. 2, British Museum, London.

12
Weddell Seal
Leptonychotes weddelli Lesson, 1826

Gerald L. Kooyman

Genus and Species

The Weddell seal (Fig. 1 (a)) was first described by Robert Jameson in James Weddell's book, "A Voyage Towards the South Pole Performed in the Years 1822-24". The description was based on six skins collected at Saddle Island, South Orkneys on 13 January 1823. Also in the book is a rather fanciful drawing of the seal by Captain Weddell which is captioned "Sea Leopard of the South Orkneys" (Fig. 1 (b)). Later, based on the illustrations and report in Weddell's book the animal was officially described for science by R. -P. Lesson in 1826 (Scheffer, 1958). The Latin binomial he gave it was *Otaria weddelli*. In 1880 the animal was renamed *Leptonychotes weddelli* by J. A. Allen and this is the accepted name at present.

Species descriptions usually require skull specimens as well, but these were not obtained for some years after Weddell's expedition.

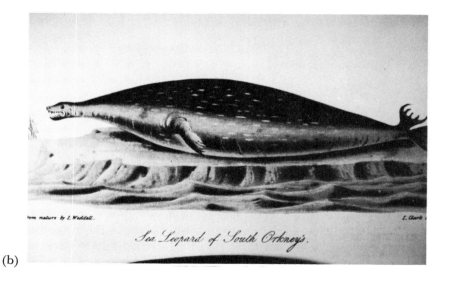

FIG. 1 (a) A Weddell seal mother and pup rest on the ice. (b) The first Figure or caricature of a Weddell seal (from Weddell's book, "A Voyage Towards the South Pole performed in the years 1822-1824").

Oddly they came from two animals found near the Santa Cruz River in Patagonia, which is near the latitude of 50° S (Gray, 1837), an exceptionally low latitude for the species.

External Characteristics and Morphology

The Weddell seal is one of the largest of all seals. The nose-to-tail-tip length of the largest female reported by Bertram (1940), was 292 cm and the largest male was 279 cm. Bruce (1913) reports a female of 329 cm and a 297 cm male. In early spring both males and females commonly weigh 400-450 kg (Kooyman *et al.*, 1973).

The head, in contrast to the rest of the body, is so small that it appears as if someone made a mistake while putting the animal together. The face of the Weddell seal is one of the most benign in the animal world. The muzzle is short, giving it the appearance of a juvenile, and the mouth seems always to have the hint of a smile. The eyes are huge, deep brown, and add a benevolent touch to the countenance. The vibrissae (moustache or whiskers) are usually short and undistinguished, which is typical of all the Lobodontini.

The fur, which is about 1 cm long, covers the entire body, except a small portion of the underside of the fore and hind flippers. Shortly after moulting the back is blue-black which grades to a silver-white spotting on the belly, hence the early name of sea leopard. As the fur ages it fades, and shortly before moulting the back is a rust brown.

Distribution

The Weddell seal occurs in greatest abundance near the coast of Antarctica (Fig. 2). For example, within a radius of 25 km from Hut Point, McMurdo Sound, 200-400 pups are born each spring. Small breeding populations occur in some of the subantarctic islands, the most northerly of which is S. Georgia Island, where in 1964, 27 pups were born (Vaughn, 1968).

Interestingly, only one animal has been sighted in the Falkland Islands, which have a latitude nearly the same as South Georgia, but due to oceanic currents the climate is different (Hamilton, 1945). The Falkland Islands are north of the Antarctic Convergence.

There are only a few other records of Weddell seals venturing north

FIG. 2 Distribution of the Weddell seal is primarily inside the Antarctic Convergence. Arrows indicate sightings of seals north of the Convergence.

of the Convergence. Four sightings have been made in New Zealand (Turbott, 1949; Ingham, 1960), one in Australia (Turbott, 1949), and one in Uruguay (Vaz Ferreira, 1956). The sighting in Uruguay (Fig. 2) was at a latitude of about 35° S which is the most northern known occurrence.

The most comprehensive survey of Weddell seal distribution near Antarctica was published in the Antarctic Map Folio Series No. 18 (Erickson and Hofman, 1974). Data are based on aerial and ship surveys of the Antartic coast from the Weddell Sea westward to the Banzare Coast. The greatest concentrations were found in the Weddell Sea.

Abundance

The abundance of Weddell seals has been estimated by several authors. All of these may be found in the report on the crabeater seal (Ch. 9). The most recent population estimates are those of Erickson and

Hofman in the Antarctic Map Folio Series No. 18 (1974). These estimates account for two-thirds of the Antarctic coast. They estimate: 52 500—Adelie, Clarie and Banzare coasts; 64 800—Oates and George V coasts; 45 600—Amundsen and Bellingshausen Seas; and intriguingly 593 700—Weddell Sea in 1968; and 92 900—Weddell Sea in 1969. The total population of Weddell seals is unknown.

Internal Anatomical Characteristics

Skeletal

The skull (Fig. 3) is distinctive. The bones are thin and light for such a large mammal. However, the canines are robust and project forward, as do the incisors. If it were not for these peculiar dental features, especially in the upper jaw, and the small narrow snout, the Weddell seal probably would not be able to remain all the year round in McMurdo Sound. These dental features result in an effective ice ream which the animal must rely on frequently. Piérard (1971) presents a detailed description of the osteology and myology of the entire animal.

Cardiovascular

The circulatory anatomy is similar to the more completely studied harbour (common) seal. The heart of the Weddell seal is in important respects the same as other mammalian hearts (Drabek, 1975). As in other Antarctic seals, its configuration is broad and flattened. A well-developed aortic bulb is present. It is interesting that one of the first descriptions of the caval sphincter, a structure peculiar of phocids, was based on observations of a Weddell seal pup (Hepburn, 1912).

Respiratory

It can be deduced from morphology that the larynx seems to be able to seal more tightly than in other pinnipeds (Piérard, 1969). It was also suggested that the vocal fold contributes little to sound production. Instead, most sounds are made at the level of the cricotracheal junction and the adjacent portion of the trachea behaves as a resonating chamber. Tracheograms of a seal hydraulically compressed to and equivalent depth of 310 m showed that, although the tracheal volume was much reduced, air still remained and resonating sound production was still possible (Kooyman et al., 1970).

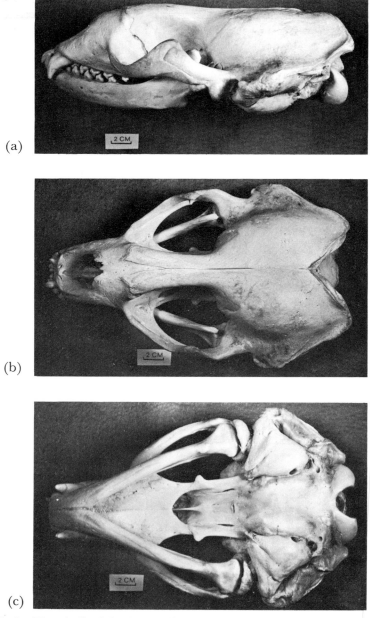

FIG. 3 The skull of the Weddell seal (a) lateral view; (b) dorsal view; (c) ventral view.

The prolonged- and deep-diving ability of the Weddell seal gives special interest to its respiratory system. A superficial study by Hepburn (1912) was one of the earliest available. More recently, a well-illustrated general discussion of some of the major features of the respiratory system, especially the airways, has been published (Boyd, 1975). In a review of the airway structures of marine mammal lungs it was noted that all seals possessed a consistent pattern of extensive cartilaginous reinforcement with a short terminal segment in which cartilage was not present (Denison and Kooyman, 1973). The Weddell seal's airway structure was consistent with that of other phocids.

Other characteristics

Hepburn (1912) describes the brain in detail and also reports on the genitourinary system. Haig (1912) reported on histological examination of the central nervous system in the same volume. Cuello (1967) describes the hypophysis, and Cuello and Tramezzani (1969) describe the pineal, which they noted was exceptionally large.

The optics of the eye were studied by Wilson (1970) and he concluded that in air vision is astigmatic. The astigmatism is corrected to some extent by orientation of the slit pupil. The pupil is in the vertical plane where the least distortion of the cornea exists.

Physiology*

Most physiological studies of Weddell seals have been done by three groups. Elsner and co-workers and Zapol and co-workers have concentrated on problems requiring conventional laboratory conditions. Kooyman *et al.* have worked with unrestrained animals at sea.

Effects of submersion

The Weddell seal has an unusually large O_2 store, which is one adaptation enabling it to make exceptionally long submersions. There are many studies of the various cardiovascular responses in seals to submersion and Weddell seals respond similarly to other phocids. In an attempt to determine the cerebral tolerance to hypoxaemia, Elsner *et al.* (1970)), simultaneously monitored arterial blood gas tensions and brain bioelectric potentials. A change in the character of the EEG,

*Since typeset the following works have been published on diving responses in Weddell seals: Hochachka, P. W. (1980), "Living Without Oxygen," Harvard University Press, Boston, MA; Kooyman *et al.* (1980), *J. Comp. Physiol.* **138**, 335-346; Murphy, B., Zapol, W. M., Hochachka, P. W. (1980). *J. Appl. Physiol.* **48**, 596-605; Zapol, W. M. *et al.* (1979). *J. Appl. Physiol.* **47**, 968-973.

indicating the onset of cerebral dysfunction, occurred at a p_{a,O_2} of about 10 mm Hg. This is believed to be a good deal lower than in terrestrial mammals. For example, it is known that man loses consciousness at a p_{a,O_2} of about 30 mm Hg (Mithoefer, 1965).

In another study the circulatory responses to asphyxia were determined in the pregnant mother and her foetus (Elsner et al., 1969-70). During asphyxia the pregnant seal reduces renal blood flow to onetenth of pre-dive levels, which is similar to that in non-pregnant seals, but uterine blood flow is maintained. The onset of bradycardia in the mother occurred immediately after asphyxiation began, but foetal bradycardia was delayed (Fig. 4). Interestingly, when pregnant females were allowed to dive voluntarily they showed no reluctance to make prolonged dives, and one remained submerged for 58 min (Elsner et al., 1970).

The heart rate varies considerably during voluntary dives, in contrast to forced dives (Kooyman and Campbell, 1972). These variations are correlated with the type of dive. There is slowing to 50% of the pre-dive rate during short, shallow dives, and a reduction of at least 75% during long submersions. The onset of this slowing is rapid.

Measurements of pulmonary function have been obtained from adult (average weight = 425 kg) free animals while they rested and after dives (Kooyman et al., 1971). Tidal volume ranged from 5.4 l at

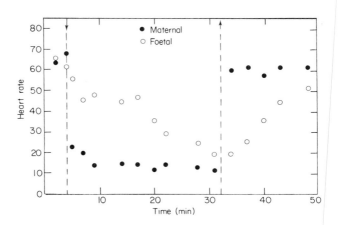

FIG. 4 Heart rates of a mother and foetal Weddell seal during a dive which is indicated by the arrows and dashed lines (from Elsner et al., 1969-70).

rest to 15.2 l after a dive. Respiration rates averaged 4 min^{-1} at rest to 15 min^{-1} after a dive. Minute volume ranged from 19.5 l min^{-1} at rest to 224 l min^{-1} after a dive (Fig. 5).

The respiratory exchange ratio of resting animals averaged 0.69, indicating a fat-based metabolism (Kooyman *et al.*, 1973). End-tidal O_2 and CO_2 determinations indicate that the seals hyperventilate before dives. The pre-dive p_{a,O_2} = 124 mm Hg and p_{a,CO_2} = 30 mm Hg. Presumably end-tidal gas samples closely approximate arterial gas tensions. If so, then even during prolonged dives CO_2 tensions never rose very high. The maximum was 38 mm Hg which is lower than during resting ventilation when the average p_{a,CO_2} = 50 mm Hg. Values of p_{a,O_2} during prolonged dives fell as low as 25 mm Hg, which, if arterial values were close to this, means that the arterial blood was at just below half saturation (Lenfant *et al.*, 1969), and well above the level at which Elsner *et al.* (1970) observed cerebral dysfunction.

Oxygen consumption (V_{O_2}) in these animals while resting in sea water at a temperature of —1.9 °C was 1.5-3 times above the predicted value for terrestrial mammals of similar size (Kleiber, 1961). This is

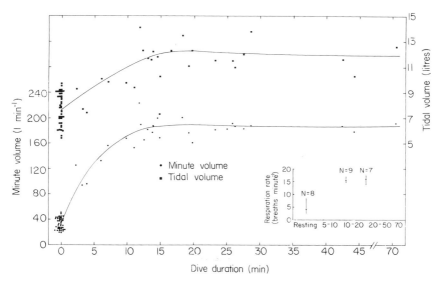

FIG. 5 Ventilation results for five Weddell seals ranging in size from 370 to 450 kg. The tidal volume, minute ventilation and respiratory rate data are averages for the first two minutes after surfacing. All volumes are litres, BTPS (from Kooyman *et al.*, 1971).

consistent with measurements obtained from other species of marine mammals (Irving, 1969). Also, V_{O_2} immediately after the dive, which should include both the needs of the animal while ventilating at the surface and the O_2 debt incurred during the dive, resulted in values lower than the resting V_{O_2} (Kooyman et al., 1973). It is clear from these results that the cost of diving is not great in the Weddell seal, and there may be a depression of metabolism, as observed in earlier work on grey seals and bladder-nosed seals (Scholander, 1940).

The diving lung-volume of Weddell seals is large enough for the animal to contract the bends were most of this gas to be absorbed during a deep dive (Kooyman, 1972). However, venous N_2 tensions of a young Weddell seal did not rise to levels considered dangerous, even after compressions to depth equivalents of 135 m (Kooyman et al., 1972).

Temperature regulation

The average body temperature of 47 sleeping adults was 36.7 °C (R = 35.6-37.8°), and that of 22 sleeping pups was 37.0 °C (R = 36.2-37.8°) (Kooyman, 1968). These pup measurements are about a degree cooler than those of active pups (Elsner et al., 1977).

Maintenance of body temperature in the neonate depends on the effectiveness of the fur coat (lanugo). For the first 24 hours there is an elevated thermogenic response. The neonates' V_{O_2} of about 490 ml min^{-1} is twice as high as that of older pups whose average V_{O_2} is about 240 ml O_2 min^{-1}. This is 1.5-2 times greater than the rate of O_2 consumption predicted for terrestrial mammals of the same body weight (Elsner et al., 1977).

Fat thickness, as estimated by abdominal skin-fold thickness, increases linearly with pup age from 0 when born to 3 cm thick at 20 days. As the pup ages the blubber becomes the most important insulator. Instead of the skin temperature being consistently high (25-34 °C) (Ray and Smith, 1968), it becomes much more variable as in the adult. The importance of the blubber as an insulator was dramatically illustrated by the comparison of the effects of immersion on an eight-hour old pup and a nine-day old animal (Fig. 6). The body temperature of the new-born dropped 2 °C in less than 20 min when immersed for 10 min, whereas there was no discomfort to the nine-day old animal (Elsner et al., 1977). However, Siniff et al. (1971) noted a two-day old pup that voluntarily entered the water and spent about one hour there. It showed no ill effects from the experience. When four days old it again entered the water, and did so regularly thereafter.

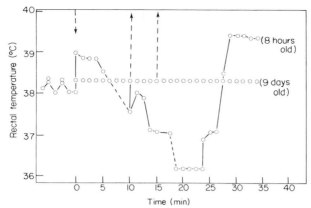

FIG. 6 Rectal temperatures of an eight-hour old and nine-day old Weddell seal pup before and after immersion in 1.8 °C sea water. Arrows indicate times of immersion and departure from the water (Elsner et al., 1977).

Water balance

Only a single study related to this problem has been conducted. It was carried out on those individuals that are under the greatest potential stress—the lactating females (Kooyman and Drabek, 1968). The average osmolality of Weddell seal blood is 316 milliosmols per litre, which is no different from that of terrestrial mammals. Average urine concentrations of 1.58 mol l^{-1} and a maximum of 2.03 mol l^{-1} is not unusually high for mammals. Man may produce a urine of 1.43 mol l^{-1} concentrations and some desert rodents such as the sand rat *(Psammonys obesus)* produce a urine of at least 6.34 mol l^{-1} (Schmidt-Nielsen, 1964). The milk produced by the females becomes more concentrated as lactation progresses. Within the first week the average amount of water in the milk is about 55%. This declines to an average of about 35% after five weeks and one sample was as low as 27%. The average milk composition from 39 samples was 42.2% fat, 14.1% non-fat solids, and 43.6% water. Unlike many species of pinnipeds the lactating mother always has fresh water available because she pups on snow and ice. It is not unusual to see the mother seals eating snow.

Behaviour

The ease with which Weddell seals can be approached and captured is well known. This fact plus its tractability has been the reason for many studies being done on this particular seal. Like some other animals in many isolated parts of the world Weddell seals do not respond appropriately with fear to man. They do not take flight at man's approach and there is little aggressiveness shown by the seals when disturbed. This phlegmatic character is not possessed by other Antarctic seals. The crabeater and the leopard seal both very actively resist intrusion. Such tractability makes the capture and manipulation of Weddell seals relatively easy. They are also abundant within a few miles of major research facilities such as McMurdo Station.

It is one of the few marine mammals that can be collected with wheeled transport such as a truck. One can simply drive up to the animals, net the desired one, place it in a sled and tow it away.

In addition, the seal normally dives beneath stable sea ice. Breathing holes are made and kept open in local areas of thin ice by using the teeth as ice reamers (Kooyman, 1969; Stirling, 1969). General natural history accounts of aspects of this ice-behaviour have been published by several authors (Wilson, 1907; Bertram, 1940; Sapin-Jaloustre, 1952). In many of those areas where breathing holes are made the sea ice is thick enough to provide a stable foundation on which to set up complex laboratories over deep water of at least 600 m in depth in which studies of the seals can be done. There are very few places anywhere in the world where oceanic stations can be established with such safety, for such extended periods, and at such low cost.

The food of seals in McMurdo Sound is almost entirely fish (Dearborn, 1965). One of the favourite prey species is the Antarctic cod *Dissostichus mawsoni*. DeVries et al. (1974) have caught these fish most consistently at depths of 400-600 m. The average size of fish caught by these investigators was 35 kg.

Diving abilities

The diving abilities of this species have been studied extensively, perhaps more so than for any other seal. The usual method for such studies was to capture a seal at a local haul-out area and transport it over the ice to a previously cut and isolated ice hole where it was released (Kooyman, 1968). At this site recorders were placed on the seals which measured depth and duration of dives (Kooyman, 1965 and 1968). These conditions probably result in some exceptionally long dives as the seal seeks other breathing holes. The maximum submersion ever measured was 73 min (Kooyman, 1981), but submersions are usually not more than 15 min (Fig. 7).

The maximum depth of dive measured was 600 m (Kooyman, 1966;

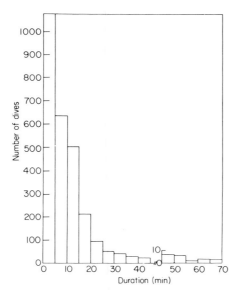

FIG. 7 Summary of the duration of dives of Weddell seals. All dives were measured at McMurdo Sound, Antarctica during the months of September to January, and in the course of several different years.

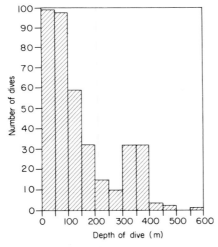

FIG. 8 Summary of the depth of dives of Weddell seals. All dives were measured during the months of October to January in 1964 and 1965 (from Kooyman, 1968).

1975) which is essentially the bottom depth of the water where these investigations were conducted. Dives to 300-400 m are common (Fig. 8) and interestingly these seem to be approximately the depths at which the Antarctic cod is most commonly caught. Dives to such depths do not normally last more than 15 min.

Bertram (1940) estimated their "normal" swimming speed by comparing it to that of a motor boat and arrived at a figure of 7 knots (13 km h^{-1}) Kooyman (1968) timed seals swimming between holes 2 km apart. The estimated speed was 4-5 knots (8-10 km h^{-1}). For seals that make hour-long dives to other breathing holes this means a possible underwater range of 9 km.

Orientation

In an initial study of under-ice orientation it was noted that seals consistently departed from and returned to the starting point in the same direction (Kooyman, 1968). A high percentage of these departures were towards the nearest shoreline. In a comparison of day and night diving behaviour it was demonstrated that the seals prefer to dive during the daylight hours (Kooyman, 1975). This is when the largest, deepest and most frequent dives are made. Many of the night hours are spent sleeping. This is a reversal from the summer period when there are 24 hours of sun. At this season the seals' peak resting period is in the afternoon (Müller-Schwarze, 1965; Smith, 1965; Stirling, 1969; Siniff *et al.*, 1971).

Sound production*

Weddell seals produce a broad array of calls, some of which are spectrographically illustrated in Fig. 9. These calls range from low-pitched buzzes to trills and whistles (1-6 kHz) (Schevill and Watkins, 1965; Kooyman, 1981; Schevill and Watkins, 1971) to chirps (25-70 kHz) (Schevill and Watkins, 1971). Based on evidence from a vertical array of three hydrophones the chirp is directional and when the seal is in its normal swimming posture the beam is directed downwards. A directional sound beam would be especially useful for echolocation; however, no evidence for this exists to date (Schevill and Watkins, 1971).

Some calls undoubtedly play an important role in social activities. Under certain circumstances the seals will respond to playbacks of calls from their own species (Watkins and Schevill, 1968). Certain calls are also used in territorial displays and other aggressive acts (Ray, 1967; Kooyman, 1968). An extensive study of the seal's under-ice behaviour in breeding areas has been conducted with the use of underwater

*Since typeset this important work on sound production by Weddell seals is available: Thomas, J. A. (1979). Ph.D. Thesis, Univ. of Minn.

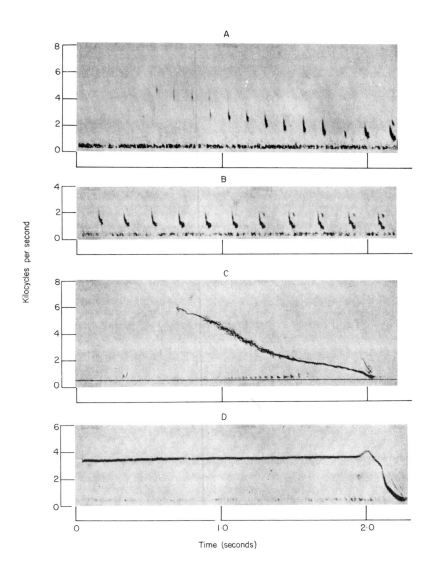

FIG. 9 Sonagrams of Weddell seal calls during agonistic behaviour. A and B are a single call which continued for approximately two seconds longer than illustrated, but the final two seconds were similar to the last five pulses of B. C is a descending trill. D is reminiscent of a shriek or scream (from Kooyman, 1968).

television cameras and hydrophones by Siniff *et al.* (1974), who have estimated the under-ice area controlled by breeding bulls to be about 20 m in diameter.

Social behaviour

The most detailed work on social behaviour is being done by Siniff's group at McMurdo Sound. Little of this work has been published yet. An interesting pattern of behaviour in mother-pup relationships has been noted (Siniff *et al.*, 1971). For a period of nearly a month a mother was radio monitored. From day 1 to day 12 she spent most of her time with the pup on the sea ice. From days 13 to 23 the seal spent about 30-40% of her time in the water, which was a marked increase over the first 13 days.

As mentioned earlier, bulls appear to set up underwater territories in the breeding areas. Females seem to be able to move freely through these areas (Siniff *et al.*, 1975). Siniff *et al.* (1975) give no description of vocalizations in this preliminary report but mention that they play an important role in the social structure of the breeding colony.

In non-pupping areas and at other times of the year the mature seals are rather aloof from each other. The adult seals do not appear to dive in groups nor to move about with one another. When hauled out they are discretely spaced and no seal touches another without causing some sort of threat display by the touched seal. Conflicts, which occur near the breathing holes (Fig. 10), are presumably over the possession of these important sites (Kooyman, 1968). However, the customary adult solitariness may not be general for young animals ranging from pups up to perhaps three-year olds. In one instance a large group of young animals was seen in a large pool near an iceberg, and much mock fighting took place. Pups collected by Kooyman and held in captivity at Scripps Institution of Oceanography and then later at Sea World of San Diego, frequently engaged in vigorous mock fights.

Seasonal movements

Some authors claim that the species is not migratory and winters in the local area (Wilson, 1907; Lindsey, 1937; Bertram, 1940; Sapin-Jaloustre, 1952; Stirling, 1969). Others dispute this claim (Smith, 1965; Kooyman, 1968, 1975). The controversy may be over a matter of definition and about where the studies were conducted. The seals may not undergo long north and south movements from southern areas

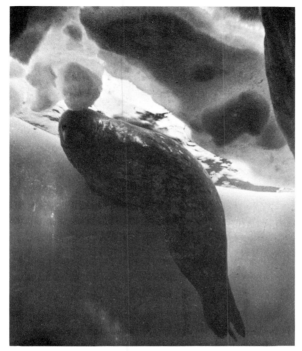

FIG. 10   Weddell seal at breathing hole.

such as McMurdo Sound to more northern regions such as the pack ice. However, in the region that we know best, most do leave the immediate area along the eastern coast of McMurdo Sound from Cape Armitage to Cape Evans. This is a sector approximately 30 km long where much pupping and breeding occurs during the summer months.

The pattern of movement into and out of McMurdo Sound is for pregnant females to begin arriving at the pupping areas in early October (Smith, 1965; Kooyman, 1968). Breeding bulls may be present at this time, but because they remain in the water they cannot be counted. Many non-pregnant females also arrive at this time. Later in November through February many more seals arrive. These include immatures and an occasional yearling. It was noted also by Siniff *et al.* (1975) that few animals less than six years old are found in the pupping areas and young animals of one to three years old are rare in the study area of Eastern McMurdo Sound (it is unclear from their report what the north-south limits of this area are but probably they are from Cape Evans to Cape Armitage). Much of this late season influx is probably a result of the sea ice breaking up in the north and many animals moving south to more secure ice (Kooyman, 1968). By late March the number of seals in the Cape Armitage area has declined considerably (Smith, 1965). One notable exception with regard to movements and other aspects of the Weddell seals' biology is the isolated population at White Island. These animals are 26 km south of the annual sea ice limit. This small group of about 30 remains in this small area all the year round (Stirling, 1972; Kooyman, 1981).

Reproduction

Copulation has been observed once. It occurred under water on 7 December and coitus lasted 5 min (Cline, Siniff and Erickson, 1971).

Ovulation occurs at the end of lactation (Bertram, 1940) and implantation occurs in January and February (Stirling, 1969). Histological examination of the testes indicate that bulls are ready for breeding from September through December (Mansfield, 1958). According to Stirling (1971b) 80% of the females three years and older are pregnant, but only 55% of the females between the ages of three to five years old are pregnant. Siniff et al. (1975) estimate that only 60% of mature females pup each year and that this non-productivity influences population regulation significantly.

Pupping begins sooner in lower latitudes. At Signy Island, South Orkneys (60° 43′ S) and at South Georgia Island (55° S) the earliest pups arrive in late August, and the peak at Signy Island is 7 September (Mansfield, 1958). In the Bay of Whales (78° S) the peak pupping is about 23 October (Lindsey, 1937).

Birth has been described several times (Mansfield, 1958; Stirling, 1969; Ray and deCamp, 1969). The delivery is rapid, and twinning is exceedingly rare. In one sample population Stirling (1971b) noted that 51.2% of the pups were males.

When born the pup's coat is a light brown to grey fur (lanugo). The hair is several times longer than the adult fur and the importance of this pelt for thermoregulation has been determined (Elsner et al., 1977). The pup begins to shed this pelt 9-21 days after birth and by an average of 30 days the moult is complete (Wilson, 1907). Adults may be found moulting any time during the summer.

When born the pup's average weight is about 29 kg and it gains from 10 to 15 kg per week. A maximum gain of 23 kg for a week was noted.

At weaning the average weight is 113 kg (Lindsey, 1937). The average duration of lactation is 45.2 days (Bertram, 1940). At birth at least the four canines are above the gums and by 34 days all teeth have pierced the gums (Lindsey, 1937). By seven weeks of age the pups will remain submerged for at least 5 min and dive to depths of 92 m (Kooyman, 1968).

Sexual maturity, based on ovarian histology and size comparisons of the animals, is reached at two years of age in females (Mansfield, 1958). By three years of age 97% have ovulated (Stirling, 1971a).

The average age of a sample taken from McMurdo Sound was 8-9 years, but 10-12-year olds were common and the maximum age was 18 years (Stirling, 1969). The oldest animal found dead in this area was a 22-year old female (Stirling and Greenwood, 1972).

References

Bertram, G. C. L. (1940). The biology of the Weddell and crabeater seals. *Sci. Rept. British Graham Land Exped.* 1934-37 **1**, 1-139.
Boyd, R. B. (1975). A gross and microscopic study of the respiratory anatomy of the Antarctic Weddell seal, *Leptonychotes weddelli. J. Morphol.* **147**, 309-336.
Bruce, W. S. (1913). Measurements and weights of antarctic seals. *Trans. Roy. Soc. Edin.* **49**, 567-577.
Cline, D. R., Siniff, D. B. and Erickson, A. W. (1971). Underwater copulation of the Weddell seal. *J. Mammal.* **51**, 204.
Cuello, A. C. (1967). Interrelaciones entre la pars intermedia y la pars nervosa en la hipofises de una foca antarctica. *Contrib. del Inst. Antartico Argent.* No. 120.
Cuello, A. C. and Tramezzani, J. H. (1969). La epifisis cerebri de la foca de Weddell, su notable tamãno of organizacion glandular. *Contrib. del Inst. Antartico Argent.* No. 121.
Dearborn, J. H. (1965). Food of Weddell seals at McMurdo Sound, Antarctica. *J. Mammal.* **46**, 37-43.
Denison, D. M. and Kooyman, G. L. (1973). The structure and function of the small airways in pinniped and sea otter lungs. *Respir. Physiol.* **17**, 1-10.
DeVries, A. L., DeVries, Y. L., Dobbs, G. H. and Raymond, J. A. (1974). Studies of the Antarctic cod, *Dissostichus mawsoni. Antarctic J.* **9**, 107-108.
Drabek, C. M. (1975). Some anatomical aspects of the cardiovascular system of antarctic seals and their possible functional significance in diving. *J, Morphol.* **145**, 85-105.
Elsner, R., Hammond, D. D. and Parker, H. R. (1969-70). Circulatory responses to asphyxia in pregnant and fetal animals: a comparative study of Weddell seals and sheep. *Yale J. Biol. Med. Dec. 1969-Feb. 1970.* 202-217.

Elsner, R., Hammond, D. D., Denison, D. M. and Wyburn, R. (1977) Temperature regulation in the newborn Weddell seal *Leptonychotes weddelli*. *In* "Adaptations within antarctic ecosystems", Proc. of third SCAR Symp. on antarctic biology, (Ed. G. A. Llano), pp. 531-40, Smithsonian Institution, Wash., D.C.

Elsner, R., Kooyman, G. L. and Drabek, C. M. (1970). Diving duration in pregnant Weddell seals. *In* "Antarctic Ecology", Vol. 1, (Ed. M. W. Holdgate), 477-482. Academic Press, New York and London.

Elsner, R., Shurley, J. T., Hammond, D. D. and Brooks, R. E. (1970). Cerebral tolerance to hypoxemia in asphyxiated Weddell seals. *Respir. Physiol.* **9,** 287-297.

Erickson, A. W. and Hofman, R. J. (1974). Antarctic Seals. Antarctic Map Folio Series (Ed. V. C. Bushnell), Folio 18, Antarctic Mammals, Amer. Geographical Society, New York.

Gray, J. E. (1837). Description of some new or little known mammalia, principally in the British Museum collection. *Mag. Nat. Hist., N.S.* **1,** 577-587 (as cited by Scheffer, 1958).

Haig, H. A. (1912). A contribution to the histology of the central nervous system of the Weddell seal. *Trans. Roy. Soc. Edinburgh* **48,** 849-886.

Hamilton, J. E. (1945). The Weddell seal in the Falkland Islands. *Proc. Zool. Soc. Lond.* **114,** 549.

Hepburn, D. (1912). Observations on the anatomy of the Weddell seal *(Leptonychotes weddelli).* II. Genito-urinary organs, 191-194. III. The respiratory system and the mechanism of respiration, 321-332. IV. The brain, 827-847. *Trans. Roy. Soc. Edinburgh* **48,** 1912.

Ingham, S. E. (1960). The status of seals (Pinnipedia) at Australian Antarctic stations. *Mammalia* **24,** 422-430.

Kleiber, M. (1961). "The Fire of Life", Wiley, New York.

Kooyman, G. L. (1965). Techniques used in measuring diving capacities of Weddell seals. *Polar Rec.* **12,** 391-394.

Kooyman, G. L. (1966). Maximum diving capacities of the Weddell seal, *Leptonychotes weddelli*. *Science (N. Y.)* **151,** 1553-1554.

Kooyman, G. L. (1968). An analysis of some behavioural and physiological characteristics related to diving in the Weddell seal. *Antarctic Res. Ser.* **11,** 227-261.

Kooyman, G. L. (1969). The Weddell Seal. *Sci,. Am.* **221,** 100-106.

Kooyman, G. L. (1972). Deep diving behaviour and effects of pressure in reptiles, birds and mammals. *In* Symposia of the Society for Exper. Biol. Vol. 26, "The Effects of Pressure on Organisms", (Eds M. A. Sleigh and A. G. Macdonald), 295-311, Cambridge University Press, Cambridge.

Kooyman, G. L. (1975). A comparison between day and night diving in the Weddell seal. *J. Mammal.* **56,** 563-574.

Kooyman, G. L. (1981). "Weddell Seal: consummate diver", Cambridge University Press, Cambridge.

Kooyman, G. L. and Campbell, W. B. (1972). Heart rates in freely diving Weddell seals, *Leptonychotes weddelli. Comp. Biochem. Physiol.* **43A,** 31-37.

Kooyman, G. L. and Drabek, C. M. (1968). Observations on milk, blood and urine constituents of the Weddell seal. *Physiol. Zool.* **41**, 187-193.

Kooyman, G. L., Hammond, D. D. and 'Schroeder, J. P. (1970). Bronchograms and tracheograms of seals under pressure. *Science (N.Y.)* **169**, 82-84.

Kooyman, G. L., Kerem, D. H., Campbell, W. B. and Wright, J. J. (1971). Pulmonary function in freely diving Weddell seals, *Leptonychotes weddelli. Respir. Physiol.* **12**, 271-282.

Kooyman, G. L., Kerem, D. H., Campbell, W. B. and Wright, J. J. (1973). Pulmonary gas exchange in freely diving Weddell seals, *Leptonychotes weddelli., Respir. Physiol.* **17**, 283-290.

Kooyman, G. L., Schroeder, J. P., Denison, D. M., Hammond, D. D., Wright, J. J. and Bergman, W. P. (1972). Blood nitrogen tensions of seals during simulated deep dives. *Am. J. Physiol.* **223**, 1016-1020.

Lenfant, C., Elsner, R., Kooyman, G. L. and Drabek, C. M. (1969). Respiratory function of blood of the adult and fetus Weddell seal *Leptonychotes weddelli. Am. J. Physiol.* **216**, 1595-1597.

Lindsey, A. A. (1937). The Weddell seal in the Bay of Whales, Antarctica. *J. Mammal.* **18**, 127-144.

Mansfield, A. W. (1958). The breeding behaviour and reproductive cycle of the Weddell seal (*Leptonychotes weddelli*, Lesson). *Falkland Islands Depend. Survey Sci. Rept.* **18**, 1-41.

Mithoefer, J. C. (1965). Breathholding. In "Handbook of Physiology, Sec. 3: Respiration", Vol. 2. (Eds W. O. Fenn and H. Rahn), 1011-1025. American Physiological Society, Washington, D.C.

Müller-Schwarze, D. (1965). Zur tagesperiodik der allgemeine aktivitat der Weddell-Robbe *(Leptonychotes weddelli)* in Hallett, Antarctica. *Z. Morph Okol Tierre* **55**, 796-803.

Piérard, J. (1969). Le larynx du phoque de Weddell (*Leptonychotes weddelli*, Lesson, 1826). *Canad. J. Zool.* **47**, 77-87.

Piérard, J. (1971). Osteology and myology of the Weddell seal *Leptonychotes weddelli* (Lesson, 1826). *Antarctic Res. Ser.* **18**, 53-108.

Ray, C. (1967). Social behavior and acoustics of the Weddell seal. *Antarctic J.* **2**, 105-106.

Ray, C, and deCamp, M. A. (1969). Watching seals at Turtle Rock. *Nat. Hist.* **78**, 26-35.

Ray, C. and Smith, M. S. R. (1968). Thermoregulation of the pup and adult Weddell seal, *Leptonychotes weddelli* (Lesson) in Antarctica. *Zoologica* **7**, 33-48.

Sapin-Jaloustre, J. (1952). Les phoques de Terre Adélie. *Mammalia* **16**, 179-212.

Scheffer, V. B. (1958). "Seals, Sea Lions and Walruses", Stanford University Press, Stanford, CA.

Schevill, W. E. and Watkins, W. A. (1965). Underwater calls of *Leptonychotes* (Weddell seal). *Zoologica* **50**, 45-46.

Schevill, W. E. and Watkins, W. A. (1971). Directionality of the sound beam in *Leptonychotes weddelli* (Mammalia: Pinnipedia). *Antarctic Res. Ser,* **18**, 163-168.

Schmidt-Nielsen, K. (1964). Terrestrial animals in dry heat: desert rodents. *In* "Handbook of Physiology Sect. 4: Adaptation to the Environment", (Ed. D. B. Dill), 493-507. American Physiological Society, Washington, D.C.

Scholander, P. F. (1940). Experimental investigations on the respiratory function in diving mammals and birds. *Hvalradets Skrifter Norske Videnskaps-Akad. Oslo* **22**, 1-131.

Siniff, D. B., DeMaster, D., Kuechle, V. B., Watson, A., Reichle, R. and Kaufman, G. (1974). Population dynamics of McMurdo Sound's Weddell seals. *Antarctic J.* **9**, 104-105.

Siniff, D. B., Stirling, I., Hofman, R., DeMaster, D., Reichle, R. and Kirby, R. (1975). Population studies of Weddell seals in eastern McMurdo Sound. *Antarctic J.* **10**, 120.

Siniff, D. B., Tester, J. R. and Kuechle, V. B. (1971). Some observations on the activity patterns of Weddell seals as recorded by telemetry. *Antarctic Res. Ser.* **18**, 173-180.

Smith, M. S. R. (1965). Seasonal movements of the Weddell seal in McMurdo Sound, Antarctica. *J. Wildlife Mgmt* **29**, 464-470.

Stirling, I. (1969). Ecology of the Weddell seal in McMurdo Sound, Antarctica. *Ecology* **50**, 573-586.

Stirling, I. (1971a). Population dynamics of the Weddell seal (*Leptonychotes weddelli*) in McMurdo Sount, Antarctica 1966-68. *Antarctic Res. Ser.* **18**, 141-161.

Stirling, I. (1971b). Variations in sex ratio of newborn Weddell seals during the pupping season. *J. Mammal.* **52**, 842-844.

Stirling, I. (1972). Regulation of numbers of an apparently isolated population of Weddell seals (*Leptonychotes weddelli*). *J. Mammal.* **53**, 107-115.

Stirling, I. and Greenwood, D. J. (1972). Observations on a stabilizing population of Weddell seals. *Aust. J. Zool.* **20**, 23-25.

Turbott, E. G. (1949). Observations on the occurrence of the Weddell seal in New Zealand. *Rec. Auckland Inst. Mus.* **3**, 377-379.

Vaughn, R. W. (1968). The status of the Weddell seal *(Leptonychotes weddelli)* at South Georgia. *Br. Antarctic Survey Bull.* No. 15, 71-74.

Vaz Ferreira, R. (1956). Caractéristics generales de las islas Uruguayas habitadas por lobos marinos. Trabajos Sobre Islas de Lobos y Marinos, No. 1, Serv. Oceanográf. Pesca, Montevideo.

Watkins, W. A. and Schevill, W. E. (1968). Underwater playback of their own sounds to *Leptonychotes* (Weddell seals). *J. Mammal.* **49**, 287-296.

Wilson, E. A. (1907). Weddell's seal, Mammalia, *In* "National Antarctic Expeditions", 1901-04, Nat. Hist. **2**, 1-66. British Museum (Natural History), London.

Wilson, G. (1970). Some comments on the optical system of Pinnipedia as a result of observations on the Weddell seal. *Br. Antarctic Survey Bull.* **23**, 57-62.

13
Southern Elephant Seal
Mirounga leonina Linnaeus, 1758

John K. Ling and M. M. Bryden

Genus and Species

Lydekker proposed three subspecies of uncertain validity (Scheffer, 1958): *falclandicus*, 1909: (Falkland Islands Dependencies); *macquariensis*, 1909 (Macquarie and (?) Chatham Islands); *crosetensis*, 1909 (Ile de Crozet and (?) Kerguelen and Heard Island). Bryden (1968) regards the different growth patterns of the Macquarie Island and Falkland Islands Dependencies (Fig. 1) populations as being phenotypically determined and not evidence for subspecific recognition. Common names include southern elephant seal and southern sea elephant.

External Characteristics and Morphology

The general appearance of the male elephant seal is shown in Fig. 2.

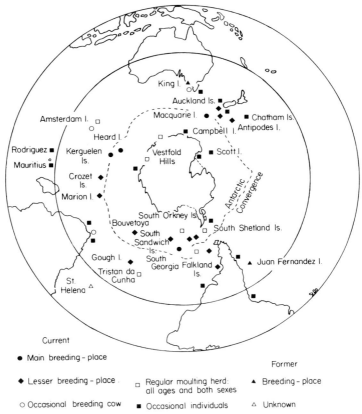

FIG. 1 Present and past distribution of the southern elephant seal, *Mirounga leonina*. (From Carrick and Ingham, 1962a).

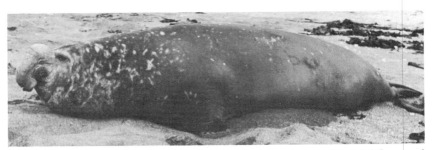

FIG. 2 Adult male southern elephant seal, *Mirounga leonina*, at Macquarie Island. This animal was branded at birth (MF BP) and photographed at 10 years of age.

Coloration

A black coat is developed before birth and, in the majority of southern elephant seals, is replaced at about three weeks of age by a short silver-grey coat which is darker above than below. A small percentage (about 1% or less) of seals born on ice and snow at the southern end of their breeding range moult or begin to moult *in utero* (Laws, 1953; Carrick *et al.*, 1962b). Successive annual moults reveal a dark slate-grey skin through which silver-grey hairs emerge and grow to form a new pelage. As seals of both sexes grow older their coats become increasingly more stained with dirt and excrement so that brownish and yellowish colours tend to predominate. However, the silver-grey colour seen in young seals certainly gives way to brown, yellowish or even cream at the time of the moult and before staining affects the coat colour in older seals.

Upon entering the water the seals assume a uniformly dark grey, almost black colour which lightens again upon drying out.

Dimensions and weights at various ages

There is a considerable volume of data on linear dimensions of elephant seals in the literature, but much less on body weight measurements. Analysis of linear measurements have shown that growth patterns and ultimate size of elephant seals at Macquarie and Heard Islands are different from those at the Falkland Islands Dependencies (Carrick *et al.*, 1962a). It has been suggested that pressures within the Macquarie and Heard Island populations retard development (and breeding) (Carrick *et al.*, 1962a; Bryden, 1968).

Growth in standard length of elephant seal populations that have been closely studied is shown in Fig. 3. As in all animal populations, there is considerable within-age variation in both standard length, and to a greater extent body weight. The figure illustrates that there is little difference in size of males and females at birth, there is a slight difference in favour of males up to three years of age and a large sexual disparity in size occurs after puberty is reached. This pattern is accentuated when body weight is used as a measure of size. The largest beachmaster seen at Macquarie Island are about 5 m standard length, whereas the largest cows are approximately 2.5 m standard length or slightly less. At Signy Island these figures are about 5.8 m and 3 m respectively.

Bryden (1969) related standard length and weight of elephant seals at Macquarie Island, and found a reasonably good correlation over the

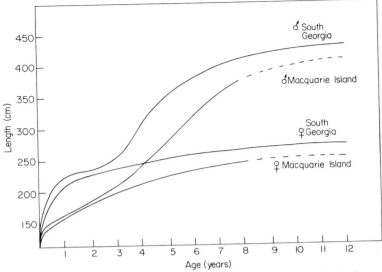

FIG. 3 Growth in length (standard length) of male and female elephant seals at South Georgia and Macquarie Island. (Modified after Carrick *et al.*, 1962a)

size range from birth to physical maturity. The relationship* can be expressed as

Males $\quad W = 131.49 + 0.00002079\,L$
Females $\quad W = 195.51 + 0.00000400\,L$

Where W is body weight in kg, and L is the cube of the standard length in cm^3. Prediction equations of equal accuracy were obtained by relating body weight and girth measurement, and body weight and total length measured in a straight line from the tip of the nose to the tip of the hind flipper.

The largest male and female animals for which both standard length and body weight data have been recorded (Bryden, unpublished data) are:

	Age (months)	Standard length (cm)	Weight (kg)
Male	191	467	3692
Female	127	250	359

It is estimated that breeding bulls are up to ten times the weight of breeding cows.

---

*Please note: There were errors in this relationship in the original paper (Bryden, 1969), which were perpetuated in a subsequent paper (Bryden, 1972a), and in a review of marine mammal growth and development (Bryden, 1972b).

Distribution and Migration

Range

Circumpolar, mainly in subantarctic waters but occurring from 16 °S at Saint Helena to 78 °S.

Scheffer (1958) and Carrick and Ingham (1962a) reported and figured all known past and present records including non-breeding and breeding sites. The latter lie between 40 and 62 °S in the South Indian and South Atlantic Oceans, but not the South Pacific other than the former breeding site on Juan Fernandez (see Fig. 1).

The chief hauling-out places for the purposes of breeding, moulting or simply "resting" are the subantarctic islands where these terrestrially cumbersome animals come ashore. They prefer to lie about on the beaches or among tussock grass (*Poa* sp.) in the hinterland. In the southern sector of their breeding range, however, they may have to lie out on snow and ice (Laws, 1953).

Movements

The pelagic phase of this species is not well known and the extent of movement from the island rookeries is still largely a conjectural matter. Of the many thousands of seals marked on Macquarie and Heard Islands a few have been sighted at more or less distant places, e.g. Campbell Island (from Macquarie) and Marion Island and the Vestfold Hills, Antarctica (from Heard). These data would indicate random dispersal rather than mass migration. There is probably only limited intermixing, if any, between seals of the three main breeding populations, Falkland Islands Dependencies, Macquarie Island, and Heard and Kerguelen Islands; hence Lydekker's inclination towards the three subspecies, already mentioned, on the basis of skull characteristics.

Life History

Pups are born during September and October throughout the geographical range with remarkable synchrony of peak pupping dates being shown at widely scattered rookeries. Most of the births take place in mid-October, lactation lasts about three weeks and mating occurs approximately 2½ weeks *post-partum*, i.e. just before weaning. The pups remain ashore for a further five to eight weeks during which time their muscles develop for an aquatic existence. Much of the blubber laid down during lactation is converted into energy and there is a loss of body weight in the pups.

At about ten weeks of age the pups go to sea having spent the latter part of the post-lactation period playing and learning to swim in the shallows off the breeding beaches.

In the unexploited population at Macquarie Island females mature at four to six years and produce a single pup each year, with perhaps a missed pregnancy after six or seven offspring. Although sexually mature at about six years, males do not participate in breeding activities until they are almost ten years old. At South Georgia, however, where several thousand elephant seal bulls were killed each year until the early 1960s, cows bred a year earlier and bulls two to three years earlier, presumably as a result of removing many of the large males from the breeding population (Carrick *et al.*, 1962a).

Lengthy pelagic phases are broken by regular haul-outs on breeding islands by different age-classes for the specific functions of breeding, moulting and "resting". Elephant seals also haul out during storms or on account of sickness.

Immature seals of both sexes, and primiparous and barren cows are ashore moulting during December and January; mature cows which gave birth to a pup in the previous breeding season moult during January and February; and breeding bulls moult in February to May (see Table 6). Each age/sex class goes to sea again; a small proportion of the yearling and one-year age group return in late March for a brief period ashore and a few older seals visit their islands during the southern winter months. The functional significance of these autumn and winter haul-outs is not known. Immature seals continue to haul out to moult during each December and January, but fewer and fewer come ashore at other times of the year until maturity is attained.

Breeding males come ashore in late August and distribute themselves over the beaches where the gravid females will haul out some weeks later to give birth. Breeding bulls may stay ashore for almost eight weeks. Each cow is ashore for about four to five weeks during her breeding season—about five days before parturition and up to 28 days *post-partum*. Mating occurs about 18 days *post-partum*, i.e. towards the end of lactation which lasts some 20 to 25 days. Breeding seals go to sea again and the last are gone by early November after which immature seals of both sexes begin arriving to moult.

SOUTHERN ELEPHANT SEAL

Females generally live 10 to 13 years and males up to 18 years but both sexes may survive to 20 years (Laws, 1953, 1960; Carrick and Ingham, 1962c).

Population dynamics

Laws (1960) estimated the world southern elephant seal population as almost 600 000 in mid-year, i.e. excluding the 50% of pups that do not survive their first winter. Of this total the main populations may be recognized as follows (Carrick and Ingham, 1962a):

South Georgia	300 000
Macquarie Island	110 000
Kerguelen	100 000
Heard Island	62 500
Marion Island	10 000

However, the Macquarie Island figure above is a provisional total for November. Stocks at Macquarie and Heard Islands are now believed to be stable after many years of protection from commercial sealing operations which drastically reduced their numbers by the end of the nineteenth century. At South Georgia and Kerguelen, where sealing was carried out in conjunction with whaling, cessation of the latter in the early 1960s brought an end to elephant seal exploitation. The more recent published assessements of the status of these populations refer to exploited stocks in the Falkland Islands Dependencies with which are compared the now stable populations at Macquarie and Heard Islands (Laws, 1960; Carrick and Ingham, 1962a). More recently, Pascal (1979) has estimated the total population at Kerguelen to be 210 000 (120 000 females and 90 000 males), while Condy (1979) estimated there to be 5538 elephant seals on Marion Island.

The sex ratios among weaned and suckling pups are 54.9% male to 45.1% female at South Georgia, 53.2% male to 46.8% female at Macquarie Island, and 52.7% male to 47.3% female at Heard Island. There may be a slightly greater pre-weaning mortality of females acting on a sex ratio at birth of 50:50 (Carrick and Ingham, 1962c).

Laws (1960) drew up provisional life tables for elephant seals at South Georgia on the basis of survival data obtained from analysis of tooth rings. He also compared the life table for exploited males with a provisional potential natural life table. Carrick and Ingham (1962b) presented similar figures based on resighting of branded seals at

Macquarie Island, but only up to eight years of age. These data are compared in Tables 1-4, reproduced unaltered or only slightly modified from the original sources.

Causes of pup mortality include abandonment and starvation, drowning, and injuries sustained in harems. During the first year of life predation and, or because of, malnutrition accounts for almost half the mortality of all seals born in the previous year. These factors operate to a lesser extent in subsequent years. The chief predator is the killer

TABLE 1 Provisional natural life table for females at South Georgia (from Laws, 1960).

Age[a]	Seals alive at beginning of year		Seals dying naturally during year	
	Number[b]	%[c]	Number	%[d]
0	46 920	100.00	18 768	40.0
1	28 152	60.00	5 630	20.0
2	22 522	48.00	2 928	13.0
3	19 594	41.76	2 547	13.0
4	17 047	36.33	2 216	13.0
5	14 831	31.61	1 928	13.0
6	12 903	27.50	1 677	13.0
7	11 226	23.92	1 459	13.0
8	9 767	20.81	1 270	13.0
9	8 497	18.11	1 105	13.0
10	7 392	15.76	961	13.0
11	6 431	13.71	1 414	22.0
12	5 017	10.69	1 154	23.0
13	3 863	8.23	966	25.0
14	2 897	6.17	869	30.0
15	2 028	4.32	852	42.0
16	1 176	2.51	564	48.0
17	612	1.31	319	52.0
18	293	0.63	176	60.0
19	117	0.25	94	80.0
20	23	0.05	23	100.0

[a] Age in years from beginning of pupping season, 20 September.
[b] Entire pup class assumed to be born on 20 September.
[c] Percentage of initial recruitment.
[d] Percentage of those alive at beginning of year (numbers to nearest whole animal).

TABLE 2 Provisional potential natural life table for males at South Georgia (from Laws, 1960)

Age[a]	Seals alive at beginning of year		Seals dying naturally during year	
	Number[b]	%[c]	Number	%[d]
0	55 080	100.00	31 234	56.7
1	23 846	43.30	7 154	30.0
2	16 692	30.31	2 837	17.0
3	13 855	25.15	2 355	17.0
4	11 500	20.88	1 955	17.0
5	9 545	17.33	1 623	17.0
6	7 922	14.38	1 347	17.0
7	6 575	11.94	1 118	17.0
8	5 457	9.90	928	17.0
9	4 529	8.22	770	17.0
10	3 759	6.82	639	17.0
11	3 120	5.66	718	23.0
12	2 402	4.36	576	24.0
13	1 826	3.31	475	26.0
14	1 351	2.45	405	30.0
15	946	1.72	397	42.0
16	549	1.00	274	50.0
17	275	0.50	165	60.0
18	110	0.20	88	80.0
19	22	0.04	20	90.0
20	2	0.004	2	100.0

[a-d] As in Table 1.

whale, *Orcinus orca*. Injuries sustained in intrasexual combat may also lead to premature mortality without being fatal at the time of fighting (Carrick and Ingham, 1962c).

Internal Anatomical Characteristics

Skeleton, skull, teeth

Other than the skull (Fig. 4), the skeleton is basically similar to that in other Phocidae. Turner (1888) described the osteology in detail. The vertebral formula usually quoted for all Phocidae is C7 T15 L5 S3 Cd11.

TABLE 3 Provisional life table for exploited males at South Georgia (from Laws, 1960)

Age[a]	Seals alive at beginning of year		Seals killed by man during year[d]	Seals alive after kill	Seals dying naturally during year	
	Number[b]	%[c]			Number	%[e]
0	55 080	100.00			31 234	56.7
1	23 846	43.30			7 154	30.0
2	16 692	30.31			2 837	17.0
3	13 855	25.15			2 355	17.0
4	11 500	20.88	5	11 495	1 954	17.0
5	9 541	17.32	169	9 372	1 593	17.0
6	7 779	14.12	992	6 787	1 144	17.0
7	5 633	10.22	2 083	3 550	603	17.0
8	2 947	5.35	1 538	1 409	240	17.0
9	1 169	2.12	675	494	84	17.0
10	410	0.74	233	177	30	17.0
11	147	0.27	79	68	16	23.0
12	52	0.09	30	22	5	24.0
13	17	0.03	10	7	2	26.0
14	5	0.01	4	1	1	

[a-c] As in Table 1
[d] Average annual kills in 1955-58, age composition from teeth samples.
[e] Percentage of those alive at beginning of year, except for age 4 and upwards. Percentage of those alive at end of kill.

In a series of 98 dissections of elephant seals, this number of cervical, thoracic, lumbar and sacral vertebrae was observed in 92 animals, the remaining six animals having C7 T14 L6 S3; the number of caudal vertebrae was not recorded (Bryden, unpublished data). Supernumerary thoracic vertebrae have been reported in other species of seals, but a reduction in the number of thoracic vertebrae to 14 is unusual in pinnipeds except in the walrus *(Odobenus)* which normally has 14 thoracic and six lumbar vertebrae.

Bryden (1971a) described the musculature, and an analysis of the relative size and growth patterns of individual muscles was reported by Bryden (1973).

TABLE 4 Survival of branded elephant seals to 8 years old, Macquarie Island (from Carrick and Ingham, 1962c). The most reliable figures are in bold type; those obviously too low, owing to inadequate searching for marked seals in some years, are in brackets.

Age (years)	Surviving males		Surviving females	
	Number	Percentage	Number	Percentage
0	—	100	—	100
1	—	—	—	—
2	234	(39.5)	226	(42·0)
3	73	(22.5)	98	(33.5)
4	101	**40.1**	115	**46.5**
5	80	31.7	107	43.1
6	67	**26.5**	99	**39.5**
7	—	—	—	—
8	39	**14.0**	51	**20.5**

The skull of South Georgia elephant seals has been studied in detail by King (1972), who observed the modifications connected with the possession of a proboscis. Growth of the skull is different in males and females: in males there is a continuous steady increase in length, width and height at least until 11 years of age, and the rate of increase of dimensions is greatest in the snout, and greater in the facial region than the cranium; in females the facial width and height increase rapidly up to 30-40 months of age, the beginning of development of the first pup, after which there is no further growth. The length and width of the cranium and the length of the face also have a rapid increase until 30-40 months, after which there is slow growth for the rest of the animal's life.

The formula for the milk dentition is

$$2 \left(I \frac{2}{1} C \frac{1}{1} PC \frac{3}{3} \right) = 22$$

Milk teeth first appear when the foetus is about 18 cm standard length and four to six weeks after implantation, and reach their maximum development about three months after implantation when the foetus is about 50 cm long. Reabsorption of the milk teeth occurs in the latter half of gestation. Eruption of permanent teeth begins about the time of birth, and is completed by about 35 days of age, after the pups become nutritionally independent of the mother.

The complete adult dentition is

$$2 \, (I \frac{2}{1} \, C \frac{1}{1} \, PM \frac{4}{4} \, M \frac{1}{1}) = 30$$

The incisors, with the exception of the second upper incisor, are very much reduced and probably have no function. The canine teeth are different in males and females: in the female they are used mainly for feeding; in males they are used during the breeding season in fights with other males, and consequently they are larger and much better developed, particularly in the upper jaw. The pulp cavity does not close and the crown apparently grows throughout life.

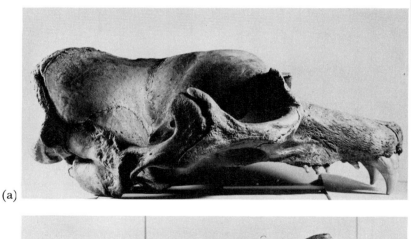
(a)

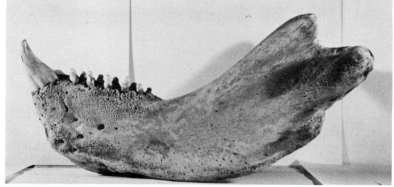
(b)

SOUTHERN ELEPHANT SEAL 309

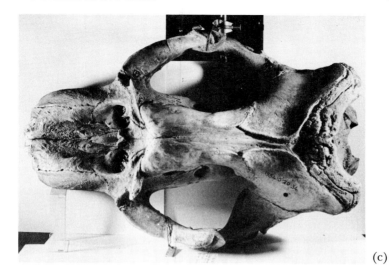

(c)

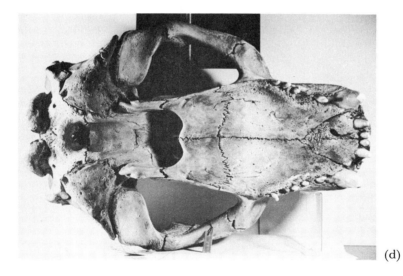

(d)

FIG. 4 Skull of an eight-year-old male taken by Dr Richard Laws at Signy Island, South Orkney Islands on 2 January 1950. The animal measured 490 cm (189 in) nose to tail. It is currently registered as 1954.5.20.35 in the British Museum and Laws station number H/354. (a) lateral view; (b) mandible; (c) dorsal view; (d) ventral view.

Some measurements of adult canine teeth (Laws, 1953; Carrick and Ingham, 1962b) are:

Dimension (mm)	Female	Male
Length enamel crown	14	29
Width enamel crown	9	22
Length root	60	110
Diameter at neck	15	32

The postcanines are small, peg-like structures which are apparently almost functionless, although they may be used to a limited extent to crush large fish. The upper fifth postcanine (first molar) is more simple than the others, and is often lacking.

Visceral anatomy

For the most part, the viscera are similar to those of other seals. Organ weights were reported by Bryden (1971b), who also analysed the growth patterns of many organs. Relative organ weights are listed in Table 5.

The stomach is simple, the small intestine is exceptionally long with a narrow lumen and there is virtually no caecum, although the ileocolic junction is obvious. The large intestine is similar in gross appearance to the small intestine, except that the lumen is a little wider.

The tracheal rings are C-shaped, not flattened as in some Antarctic phocid seals. Detailed anatomy of the respiratory tract has not been reported.

The kidneys are renulate as in other seals. In the male, the testes are held in inguinal pouches, outside the abdominal cavity but deep to the blubber. Although there is no scrotum, the testicular temperature is about 5 °C lower than body temperature. Enlargement of the testes occurs in the late foetal stage, but by birth this enlargement is no longer noticeable; the prostate declines sharply in weight from birth to about two weeks of age (Bonner, 1955). The cranial part of the corpus cavernosum penis is ossified, forming the os penis, as in other pinnipeds. The length of the os penis increases markedly about the time of puberty (between four and five years in the South Georgia population). Lengths are approximately 10 cm at birth, increasing to 23 cm at four years, 31 cm at five years and 34 cm at eight years (Laws, 1956a). The ovaries are almost completely enclosed in an ovarian bursa, which opens by a narrow dorsomedial slit into the peritoneal cavity. Mesonephric remnants are common. The uterus is bicornuate,

TABLE 5 Mean percentage of total body weight represented by the weight of each organ in nutritionally independent elephant seals.

Organ	Relative weight (%)	
	Males	Females
Oesophagus	0.18	0.18
Stomach	0.54	0.54
Small intestine	1.75	1.49
Large intestine	0.19	0.19
Liver	2.08	2.12
Pancreas	0.06	0.07
Trachea	0.16	0.16
Lungs	1.22	1.34
Kidneys	0.37	0.40
Bladder	0.03	0.03
Reproductive tract	0.09	0.21
Gonads	0.04	0.004
Thymus	0.06	0.04
Spleen	0.84	1.13
Lymph nodes	0.31	0.26
Heart	0.65	0.69
Brain	0.35	0.36
Adrenals	0.01	0.01
Thyroid	0.01	0.01
Eyes	0.12	0.13

NOTE: The relative weights of some organs change with increasing absolute body weight. For a complete discussion of this and growth patterns of these organs, see Bryden (1971b).

with a short body and single cervix approximately 3 cm long. There is a conspicuous hymen between the vagina and vestibule. The corpus cavernosum clitoridis is usually partially chondrified, or ossified forming an os clitoridis, which may reach 2 cm in length.

The cardiovascular system follows the general mammalian plan, with the modifications of the venous system described for other seals (Harrison and Tomlinson, 1956). There is a well-developed caval sphincter and a large hepatic sinus associated with the caudal vena cava.

The spleen is large as in other pinnipeds, and in contrast to the very small spleen in Cetacea. Blessing et al. (1972) described the morphology of the spleen of several marine mammals, including M.

leonina. The shape is an elongated oval, and the organ is smooth, not lobed. Its function in marine mammals is not fully understood, but it is thought to be related not so much to storage or metabolism as to a function more specifically related to aquatic adaptation.

Central nervous system

The central nervous system has not been studied in detail, but is apparently similar to that in other pinnipeds. Brain weights are greater in males than females, and growth of the brain is very rapid during suckling. At birth, brain weight is about 540 g in males and 500 g in females, and at one month of age (just after weaning) these figures have increased to about 620 g in males and 560 g in females. Brain growth is very slow after weaning, but the sex difference remains so that the adult brain weights are approximately 1350 g in males and 900 g in females (Bryden, 1971b).

Skin, moult

The skin (hide or pelt), including the dermis and epidermis (Ling, 1974), varies from a few millimetres in thickness in the axillary regions to more than 4 cm in the heavily bossalated throat region, particularly of large breeding bulls. The epidermis is up to 1 mm in depth of which the horny layer may form more than 90%. This is characterized by its forming continuous sheets to which the hair roots also become attached, so that the moult involves the loss of large ragged patches of cornified epidermis and old hairs (Ling and Thomas, 1967; Ling, 1968).

Skin pigmentation is very dense between the hair follicles and the follicle bulbs are richly pigmented; but the follicle walls lack pigment. Granular and clear layers are absent except during a hair growth phase, when a rudimentary granular layer may be present.

The dermis is well vascularized; dense connective tissue is more or less continuous with the underlying hypodermis (blubber) layer. The dermis is laid down on a rhombic framework between the elements of which lie the hair follicles surrounded by loose fat-charged cells.

The hair follicles are simple and a single hair is formed; i.e. there is no underfur. In common with the skin of all other pinnipeds studied, *M. leonina* does not have erector muscles serving the hair follicles. Also, like all other seals, this species moults annually at varying times according to age, sex and reproductive condition. Hair shedding begins in the axillae, between the base of the tail and hind flippers, and around the eyes and old scars. It spreads dorsad over the shoulders and

TABLE 6 Summary of moulting seasons and duration of moult stages at Macquarie Island 1962-64

Sex and Age[a] Categories	Moulting season Range[c]	Peak[d]	Average duration of moult haul-out (days)[b]			
			Pre-moult	Moult	Post-moult	Total
Immature males (second to fourth years)	15 Nov-24 Feb	18 Dec	5.0	21.0 (11-33)	9.3 (3-18)	35.3
Immature females (second and third years)	15 Nov-24 Feb	18 Dec	6.0	23.2 (16-37)	15.0 (5-19)	44.2
Subadult males (fifth to ninth years)	20 Nov-22 Mar	12 Jan	7.0	21.2 (10-32)	14.0 (5-24)	42.2
Adult males (ages not all known)	31 Jan-3 June	23 Feb	6.0	22.6 (13-39)	11.5 (5-23)	40.1
Mature females (fourth year)	9 Dec-20 Feb	13 Jan	6.0	16.0 (15-19)	7.0	29.0
Mature females (fifth to thirteenth years)	13 Dec-1 Mar	20 Jan	5.0	16.2 (10-26)	6.0 (2-16)	27.2

[a] Based on branding at about one month of age.
[b] Range in parenthesis.
[c] Earliest and latest date of sighting branded seals.
[d] Modal date of commencement of moult.

rump until the moulting areas meet on the back. Shedding then extends over the sides to the abdomen which is the last area to moult. Details of timing for the three stages of the moult haul-out in immature and mature seals are shown in Table 6.

Behaviour

Diving and swimming capabilities

Experiments have not been performed to test the diving and swimming capabilities. However, dives of more than 30 min have been observed (cited by Matthews, 1952) which may not be maximal. Indirect evidence suggests that elephant seals can make deep dives and of long duration. There is a large blood volume (about 15% of body weight) and the blood oxygen capacity (40 vol.%) is greater than any other marine mammals recorded (Scholander, 1940; Bryden and Lim, 1969).

Even very young pups (one to two days old) can swim, although they do not usually enter the water until they are more than five weeks old. Younger pups that are washed into the sea, or fall into deep water when the fast-ice breaks up, can keep afloat for a time, but later weaken and drown. The mother does not teach her pup to swim and dive; she has left the pup before these activities begin, under normal circumstances. Swimming is by means of the caudal part of the body and hind flippers, with the fore flippers pressed to the sides or used as rudders and possibly to assist in rapid descent to deep water. Aquatic locomotion has not been studied in detail, although Bryden (1973) has discussed it in relation to growth patterns of individual muscles. The maximum speed of free swimming adults is about 20-25 km h^{-1}. They have never been seen to leap clear of the water, and they do not display the great agility in the water which is characteristic of some other pinnipeds.

Social behaviour

Detailed accounts of behaviour in *M. leonina* have been given by Angot (1954) and Paulian (1953)—Kerguelen, Amsterdam and St. Paul; Laws (1953)—South Georgia; and Carrick *et al.* (1962a, b)—Macquarie and Heard Islands. These studies have revealed similar patterns to those described in the northern species, *M. angustirostris*, by Bartholomew (1952). Elephant seals are generally gregarious, have well-defined seasonal cycles and as a result demonstrate a rich array of social behavioural traits according to age, sex and season. The young stay close to the cow in the early days of life and move with her only as far as she does during the breeding fast until weaning. They then move away from the harem up the beach where they continue to rest using their body reserves for energy. After about four weeks of fasting the pups venture to the water's edge where they splash and play and eventually learn to swim. The mother takes no part in training, having deserted her pup at weaning. While cows can recognize their own pups—probably by smell—pups apparently do not discriminate between cows and often attempt to suckle the wrong one. They go to sea for the first time at about 8-10 weeks of age.

A proportion of young seals haul out during the following autumn-winter months of April-July, but the main haul-out for this category does not take place until the next summer—November, December and January—to moult. The yearling seals generally choose the tussock grass (*Poa* sp.) above the beaches where the sand is dry. The moulting seals lie wriggling and scratching as the outer epidermis peels off with the old hairs attached. After the moult which takes about 21-23 days (Table 6) the yearlings move to the beaches again, where they remain for a week or more, often going into the shallows and engaging in play and mock-fighting among the young males. Older age groups of immature seals also moult in the tussock, where males up to three years old engage in sparring exercises.

Moulting mature females generally select deep muddy wallows in which to shed their hair. Here they lie packed together in a steaming mud bath, wriggling, shoving and scratching for approximately 16 days while the old coat is lost and the new hairs emerge. Reasons advanced for this choice of habitat have previously related it to the relief it may afford to the apparently intense irritation of the moult (Carrick *et al.*, 1962b). However, work on other species, *Odobenus rosmarus* and *Phoca vitulina*, suggests that the maintenance of high skin temperatures may be important in effecting the moult (Ling *et al.*, 1974). The high temperatures in the moulting wallows may therefore hasten the process, thereby enabling the moult haul-out to be reduced to a minimum for the pregnant cows which have the shortest moult period—even shorter than small immature seals (see Table 6). Thus size alone does not appear to be the crucial factor in determining the duration of the moult haul-out which is accompanied by a fast. In pregnant cows the need to conserve energy seems to be paramount.

Elephant seals sometimes scoop sand over themselves with their fore flippers. This has recently been shown by White and Odell (1971) to assist in thermoregulation in northern elephant seals by lowering the peripheral temperature.

Seals are often seen to stretch their hind flippers and rub them together both when the hairs are being shed and also before and after, apparently manifesting some irritation which may be caused by moulting or the presence of ectoparasites. The older immature (and mature) animals spend some time in the sea cruising along the shore and often stopping to rest with the hind flippers and head out of the water and only the strongly arched body submerged.

Terrestrial locomotion is by means of an abdominal hitching movement with the aid of the fore flippers which are rotated forwards and backwards, thereby pushing the seal along. The head bobs vigorously and undulations of the skin and underlying blubber pass down the body.

Laws (1953) was the first to describe the behaviour of *M. leonina* in relation to ice. Elephant seals make every effort to climb out onto the ice to rest wherever it will support their weight. Breathing holes about 60 cm in diameter are made simply by butting the ice until it breaks, rather than by chewing as in the case of *Leptonychotes*. Elephant seals may travel quite long distances (up to 20 km) under ice which means they would have to utilize pockets of air to survive.

A few mature seals of both sexes, but chiefly females, haul out during the winter months. Most of the cows appear to be in some sort of distress and the number of still-births late in pregnancy may explain this behaviour. At other times of the year seals of all ages and of either sex come ashore occasionally with wounds of varying severity caused by the attacks of killer whales *Orcinus orca*.

At the beginning of the breeding season, which lasts from early August to early December, sexually and socially mature bulls haul out along the beaches around the island rookeries. The oldest and strongest, but not necessarily the largest beachmasters arrive first. The bachelor bulls do not haul out until mid-August. Agonistic behaviour is intense and involves vocal exchanges, much postering and some actual combat. The cows begin arriving at the end of August and form large pods, which are kept herded together by breeding bulls. Mature males continue to arrive until mid-September and their numbers remain high until late October. The biggest males are ashore for the longest time.

Carrick *et al.* (1962a, p. 163) have defined the various categories of elephant seal males which participate in breeding season activities as follows:

Breeding bull—A large bull 4.2 m or more in length, with a fully developed proboscis and many scars on the neck and shoulders; 14 years old or more.

Beachmaster—A breeding bull in sole possession of a small harem, or the dominant bull in a large harem.

Challenger—A breeding bull temporarily without a harem, but fighting with beachmasters for one.

Assistant beachmasters—A breeding bull in charge of part of a harem which is too large for a single beachmaster.

Bachelor—A smaller bull 3.3-4.2 m long, whose proboscis lacks the deep transverse cleft of the beachmaster, whose neck and shoulders are lightly scarred, and who does not fight with breeding bulls; from six to about 13 years old (Fig. 2).

The terms "breeding bull" and "bachelor" correspond to the "dominant bull" and "subordinate bull" of Laws (1956b).

Cows aggregate into small peaceful groups when they start coming ashore in late August. If a breeding bull takes charge such a group will form the nucleus of a harem which will attract more cows and grow quickly, unless it is disrupted by the attempts of bachelor bulls to copulate with the pregnant cows at this stage.

Thus harems develop spontaneously and true territorial behaviour is not a feature of this species; boundaries are neither defined nor defended. The fighting between males is for dominance over a large number of females which are controlled in a loosely defined area by one or more males in a hierarchy. Combat between bulls rarely takes any other form than threat and posture, although bulls may come to blows, often with bloody results. Usually, however, one or the other combatant backs down. This is often followed by the victor copulating with the nearest cow; though copulation does not necessarily have to be preceded by a fight.

The main duties of beachmasters and breeding bulls are maintenance of the harem, i.e. stopping the escape of cows, impregnating them and preventing subordinate males from doing likewise.

Birth-site tenacity is fairly high (60-70%) over an age span of four to 11 years, except for the purpose of moulting which requires a different habitat from the birth-site (Nicholls, 1970).

Feeding habits

Southern elephant seals feed only at sea, and fast for long periods when on land. After weaning, the pups fast for up to six weeks before going to sea. The nutritionally independent animal feeds exclusively on cephalopods and fish. The stomach usually contains squid beaks, and sometimes fresh food. Possibly these seals feed primarily on fish (mainly *Notothenia* spp.) in inshore waters and on cephalopods elsewhere.

Adult bulls spend over eight weeks on land during the breeding season, and a shorter time when moulting. Cows spend only about four weeks ashore during the breeding season, but they lose about 135 kg body weight while suckling the pup. For much of these long periods when the seals are ashore they fast completely, although occasionally, and particularly towards the end of the breeding season or the moult, the males go into the water and feed.

While on land, many seals fill the stomach with sand and small pebbles. This is certainly done deliberately, as many animals at Macquarie Island, particularly pups soon after weaning, have been seen eating beach material. One large male had 35 kg of pebbles in his stomach.

Sound production

Nothing has been reported of sound production in the water by southern elephant seals, but it would be expected to occur as it does in other species, e.g. bearded seal, Weddell seal. Many sounds are produced on land and have been described by Laws (1956b) from whose work these descriptions are largely drawn.

A wide variety of snorts, sneezes, whistles, grunts, yawns, and internal noises which are produced have no apparent significance. Nearly all the socially significant sounds are concerned with territorial protests or threats, apart from communication between cow and pup, and there is much less vocalizing in the non-breeding season than the breeding season when social relations are more complex.

The pup emits a sharp bark or yap, which is the response to most new situations, and only occasionally receives a response from the mother.

Cows produce a high-pitched moaning or "yodelling" sound after the birth of the pup, which seems to be produced in response to a situation, although it is apparently related to the presence of the pup.

Bulls produce a bubbling roar which can carry for miles on a calm day. The sound appears to be produced in the throat as in a gargling action, that results in a harsh rattling sound of great power. The proboscis may act as a resonating chamber to amplify the sound. The roar may be used as a threat to a specific individual, as a generalized response to a situation, or merely spontaneously. Vocalization is used extensively by beachmaster bulls in the establishment and maintenance of harem sites.

Captivity

Southern elephant seals have been kept successfully in many zoos and oceanaria around the world, but little has been written about this species in captivity. Bullier (1954) described the problems involved in getting recently weaned elephant seals to accept food, and stressed the advantages of offering live, fast-swimming fish in the early acclimatization period. Two young elephant seals (about 15 months old) taken to the Sydney Zoo in 1964 were kept in apparently good health for about four months, then died from unknown causes. Difficulty was experienced in getting those animals to accept food in captivity, and finally they were offered live squid which they accepted readily and could soon be enticed to accept live fish, then dead fish. An emaciated and badly injured subadult female elephant seal which came ashore at Coffs Harbour, New South Wales in 1971 was placed in a local oceanarium, readily accepted dead fish and has been in excellent health ever since apart from a temporary problem with corneal ulcers. It appears that this species can be kept in captivity quite well, and that a suitable diet (fresh small fish or squid) is most important. The husbandry principles used for other seals kept in captivity apply equally to southern elephant seals.

Reproduction

Gestation

Mating and conception take place in November, about 18 days after birth of the pup. Thus gestation is a little over eleven months' duration, although development almost, or completely, ceases for about 16 weeks after the attainment of the blastocyst stage. Blastocysts have been observed lying free in the lumen of the uterine cornu up to the end of February, and implantation occurs in late February or early March. Thus gestation from the time of implantation to parturition is about eight months.

Because pregnant females remain at sea during the winter months, very few embryos have ever been collected. Laws (1953) and Bryden (1969) each gave body length data for seven foetuses (Table 7). The placenta is of zonary type. Usually there is a single foetus; Bryden (1966) described one female with twin foetuses, but only one twin was alive. Laws (1953) reported siamese twins born at South Georgia in September, 1951.

Maturity

Females mature at two years of age at South Georgia and at about three years on Heard and Macquarie Islands. All third-year cows at South Georgia are pregnant; one-third of Macquarie Island cows are pregnant in their fourth year but a quarter are still not pregnant until their seventh year (Laws, 1953; Carrick et al., 1962a). The average expectation of life after the first pregnancy is about seven years at South Georgia.

TABLE 7 Length of elephant seal foetuses at South Georgia (SG) and Macquarie Island (MI). Measurements taken in a straight line from nose to tail. Age of foetus in months post-implantation, assuming that implantation occurs at the end of February.

Location	Sex	Approx. age (months)	Standard length (cm)
SG	—	1.0-1.5	18.6
SG	—	1.0-1.5	19.2
SG	—	1.5-2.0	24.8
SG	—	1.75-2.5	28.2
SG	—	2.25-3.25	41.0
SG	—	2.5-3.5	49.0
SG	—	5.0-6.0	89.0
MI	Male	2.0-2.5	38.2
MI	Male	2.0-2.5	44.0
MI	Female	3.25-3.75	72.5
MI	Female	3.25-3.75	67.0
MI	Female	3.75-4.25	82.0
MI	Male	6.5-7.0	96.5
MI	Female	7.0-8.0	94.0

Males reach puberty at four years at South Georgia and six years on Macquarie Island. However, full breeding ("Beachmaster") status may not be attained until 12 or 13 years of age (Carrick *et al.*, 1962a).

The accelerated maturity in the Falkland Islands Dependencies population relative to Macquarie Island is believed to be due to a cropping programme which was carried out at the former until the early 1960s. Removal of older (larger) bulls is believed to have allowed younger seals of both sexes to exploit the food resources leading to earlier maturity. Conversely, there is probably more intraspecific competition for food resources in the unexploited Macquarie Island population, in which the older seals are more successful. As a result, intrasexual strife delays the attainment of sexual maturity in aspiring breeders (Carrick *et al.*, 1962a).

Reproductive physiology

Spermatogenesis occurs throughout the breeding season from August to November, but by December the testes have regressed to an anoestrous state in which they remain until the following August. At South Georgia there is no sign of any decline in male potency up to at least 11 years (Laws, 1956a). Enlargement of the testes during sexual activity is caused chiefly by increased tubule size; while interstitial tissue volume varies only slightly throughout the year. The epididymis also undergoes enlargement during the breeding season but somewhat later than the seminiferous tubules. Males at South Georgia are sexually mature at 47 months on the basis of histological evidence.

Females are mature at 22 months, based on signs of first ovulation; one has been noted to be mature at 12 months and a few at 36 months on South Georgia (Laws, 1956a). Ovarian evidence and data on branded seals at Heard and Macquarie Islands indicate a minimum age of maturity of 36 months with some cows not maturing until their seventh, and giving birth in their eighth year (Carrick *et al.*, 1962a).

Cows usually ovulate only once a season, but a second ovulation may occur in a very early breeder which misses impregnation at the first oestrus (Laws, 1956b). Ovaries function alternately each year, and the right ovary is almost always the first to ovulate upon the attainment of sexual maturity. Following ovulation there is a great spate of follicular activity with as many as 16 and 25 follicles becoming atretic in the "active" and "inactive" ovary respectively. It is not known whether ovulation is spontaneous or induced.

The corpus luteum forms in the usual way, but secretory activity is low in the period immediately following ovulation and successful impregnation. From mid-February secretory activity increases as the time for implantation of the blastocyst approaches.

The approximately 16-week free blastocyst stage of the reproductive cycle ends around late February or early March when implantation occurs. Laws (1956a) believed that the end of the annual moult in cows signals implantation and gestation to begin, but cows have been observed with quite advanced embryos while still moulting (Ling, 1970). Thus the end of the moult and implantation do not seem to be directly related. The process of delayed implantation enables successful mating to occur when the population of sexually mature seals is at its maximum, thereby affording high reproductive efficiency and gene flow.

There appears to be a retardation by as much as a week each year in the onset of ovulation in females at South Georgia which results in a missed pregnancy after about seven years (= 16%) when ovulation is too late for successful impregnation to be accomplished (Laws, 1956b). At Macquarie Island retardation, if it occurs at all (some cows are known to have pupped earlier in a subsequent season), is not sufficient to cause pregnancies to be missed in the lifetime of a breeding cow (Carrick et al., 1962a).

Breeding season

A bull may spend up to eight weeks ashore during the breeding season. The last of the breeding bulls do not leave the rookeries until early December, returning again during February to April or even May to moult.

Each cow is ashore for about 28 days during the breeding season between September and October. The majority of cows haul out over a three-week period around the middle of October throughout the species' range. Gravid cows land about five days before the pup is born, lactation lasts 20-25 days with the *post-partum* oestrus occuring on about the eighteenth day of lactation.

The location of breeding sites is dictated largely by their ease of access to gravid cows as they come ashore to give birth to their pups. Thus favourable beaches quickly become crowded with up to thousands of cows towards the peak pupping time. Harems are presided over by beachmaster bulls along with varying numbers of assistant beachmasters, according to harem size. Only the most dominant bulls retain ownership for any length of time; otherwise the ownership changes frequently among beachmasters and challengers.

While harems can become quite large and still be guarded by a single bull, the average number of cows per bull over longer periods and when harems reach maximum size varies between 28 and 53 (Carrick *et al.*, 1962a). The lower numbers of cows per bull are due to spreading of the harems as pups are born and the rapidly increasing numbers of cows in *post-partum* oestrus.

Mating

The vast majority of cows which pup on land are re-impregnated in harems whilst ashore. A few may escape the attention of beachmasters only, it is believed, to be mated at sea by patrolling subordinate bulls which have not yet reached sufficient social status to participate in harem activity. Virgin and possibly non-parous cows are also believed to be impregnated at sea by these males (Laws, 1953; Carrick *et al.*, 1962a). Angot (1954) however, while agreeing that the young cows are usually fertilized away from the harems, was not sure that impregnation did not occur on land in remoter parts of the breeding island (Kerguelen). Laws (1956a) has observed mating in shallow water, if not farther out to sea.

Laws (1956a) and Carrick, Csordas and Ingham (1962a) have described and photographed terrestrial mating in this species. The breeding bull—considerably larger and heavier than the cow—pins her to the ground by throwing his weight across her neck and shoulders. He then grips her neck in his mouth—hence this very scarred region in older cows—places a fore flipper over the back and draws her towards his belly which he exposes by rolling over on to a side. The penis is then protruded and inserted into the now adjacent vulva.

Bulls may mate or attempt to mate with non-oestrous cows as well as those previously served. They even try to seduce immature males, apparently being deceived by the small size which Carrick *et al.*, (1962a) regard as the first clue to the presence of a potential partner. This is likely to occur most frequently with younger inexperienced bulls which must learn to distinguish females from small males, and oestrous from non-oestrous cows.

Birth and lactation

Birth and lactation almost always occur on land, in the harems. A few births have been observed away from the harems, but in most such cases the pup is abandoned by the cow who moves to a harem soon after her pup is born.

At the onset of parturition the cow becomes irritable and moves to the edge of the harem, or at least to a less crowded part of it. At first, straining is accompanied by opening and closing of the urogenital orifice, which probably lasts for some hours. Later on, contractions become more frequent, and the foetal membranes appear at the vulva. The membranes soon break, revealing either the nose or the hind flippers of the pup. Actual birth occurs quite rapidly (2-15 min), there being no problems associated with caudal presentations because of the fusiform shape of the animal. Approximately 50% of births are head presentations, although tail presentations may be slightly more frequent 56% of 196 observed births (Carrick et al., 1962a). The placenta is normally presented 15-60 min after the pup is born. The umbilical cord is usually broken soon after the pup is born by movements of the cow, but sometimes the cord remains intact for some time so that occasionally new-born pups are seen with the placenta still attached by the umbilical cord. The placenta is usually consumed by scavenging birds, and is not eaten by the cow. The various phases of parturition have been described more fully, and well illustrated, by Carrick et al., (1962a).

Twin births are very rare, only one undoubted twin birth having been reported, (Carrick et al., 1962a).

As in all Phocidae, lactation is characteristically of short duration, and is associated with rapid growth of the pup. Lactation lasts an average of 23 days, when pups at Macquarie Island increase their birth weight of about 45 kg to a weaning weight of about 110 kg and pups at Falkland Islands Dependencies increase from about 50 kg to about 160 kg during the suckling period. The lesser birth weight and slower growth of Macquarie Island pups may be due to greater competition for food among pregnant cows particularly near the end of pregnancy, and possibly more crowding and disturbance in the harems, in that population. The milk is high in fat and protein content, with negligible carbohydrate, and a striking difference in composition of the milk at different stages of lactation has been demonstrated (Bryden, 1968).

Sucking is initiated by the pup when it vocalizes and nuzzles its mother, who then rolls on her side to expose the two nipples. The cow does not assist the pup to suck, so that often in the early stages of lactation a lot of time is spent by the pup in searching for the nipples. Pups suck about ten times per day at the beginning of lactation, total suckling time being about 25 min per day. These figures increase to about 20 times per day and 180 min total suckling time at 21 days of age at Macquarie Island (Bryden, 1968). These suckling frequency figures are probably considerably lower in the Falkland Islands Dependencies population (Laws, 1953) although total suckling time may not differ greatly.

Size of neonates

Length at birth is approximately 130 cm; weight is about 45 kg in Macquarie Island seals and almost 50 kg in those at the Falkland Islands Dependencies. Therefore pups born at the latter location are heavier, but not longer, than those at Macquarie Island.

References

Angot, M. (1954). Observation sur les mammifères marins de L'Archipel de Kerguelen avec une étude detaillée de l'elephant de mer, *Mirounga leonina* (L.). *Mammalia* **18**, 1-111.
Bartholomew, G. A. (1952). Reproductive and social behavior of the northern elephant seal. *Univ. Calif. Publ. Zool.* **47**, 369-472.
Blessing, M. H., Ligensa, K. and Winner, R. (1972). On the morphology of the spleen of some water mammals. *J. Wiss. Zool.* **184**, 164-204.
Bonner, W. N. (1955). Reproductive organs of foetal and juvenile elephant seals. *Nature (London)* **176**, 982.
Bryden, M. M. (1966). Twin foetuses in the southern elephant seal, *Mirounga leonina* (L.). *Pap. Proc. R. Soc. Tamania* **100**, 89-90.
Bryden, M. M. (1968). Lactation and suckling in relation to early growth of the southern elephant seal, *Mirounga leonina* (L.). *Austr. J. Zool.* **16**, 739-748.
Bryden, M. M. (1969). Regulation of relative growth by functional demand: Its importance in animal production. *Growth* **33**, 69-82.
Bryden, M. M. (1971a). Myology of the southern elephant seal *Mirounga leonina* (L.). *Antarctic Res. Ser.* **18**, 109-140.
Bryden, M. M. (1971b). Size and growth of viscera in the southern elephant seal *Mirounga leonina* (L.). *Austr. J. Zool.* **19**, 103-120.
Bryden, M. M. (1972a). Body size and composition of elephant seals (*Mirounga leonina*): absolute measurements and estimates from bone dimensions. *J. Zool.* **167**, 265-276.
Bryden, M. M. (1972b). Growth and development of marine mammals. *In* 'Functional Anatomy of Marine Mammals'', Vol. 1, (Ed. R. J. Harrison) Academic Press, London and New York. pp. 1-79.
Bryden, M. M. (1973). Growth patterns of individual muscles of the elephant seal, *Mirounga leonina* (L.). *J. Anat.* **116**, 121-133.

Bryden, M. M. and Lim, G. H. K. (1969). Blood parameters of the southern elephant seal (*Mirounga leonina*), in relation to diving. *Comp. Biochem. Physiol.* **28**, 139-148.

Bullier, P. (1954). Alimentation et acclimation d'éléphants de mer (*Mirounga leonina* L.) en captivité, au parc zoölogique du Bois de Vincennes. *Mammalia (Paris)* **18**, 272-276.

Carrick, R. and Ingham, S. E. (1962a). Studies on the southern elephant seal, *Mirounga leonina* (L.). I. Introduction to the series. *CSIRO Wildl. Res.* **7**, 89-101.

Carrick, R. and Ingham, S. E. (1962b). Studies on the southern elephant seal, *Mirounga leonina* (L.). II. Canine tooth structure in relation to function and age determination. *CSIRO Wildl. Res.* **7**, 102-118.

Carrick, R. and Ingham, S. E. (1962c). Studies on the southern elephant seal, *Mirounga leonina* (L.). V. Population dynamics and utlization. *CSIRO Wildl. Res.* **7**, 198-206.

Carrick, R., Csordas, S. E. and Ingham, S. E. (1962a). Studies on the southern elephant seal, *Mirounga leonina* (L.). IV. Breeding and development.*CSIRO Wildl. Res.* **7**, 161-197.

Carrick, R., Csordas, S. E., Ingham, S. E. and Keith, K. (1962b). Studies on the southern elephant seal, *Mirounga leonina* (L.). III. The annual cycle in relation to age and sex. *CSIRO Wildl. Res.* **7**, 119-160.

Condy, P. R. (1979). Annual cycle of the southern elephant seal *Mirounga leonina* (Linn.) at Marion Island. *S. Afr. J. Zool.* **14**, 95-102.

Harrison, R. J. and Tomlinson, J. D. W. (1965). Observations on the venous systems in certain Pinnipedia and Cetacea. *Proc. Zool. Soc. Lond.* **126**, 205-233.

King, J. E. (1972). Observations on phocid skulls. *In* "Functional Anatomy of Marine Mammals", Vol. 1, (Ed. R. J. Harrison), Academic Press, London and New York. pp. 81-115.

Laws, R. M. (1953). The elephant seal (*Mirounga leonina* Linn.). I. Growth and age. *Falk. Is. Depend. Surv., Sci. Rept.* No. 8.

Laws, R. M. (1956a). The elephant seal (*Mirounga leonina* Linn.). III. The physiology of reproduction. *Falk. Is. Depend. Surv., Sci. Rept.* No. 15.

Laws, R. M. (1956b). The elephant seal (*Mirounga leonina Linn.*). II. General, social and reproductive behaviour. *Falk. Is. Depend. Surv., Sci. Rept.* No. 13.

Laws, R. M. (1960). The southern elephant seal (*Mirounga leonina* Linn.) at South Georgia. *Norsk Hvalfangst-tid.* **49**, 466-476, 520-542.

Ling, J. K. (1968). The skin and hair of the southern elephant seal, *Mirounga leonina* (L.). III. Morphology of the adult integument. *Austr. J. Zool.* **16**, 629-645.

Ling, J. K. (1970). Pelage and molting in wild mammals with special reference to aquatic forms. *Quart. Rev. Biol.* **45**, 16-54.

Ling, J. K. (1974). The integument of marine mammals. *In* "Functional Anatomy of Marine Mammals", Vol. 2, (Ed. R. J. Harrison), Academic Press, London and New York. pp. 1-44.

Ling, J. K., Button, C. H. and Ebsary, B. A. (1974). A preliminary account of gray seals and harbor seals at Saint-Pierre and Miquelon. *Can. Field-Nat.* **88**, 461-468.

Ling, J. K. and Thomas, C. D. B. (1967). The skin and hair of the southern elephant seal, *Mirounga leonina* (L.). II. Pre-natal and early post-natal development and moulting. *Austr. J. Zool.* **15**, 349-365.

Mathews, L. H. (1952). "British Mammals", Collins, London.

Nicholls, D. G. (1970). Dispersal and dispersion in relation to the birthsite of the southern elephant seal, *Mirounga leonina* (L.), of Macquarie Island. *Mammalia (Paris)*, **34**, 598-616.

Pascal, M. (1979). Essai de dénombrement de la population d'éléphants de mer (*Mirounga leonina* (L.)), des îles Kerguelen (49 °S, 69 °E). *Mammalia (Paris)*, **43**, 147-159.

Paulian, P. (1953). Pinnipèdes, cétaces, oiseaux des Iles Kerguelen et Amsterdam. Mission Kerguelen, 1951. *Mem. Inst. Sci. Madagascar* **8**, 111-234.

Scheffer, V. B. (1958). "Seals, Sea Lions and Walruses. A Review of the Pinnipedia", Stanford University Press, Stanford, CA.

Scholander, P. F. (1940). Experimental observations on the respiratory function in diving mammals and birds. *Hvalrådets Skrifter, Norske Videnskaps-Akad., Oslo* No. 22.

Turner, W. (1887-88). Report on the seals collected during the voyage of H.M.S. Challenger in the years 1873-76. **26**, 1-240. Rept. Sci. Results Voy. H. M. S. "Challenger" in the years 1873-76, Challenger Office, Edinburgh.

White, F. N. and Odell, D. K. (1971). Thermoregulatory behavior of the northern elephant seal, *Mirounga angustirostris. J. Mammal.* **52**, 758-774.

14
Northern Elephant Seal
Mirounga angustirostris Gill, 1866

Samuel M. McGinnis and Ronald J. Schusterman

Genus and Species

The Northern elephant seal, *Mirounga angustirostris*, is one continuous species throughout its range with no subspecies or races currently recognized. Bonnell and Selander (1974) surveyed five breeding colonies electrophoretically and found no polymorphisms among 21 proteins encoded by 24 loci. It was originally described by Gill (1866) A more recent discussion of the taxonomy is found in Scheffer (1958). The northern elephant seal is the largest pinniped in the northern hemisphere and second only in size to the southern elephant seal, *Mirounga leonina*. There are the only two species in the genus *Mirounga*.

External Characteristics and Morphology

At birth northern elephant seal pups weigh from 30-34 kg and have a mean nose-to-tail length of 127 cm. The pup's weight increases to a mean of 158 kg at weaning (LeBoeuf *et al.*, 1972). They are born with a moderately thick, black pelage and essentially no subcutaneous fat (Fig. 1). After the one-month nursing period they undergo a moult in which the black fur is replaced with a silvery hair coat (Fig. 2). By this time they have also acquired a thick subcutaneous fat layer which functions as body insulation throughout life.

Adult cows (Fig. 4) rarely exceed 900 kg and are about 3 m in length. They exhibit the standard phocid body form with only a slight elongation in the nasal region. Adult bulls reach a maximum size of about 2700 kg and 6 m total length (Scheffer, 1958). In addition to its massive size, the male northern elephant seal develops a greatly enlarged and elongated nasal chamber which, as the species name suggests, is elephantine (Fig. 3). They also develop at maturity, a pink and grey mottled cornified skin layer on the ventral neck and chest regions. Both sexes have a light brown hair coat with little shading. Both the hair and the surface epidermal layer is replaced in an annual moult. Even though the hair coat has essentially no insulative value compared to the thick subcutaneous fat layer (McGinnis, 1973),

FIG. 1 Nursing black furred pup, approximately three to four weeks old Note the contrast of pelt colour and texture between the pup and the buff-coloured hair coat of the adult.

FIG. 2　Weaned pups in various stages of moulting. The lighter animal on the right has completed the process and now possesses a silvery hair coat.

FIG. 3　Typical breeding bull posture while emitting the "clap-threat" vocalization. Note the elevated posture and sand layer deposited on back by backward scooping motion of fore flipper.

moulting apparently must take place on land. Also, it appears that moulting and reproductive activity are not compatible since only weaned pups moult on the breeding rookery. All others return at various times according to their age and sex group throughout the remainder of the year to moult at the rookery site (Orr and Poulter, 1965).

The hind limbs on the northern elephant seal are similar in form and function to those of other phocids. However, the forelimbs differ with respect to their extreme digital dexterity (Fig. 4). This ability permits precise scratching of most portions of the body including the back and dorsal neck and head regions. It also makes possible the scooping action by which the thermoregulatory and displacement behaviour of sand flipping are accomplished.

FIG. 4 Female northern elephant seal exhibiting the high degree of manipulation of the fore limb.

Distribution and Migration

The present range of the northern elephant seal is from Cedros Island off the Pacific coast of central Baja California, Mexico, north to the Farallon Islands, due west of San Francisco, California, USA. Non-breeding individuals are often seen offshore along the west coast of North America up to Vancouver Island, Canada (Scheffer, 1958). The most northerly sighting was by Willett, 1943, who recovered a fresh carcass of a subadult male on Prince of Wales Island, Alaska, nearly 3000 km from the most northerly breeding sight. Except for such occasional marine sightings and carcass recoveries, we have no knowledge of this species' activities and movements once it leaves the breeding or moulting grounds. Tagging operations conducted by LeBoeuf and his associates at the University of California, Santa Cruz, clearly show an annual homing to the birth sight each year for breeding and/or moulting. In one sense, then, the northern elephant seal may be considered a migratory species in that there are yearly movements from unknown feeding grounds to specific rookery sights.

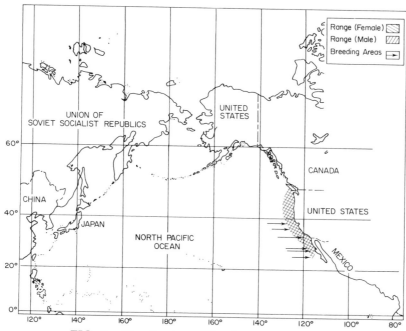

FIG. 5   Distribution of Northern elephant seal.

Abundance and Life History

As a brief sketch of the life history of the northern elephant seal we present the annual cycle as observed on Año Nuevo Island off the coast of Central California as viewed by these authors over the past decade. Adult bulls haul out on the island to begin the breeding season in late November and remain until March. Although some bulls enter the water during this time, it is doubtful that they feed. In mid-December adult females arrive, give birth, nurse their pups for about one month, come into oestrus, copulate, and depart. They rarely enter the water during this period and, like the bulls, apparently metabolize subcutaneous fat stores. Both cows and bulls appear considerably thinner at the time of departure.

After weaning, the pups, now referred to as "weaners", gather in pods on the beach where they remain for about one month. During this time they do not feed but instead also metabolize fat reserves laid in during the nursing period. Weaners make short excursions into tide pools around the breeding beach during this time and appear to be learning to swim and dive. They leave the rookery in small groups unaccompanied by any adults. Apart from reproduction, the only other apparent reason for returning to the land is to moult.

The recent population dynamics of the northern elephant seal have indeed been just that, dynamic! It was the prime target for commercial sealing because of its extensive subcutaneous fat layer which from a bull four m long yielded about 325 l of oil (Scammon, 1874). Exploitation of this species began about 1818 and continued until 1869 when the species was considered commercially extinct. By 1890 only one herd of about 100 northern elephant seals on Guadaloupe Island was reported (Bartholomew and Hubbs, 1960). This was perhaps a count of a breeding colony and did not take into account juveniles which were out at sea. Yet the fact remains that the species was at a dangerously low level. From this nucleus population on Guadaloupe Island, Mexico (29 °N), the species rapidly increased and recolonized its former breeding islands and beaches. In 1918 the San Benito Islands (lat. 28.5 °N), closest to Guadaloupe, were recolonized (Rice *et al.*, 1965). This first step in recovery was aided by a law affording the northern elephant seal complete protection passed by the Mexican Government in 1922.

The United States soon followed suit. The re-occurrence of the species on other islands followed in this order: San Miguel Island (34 °N) 1925; The Coronado Islands (32 5 °N) 1948; Santa Barbara

Island (33.5 °N) 1948; San Nicolas Island (33.3 °N) 1949; Año Nuevo Island (37 °N) 1955; Anacapa Island (28.5 °N) 1958; the Farallon Islands (37.6 °N) 1959; Cedros Island (28.5 °N) 1965; and Santa Rosa Island (34 °N) 1965.

In their publication, Bartholomew and Hubbs (1960) estimated a total population size of 13 000 animals in 1957. LeBoeuf (in press) estimated that in 1976 there were 47 684 northern elephant seals in existence. Thus within 20 years the population has more than tripled. LeBoeuf has summarized the initially observed locations, the initiation of breeding, Colony size, number of breeding females, and total colony numbers (Table 1).

TABLE 1 Northern elephant seal breeding colonies, the time of colonization, and estimates of colony sizes.[a]

Colony	Seals initially observed	Breeding began	Colony size	
			Number of breeding females at peak season	Estimated total animals
Isle of Guadaloupe	—	—	4 652	18 596
Islas San Benito	1918	1930s	2 382	9 238
San Miguel Island	1925	1930s	3 842	13 980
Los Coronados	1948	1950s or later	44	152
Santa Barbara Island	1948	late 1950s	68	252
San Nicolas Island	1949	late 1940s	616	2 214
Ano Nuevo Island	1955	1961	687	2 718
South-east Farallon Island	1959	1972	60	260
Isla Cedros	1965	1960s(?)	63	274
Population totals			12 414	47 684

[a]The last are based on censuses taken in late January during the 1976 breeding season on all rookeries except the Isle of Guadaloupe (1975), Islas San Benito (1970), and Isla Cedros (1970). The number of breeding females was obtained from peak censuses taken in late January. Approximately 10-15% of the breeding females in each colony are still at sea at this time. Estimate of total animals in each colony was calculated by simply doubling the censuses in late January (LeBoeuf, 1977).

As may be seen from the latitude sequence of island re-occupancy, colonization of the former range has not been a direct progression northward. Instead it appears that perhaps the most suitable breeding sights, i.e. those with good sand or gravel beaches remote from human activity, were occupied first, and the less suitable rookery sights were occupied sometime thereafter. The pattern of colonization which these authors witness over the past decade on Año Nuevo Island and which has been reported by LeBoeuf *et al.* (1974) for the Farallon Islands had allowed the following synthesis of the process:

1. The closest colonized breeding islands supply the initial stock to a new sight. In the case of the Farallon Islands there were Año Nuevo, San Miguel and San Nicolas Islands (LeBoeuf *et al.*, 1974).

2. The colonizers are young of the year which use the new sight as a hauling out place for the summer or fall moult.

3. The first breeding at the new site involves subadult males (six to seven years) and adult females. Subadult females do not usually participate in the early stage of colonization. One may view these first breeders as wanderers from an established colony with a large established population. The first incidence of breeding on the Farallons in 1972 was characterized by the arrival first of several cows followed by the subadult bulls (LeBoeuf *et al.*, 1974). This is the reverse of the sequence in an established colony.

4. Once established the new colony grows rapidly and the harem-type breeding structure gradually changes. In the early 1960s the harems on Año Nuevo Island were for the most part successfully defended by an alpha bull, and most of the breeding was accomplished by relatively few bulls (LeBoeuf and Peterson, 1969). By the late 1960s, however, the breeding beaches were literally covered with adult bulls and females with no possibility of a discrete, well-defended harem existing. The result was breeding by a greater number of bulls and increased pup mortality. The latter was studied on Año Nuevo Island by LeBoeuf and Briss (in press) between 1968 and 1976. They found that throughout this period pup mortality ranged from 13% to 26% with the two main sources of pup death being aggression of females towards foreign pups and trampling by bulls. These are the same sort of conditions which exist apparently on a permanent basis on Guadaloupe Island. It is perhaps this "full house" condition which stimulates the roving of breeding animals to a new rookery site. This idea is supported by the fact that breeding occurred for a decade on Año Nuevo Island before adult females ventured less than 200 km north to the Farallons. The increase in pup mortality in the rookery with the increase in size of the breeding population is also an excellent example of intra-population regulation of animal numbers, since the factors causing increased pup mortality in the full beach condition are all density dependent.

Population kinetics as studied on Año Nuevo and south-east Farallon Islands by LeBoeuf *et al.* (1974) and on San Nicolas Island by Odell (1974) show a similar pattern in all three even though the

Farallons have only recently been colonized. The breeding populations all peak in February, juvenile peaks occur in April-May and again in November. The beach populations reach their lowest points in August.

Perhaps the most significant colonization event to date took place on Año Nuevo Point Beach, the mainland shore opposite the island, in 1975. For more than a decade, adult and subadult bulls from the island have been swimming to the mainland and spending days and weeks on the beach and in the surrounding sand dunes. In early January, 1975, a pregnant female hauled out on the beach and gave birth to a pup. She was chaperoned from the start by the dominant beach bull with which she later bred. The following season a total of seven cows with pups were ashore at this locality. Mainland colonization at Año Nuevo Point Beach has continued to grow rapidly during the past three years, and in 1979 there were approximately 90 females, 80 pups and 260 bulls. This is the first known instance of a mainland breeding colony since the species' recovery from near extinction. It now appears that a full recolonization of the original range via the establishment of a breeding colony at the Point Reyes National Seashore (38 °N) is most probable.

Internal Characteristics and Morphology

Adult bull skulls at the museum of Vertebrate Zoology, University of California, Berkeley, average 50 cm in total length and 27 cm in width at the zygomatic arch. In contrast, the average nursing pup skull in this collection is 20 cm long and 13 cm wide (Fig. 6). The adult dental formula is

$$I\frac{2}{2} \, C\frac{2}{2} \, PC\frac{5}{5}$$

and the deciduous formula is (Harrison, 1974)

$$I\frac{2}{2} \, C\frac{1}{1} \, PC\frac{3}{3}$$

Total upper canine length of an adult bull when removed from the skull ranges from 5-6 cm. However, only a little more than half this length protrudes beyond the gum. These are the teeth which inflict most of the wounds during bull combat, but their functional length is less than half the skin-fat layer on most parts of the body. The implication that

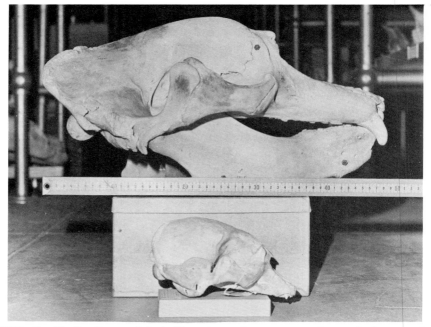

FIG. 6 Skull of an adult bull and nursing pup northern elephant seal, photographed at the Museum of Vertebrate Zoology, University of California, Berkeley.

relatively short canine length is a selective feature which prevents high bull mortality is born out by the results of an autopsy of an adult bull by Robert Jones, Museum of Vertebrate Zoology, Berkeley. This animal died as the result of an eight-hour running fight with another bull on Año Nuevo Point, California. Although there were numerous cuts in the neck shield area, no deep artery or vein was severed. Instead death was due to a shattered scapula which punctured the thoracic cavity.

Subcutaneous fat forms a major part of the mass of the northern elephant seal. In a recent disection of a two-year-old female we found that the skin and fat layer weighed slightly more than the remainder of the body. Fat-layer thickness in this animal ranged from 87 mm in the dorsal thoracic region to only a few millimetres near the nose and naked portions of the hind flippers.

As for other aspects of internal anatomy, the literature is almost completely lacking with respect to the northern elephant seal, except for a few minor observations (Ridgway, 1972). This is in sharp contrast to the southern elephant seal, *Mirounga leonina*, for which much information can be found. This is no doubt due to the unavailability of fresh wild specimens of the northern elephant seal for anatomical study due to its complete protection by the United States and Mexico. Many of the findings for the southern elephant seal probably apply to the northern species as well, although this must still remain speculative.

Behaviour

Thermoregulation

Initial interest in the body temperature of the northern elephant seal was stimulted by Bartholomew (1954) when he obtained rectal temperatures of sleeping animals in a rookery and found a surprisingly wide range of body temperatures. McGinnis and Southworth (1967) obtained similar results from a captive juvenile which led to a more detailed telemetry study of both wild and captive animals of various age groups (McGinnis and Southworth, 1971). In this work they found that most wild northern elephant seals were very stenothermic with the exception of undernourished pups and bulls at the end of the breeding season. However, numerous long-term captive animals all exhibitied marked eurythermy as previously discussed.

Insulation in the adult and weaned young is entirely by subcutaneous fat deposits. New-born pups, however, possess no subcutaneous fat. Instead they rely to a great extent on an abnormally high metabolic rate to offset excessive body heat loss until the fat insulation can be acquired (Heath *et al.*, 1977). The black pelage of nursing pups, though a better insulator than the adult hair coat, (McGinnis, 1973) does not make up for the lack of subcutaneous fat. Body heat dissipation is primarily by vasodilation of blood vessels in the hind flippers (McGinnis *et al.*, 1970). Sand flipping also provides an avenue for body heat dissipation, especially if the sand is moist when flipped on the back. Excessive activity on land, especially on warm days, results in elevated body temperature (McGinnis and Southworth, 1971) and apparently cannot be tolerated for very long.

Diving

Both direct and indirect evidence presented here under feeding behaviour indicate that the northern elephant seal is capable of long, deep dives. Various physiological blood properties associated with diving presented by Lenfant (1969) and Simpson *et al.* (1970) suggest that this species is far more similar in diving ability to other deep diving species of pinnipeds and cetaceans than to the surface-feeding, shallow-diving forms. Elsner *et al.* (1964) demonstrated that the inferior vena cava of *M. angustirostris* may expand to such an extent during a dive as to contain one-fifth of the total blood volume. In this manner it acts as a storage compartment for de-oxygenated blood during a dive. Marked bradycardia during simulated diving has been reported both for adults (Van Citters *et al.*, 1965) and nursing pups (Hammond *et al.*, 1969).

Because of the large size of adults in this species many direct tests in laboratory chambers, now so prevalent for smaller species, are lacking here. However, there appears to be ample indirect evidence that long-term, deep diving is a major feature of the physiological niche of the northern elephant seal.

Feeding behaviour

In a stomach content analysis of a 2.5 ton male, Huey (1929) found rat fishes, dogfish, shark, puffer, shark-skates and squid. All of these prey are deep water inhabitants. They are found at depths of approximately 22 m. Scheffer (1964), reported that northern elephant seals were hooked on fishing lines at depths of 183 m. Although sea lions can dive to 250 m (Ridgway, 1972), elephant seals probably dive much deeper permiting this species to exploit a food niche which is not available to other pinnipeds on the west coast of North America.

Census data an Año Nuevo Island and Southeast Farallon Island correlates with fluctuations in salinity, water temperature, coastal upwellings and possible elephant seal food (LeBoeuf *et al.*, 1974). This suggests that abundant food prey is most likely to be available from May to September.

Sensory perception

Virtually nothing is known about the sensory systems of elephant seals. It is likely that acoustic, visual, olfactory and tactile cues account for various aspects of social orientation, feeding orientation and migrating behaviour.

Comparative histology of retinas from four pinniped species revealed that the northern elephant seal had the highest ratio of receptor to ganglion cells (Landau and Dawson, 1970). This suggests that although *Mirounga* may have relatively poor visual acuity, its sensitivity to light may be greater than other pinniped species.

Vocalizations

The most detailed structural and functional analysis of northern elephant seal vocalizations was originally given for the breeding population on San Nicolas Island, California, by Bartholomew and Collias (1962). These investigators presented excellent spectrographs of the major types of phonations emitted by males and females at different ages. Later work, primarily of a comparative nature was done at Año Nuevo Island, California (LeBoeuf and Peterson, 1969b; LeBoeuf and Petrinovich, 1974a; LeBoeuf and Petrinovich, 1974b; Petrinovich, 1974; Sandegren, 1976).

Although Sandegren (1976) distinguishes between four different adult male vocalizations, the two most commonly heard during the breeding season are the "snort" and the "clap-threat". Both are threat vocalizations with the former often given from a prone position or in water. The clap-threat is always associated with other elements of threat behaviour and is clearly related to aggression and dominance. The clap-threat consists of a series of pulses which are guttural and low-pitched with most of the energy concentrated below 2500 Hz. Sandegren (1976) found that the spectral characteristics of these pulses are different for individuals and could be used for individual recognition. LeBoeuf and Peterson (1969b) and LeBoeuf and Petrinovich (1974a) found that the pulse repetition rate of the clap-threat varied significantly from one colony to another, indicating the presence of a dialect. Finally, LeBoeuf and Petrinovich, 1974b, found species differences in both temporal and pitch parameters of the clap-threat emitted by *M. angustirostris* and *M. leonina*.

When a pup is ready to suckle or has been separated from its mother, or has in some manner been stressed by another adult, it emits a shrill distress cry of 0.3 s duration, repeated several times at intervals of about one second. The fundamental frequency is about 1000-1500 Hz with harmonics above 5000 Hz. Adult females frequently respond to their own pup's distress calls with a pup-attraction call having a fundamental frequency between 600 and 1000 Hz with rises and falls in pitch five to six times per second. Petrinovich (1974) has experimentally demonstrated individual recognition of pup distress calls by their mothers. Females also have a threat vocalization which is loud and prolonged, having most of its sound spectral energy below 700 Hz.

Male dominance hierarchies

Studies of social behaviour in the northern elephant seal began in earnest with the work of George Bartholomew (1952). Brief and relatively inaccurate statements regarding their reproductive behaviour were first made by Scammon (1874) and later by Townsend (1912). Since 1968 LeBoeuf and his associates at the University of California, Santa Cruz (LeBoeuf, 1974; LeBoeuf and Briggs, 1977), have studied reproductive behaviour of elephant seals on Año Nuevo Island, California in relation to demographics in remarkable detail.

Soon after adult males (eight years old) arrive on the breeding grounds they begin to fight for dominance status within a social hierarchy, and with repeated encounters among several males the emerging alpha male acquires greater access to females than lower-ranking males (Bartholomew, 1952; LeBoeuf and Peterson, 1969a; LeBoeuf, 1974). High-ranking males are involved in more aggressive interactions than lower-ranking males, but overt fighting is usually infrequent (6% or less) relative to the total number of aggressive encounters (LeBoeuf and Peterson, 1969a; Sandegren, 1976). Threat exchanges among participants include such motor elements as an elevated posture of the neck and/or body to about a 90-degree angle with the substrate, an inflated proboscis, a head toss, a slamming of the body to the ground (Fig. 3), a "snort" vocalization and a series of low-pitched gutteral pulses (Sandegren, 1976). Defeated males retreat backwards while facing their foe, taking a low body posture. The most frequent behaviour of submissive males when threatened at some distance by a dominant male is to turn and move away. Submission by a beta individual when alpha is within striking distance includes one or more of the following: lateral swimming motions with the hindquarters, wide-eyed scared expression, proboscis retraction, relaxed open mouth, nip-biting (Sandegren, 1976). When a bull is in a fight-or-flight conflict he uses his fore flippers to flip sand on the back (Heath and Schusterman, 1975).

Reproduction

As already mentioned, the complete protection afforded this species during most of its population recovery has precluded in-depth investigation of all aspects of internal anatomy, particularly those pertaining to reproductive biology. Therefore, such suspected phenomenon as delayed implantation has yet to be documented for this species. Harrison (1969) presents numerous data on the reproductive physiology of the southern elephant seal, *M. leonina*, much of which probably applies to the northern elephant seal as well, but with no direct field data this can only be speculative.

We do know, however, that the females come into oestrus just at or about the weaning period which is quite different from western otariids. The bulls, like all phocids, have internal testis, which poses the question of sperm viability and body temperature. Bryden (1967) found that the testicular temperature of two immobilized southern elephant seal bulls was about 60 °C below deep body temperature but does not postulate a mechanism to account for this. The shunting of cool venous blood from the hind flippers to a plexus surrounding the testis could account for such a thermal differential.

Perinatal, maternal and pup behaviour

Perinatal behaviour of *M. angustirostris* females and their young has been studied during four breeding seasons by LeBoeuf *et al.* (1972) at Año Nuevo Island, California. Although females can give birth during their third year, the majority start giving birth during their fourth or fifth year of life. Arrival time of females to time of parturition is six to seven days, with a nursing period of about 28 days. An oestrus of three to five days occurs during the final days of nursing and females wean their young and depart approximately 34 days after arriving. An oestrus female mates promiscuously. After departing the breeding grounds females feed at sea for a few weeks, return to the rookery to moult and rest for nearly a month, and return to sea for the next eight to nine months before the next reproductive season.

The peak period for births throughout the northern elephant seal's range is 20 January to 1 February. Generally, most births occur at night or during the coolest hours of the day while the female is in a "harem" or clumped together with other females. Individual distance among females varies from light contact to one or two m (LeBoeuf and Briggs, 1977). A detailed description of a birth is given by LeBoeuf *et al.* (1972). In the final phase of parturition, the female assumes a "U"-shaped posture. Immediately after birth, the mother turns, faces the newborn (thereby breaking the umbilical cord), vocalizes towards the pup and appears to smell the pup. The precocial pup vocalizes in return and this behavioural sequence in which mother and pup learn to recognize one another continues for about an hour. Females become considerably more aggressive after giving birth and attempt to keep nearby females, who are frequently attracted to new-borns, away by vocal and postural threats. Small and less aggressive mothers (probably primiparous) often have their pups bitten by larger and more aggressive nearby females almost as soon as the pup is expelled. Pups of subordinate females were much more likely to be bitten and eventually become orphaned than pups of dominant, aggressive and probably older cows (Christenson, 1974).

Duration of parturition from the time the foetus is barely visible ranges from approximately 1 to 30 min with a mean of about 7.3 min. LeBoeuf, *et al.* (1972) observed 29 presentations of which 62% were cephalic and 38% were caudal. In the majority of births, delivery of the placenta, which weighs about (10 lb) 4.5 kg occurred within two min of birth. Usually gulls consume the entire afterbirth within 24 h.

A new-born pup's eyes are open, it can scratch, flip sand on its back and move forward as well as vocalize and hold its head somewhat erect. Suckling, which occurs within 15 min to one h after birth, takes place two to four times per day and increases in frequency and duration with age (Fig. 1). The close filial relationship during the nursing period of four weeks not only ensure nurturance of the pup, but also its protection by its mother from injury by other females. This close association between mother and pup ends abruptly when the female leaves the harem and returns to sea, leaving the pup on the rookery and no longer interacting with it on land. The weaned pups or "weaners" as they are called, leave the female harem and form weaner pods or aggregates. Recent observations suggest that some weaners, mainly males, re-enter the harem and try to obtain additional nourishment by suckling nursing females (Reiter, 1975).

Generally, females nurse their own pups (36 of 50 marked pairs) and reject suckling attempts of alien pups. In many of those cases where a female suckled an alien pup or weaner, the female was asleep and did not discriminate between her offspring and the alien, or the female had lost her own pup and was probably physiologically primed to respond maternally (LeBoeuf *et al.*, 1972). These findings contradict those of Klopfer and Gilbert (1967).

Swimming and diving skills of weaners improve gradually as they spend increasing amounts of time in the water (Reiter, 1975). "Mock fighting" by weaners resembles adult agonistic interactions and is sexually dimorphic, both quantitatively as well as qualitatively. Male mock-fight more frequently and for longer periods of time than do females. Male weaners had long bouts of mock-fighting interactions including throat-pressing, neck-biting, rear and slam, etc., while female weaners briefly faced each other with open mouths emitting loud vocalizations. Detailed descriptions of the motor patterns involved in self- and object-manipulation by weaners is given by Rasa (1971) who also discusses the adaptive significance of these early behaviour patterns. Weaners departing from the rookery in mid-April and early May, a period of coastal updwellings and phytoplankton blooms, are about 14 weeks old (Reiter, 1975).

Copulatory behaviour

As indicated earlier, status within the male social hierarchy determines access to the female harem or aggregate, and since high-status males prevent subordinates from approaching, mounting or copulating with females, there is a high positive correlation between social status and the estimated number of females inseminated (LeBoeuf, 1974). The high positive correlation between status and reproductive success among males depends to some extent on the number of oestrus females, the number of competing males and the location of the female harem. The more difficult it is to defend a harem, either because of its size or location, the more likely younger males will begin copulating (LeBoeuf, 1972; LeBoeuf, 1974). Generally, males begin copulating at eight or nine years of age and highly successful males may breed for three or four successive years with as many as approximately 250 females. However, the reproductive success of most males is zero or low because many succumb before attaining reproductive status, and those that do are prevented from mating by high-status males. According to LeBoeuf (1974) the mortality of males may be as high as 97% prior to reaching age eight and as high as 86-93% before reaching six or seven years of age.

Copulating activity is highest during the third week in February. Copulation is initiated by the male without any preliminary courtship (LeBoeuf, 1972). Mounting is from the side with the male using the great weight of his head and neck, and fore flipper clasp or a neck-bite to restrain the female. All females are mounted, those which are pregnant as well as those in oestrus, although the latter are mounted more frequently. A cooperative female is passive and may facilitate intromission by assuming the lordotic posture and spreading her hind flippers. Intromission last about five min on land and about six min in water. Males usually terminate intromission.

When a male attempts to mount an unreceptive female, she continually emits vocal threats, swings her hindquarters from side to side, vigorously flips sand (Heath and Schusterman, 1975) with her fore flippers back into the male's face, and struggles to get away. Many of these protested mounts occur during early oestrus and result in either the male stopping his attempted copulation and the female moving away, or the mounter being interrupted by the threatening actions of a more dominant male or a completed copulation. Cox and LeBoeuf (1976) found that the loud protests of mounted females are likely to result in the interruption of mounts by low-ranking males and not high-ranking males. Thus, females during early oestrus incite male competition with the result that they mate primarily with high-status males, thus increasing their own inclusive fitness.

Captivity

Northern elephant seals have been kept successfully in several zoos in the United States including St Louis, San Diego, Chicago, Washington D.C. and the New York Aquarium. As is true with other pinniped species, they are difficult to breed in captivity and there have been no known successful breeding programmes. The activity patterns or time budgets of two yearlings were quantified by Schusterman (1968) who gives the percentage time that each animal spent in mock-fighting, grooming, environmental manipulations, etc. McGinnis and Southworth (1971) found that long-term captive northern elephant seals were strikingly eurythermal compared to stenothermic wild specimens. They suggest that the captive diet of frozen herring does not permit a constant resting metabolism. Lenfant (1969) reports a decrease in blood haemoglobin content in captivity which also suggests dietary deficiencies.

In recent years at least two oceanaria, Sea World in San Diego and Marineland near Los Angeles, have made considerable progress in rehabilitating sick pups that come to mainland beaches. Both organizations currently maintain captive colonies of elephant seals.

Diseases

Lists and discussions of major pinniped diseases, some of which may be found in the northern elephant seal, are presented by Ridgway (1972) and Hubbard (1968). The most complete case history of a disease in this species is that of an incidence of obstructive emphysema by Saunders and Hubbard (1966). Given that the literature on disease in *M. angustirostris* is very sparse, three separate reports on emphysema as a direct cause of death make this malady the most prevalent to date.

Acknowledgement

In the preparation of this manuscript R. J. Schusterman was supported by Contract N00014-72-C-0186 from the Office of Naval Research.

References

Bartholomew, G. A. (1952). Reproductive and social behavior of the northern elephant seal. *University of California Publ. Zool.* **47**, 369-372.

Bartholomew, G. A. and Hubbs, C. L. (1952). Winter population of pinnipeds about Guadalupe, San Benito, and Cedros Islands, Baja California, *J. Mammal.* **33**, 160-171.

Bartholomew, G. A. (1954). Body temperature and respiratory and heart rates in the northern elephant seal. *Mammalia (Paris)* **35**, 211-218.

Bartholomew, G. A. and Hubbs, C. L. (1960). Population growth and seasonal movements of the northern elephant seal, *Mirounga angustirostris. Mammalia (Paris)* **24**, 313-324.

Bartholomew, G. A. and Collias, N. E. (1962). The role of vocalization in the social behavior of the northern elephant seal. *Anim. Behav.* **10**, 7-14.

Bonnell, M. L. and Selander, R. K. (1974). Elephant seals: genetic variation and near extinction. *Science (N. Y.)* **184**, 908-909.

Bryden, M. M. (1967). Testicular temperatures in the southern elephant seal, *M. leonina. J. Reprod. Fert.* **13**, 583-584.

Christenson, T. E. (1974). Aggressive and maternal behaviour of the female northern elephant seal. Ph.D. Dissertation, University of California, Berkeley.

Cox, C. R. and LeBoeuf, B. J. (1972). Female incitation of male competition: A mechanism in sexual selection. *Am. nat.* **111**, 317-335.

Elsner, R. W., Scholander, P. F., Craig, A. B., Diamond, E. G, Irving, L., Pilson, M., Johansen, K. and Bradstreet, E. (1964). Venous oxygen reservoir in the diving elephant seal. *Physiologist* **7**, 124.

Gill, T. (1866). Prodrome of a monograph of the pinnipeds. Proc. Esser. Inst. Salem. *Communications* **5**, 3-13.

Hammond, D. D., Elsner, R., Simison, G., and Hubbard, R. (1969). Submersion bradycardia in the newborn elephant seal, *Mirounga angustirostris. Am. J. Physiol.* **216**, 220-222.

Harrison, R. J. (1969). Reproduction and reproductive anatomy. *In* "The Biology of Marine Mammals", (Ed. H. T. Andersen), Academic Press, New York and London.

Harrison, R. J. (1974). "Functional Anatomy of Marine Mammals", Vol. 2, Academic Press, London and New York.

Heath, M. A., McGinnis, S. M. and Alcorn, D. (1977). Comparative thermoregulation in nursing and weaned pups of the northern elephant seal, *Mirounga angustirostris. Comp. Biochem. Physiol.* **57a**, 203-206.

Heath, M. A. and Schusterman, R. J. (1975). "Displacement" sand flipping in the northern elephant seal *(Mirounga angustirostris). Behav. Biol.* **14**, 379-385.

Hubbard, R. C. (1968). Husbandry and laboratory care of pinnipeds. *In* "The Behavior and Physiology of Pinnipeds", (Eds R. J. Harrison, R. C. Hubbard, C. E. Rice, and R. J. Schusterman), Appleton-Century-Crofts, New York.

Huey, L. M. (1930). Capture of an elephant seal off San Diego, CA and notes on stomach contents. *J. Mammal.* **11**, 229-231.

King, J. E. (1964). "Seals of the World", Trustees of the British Museum, London.

Klopfer, P. H. and Gilbert, B. K. (1967). A note on retrieval and recognition of young in the elephant seal, *Mirounga angustirostris. Z. Tierpsych.* **6**, 757-760.

Landau, D. and Dawson, W. W. (1970). The histology of retinas from the pinnipedia. *Vis. Res.* **10**, 691-702.

LeBoeuf, B. J. (1972). Sexual behavior in the northern elephant seal, *Mirounga angustirostris. Behavior* **41**, 1-26.

LeBoeuf, B. J. (1974). Male-male competition and reproductive success in elephant seals. *Am. Zool.* **14**, 163-176.

LeBoeuf, B. J. (1977). Back from extinction? *Pacific Discovery*, **30**, 1-9.

LeBoeuf, B. J. and Briggs, K. T. (1977). The cost of living in a seal harem. *Mammalia (Paris)* **41**, 167-195.

LeBoeuf, B. J. and Petrinovich, L. F. (1974a). Dialects of northern elephant seals, *Mirounga angustirostris:* origin and reliability. *Anim. Behav.* **22**, 656-663.

LeBoeuf, B. J. and Petrinovich, L. F. (1974b). Elephant seals: interspecific comparisons of vocal and reproductive behaviour. *Mammalia (Paris)* **33**, 16-32.

LeBoeuf, B. J., Ainley, D. G. and Lewis, T. J. (1974). Elephant seals on the Farallones: population structure of an incipient breeding colony. *J. Mammal.* **55**, 370-385.

LeBoeuf, B. J., Whiting, R. J. and Gantt, R. F. (1972). Perinatal behavior of northern elephant seal females and their young. *Behavior* **34**, 121-156.

LeBoeuf, B. J. and Peterson, R. S. (1969a). Social status and mating activity in elephant seals. *Science (N. Y.)* **163**, 91-93.

LeBoeuf, B. J. and Peterson, R. S. (1969b). Dialects in elephant seals. *Science (N. Y.)* **166**, 1654-1656.

Lenfant, C. (1969). Physiological properties of blood. *In* "The Biology of Marine Mammals", (Ed. H. T. Andersen), Academic Press, New York and London.

McGinnis, S. M. (1973). Comparative thermoregulation of the pup and adult pelage of the northern elephant seal, *Mirounga angustirostris. Proc. Tenth Ann. Conf. Biol. Sonar and Diving Mammals,* Menlo Park, CA.

McGinnis, S. M. and Southworth, T. P. (1967). Body temperature fluctuations in the northern elephant seal. *J. Mammal.* **48**, 484-485.

McGinnis, S. M. and Southworth, T. P. (1971). Thermoregulation in the northern elephant seal, *Mirounga angustirostris*. *Comp. Biochem. Physiol.* **40**, 893-901.

McGinnis, S. M., Cryer, E. T. and Hubbard, R. C. (1970). Peripheral heat dissipation in pinnipeds. *Seventh Annual Conf. on Biol. Sonar and Diving Mammal*, Stanford Research Institute, Menlo Park, CA.

Odell, D. K. (1974). Seasonal occurrence of the northern elephant seal, *Mirounga angustirostris*, on San Nicolas Island, CA. *J. Mammal.* **55**. 81-95.

Orr, R. T. and Poulter, T. C. (1965). The pinniped populations of Ano Nuevo Island, CA. *Proc. Calif. Acad. Sci.* **22**, 377-404.

Petrinovich, L. F. (1974). Individual recognition of pup vocalization by northern elephant seal mothers. *Z. Tierpsychol.* **34**, 308-312.

Rasa, O. A. (1971). Social interaction and object manipulation in weaned pups of the northern elephant seal, *Mirounga angustirostris*. *Z. Tierpsychol.* **29**, 82-102.

Reiter, J. (1975). Behavior of newly weaned elephant seal pups. Paper presented at the *Conf. Biol. Conserv. Marine Mamm.* Dec. 4-7.

Rice, D. W., Kenyon, K. W. and Lluch, D. B. (1965). Pinniped population at Islas Guadalupe, San Benito, and Cedros, Baja California, in 1965. *Trans. San Diego. Soc. Nat. Hist.* **14**, 73-84.

Ridgway, S. (Ed.) (1972). "Mammals of the Sea", Thomas, Springfield, Ill.

Sandegren, F. E. (1976). Agonistic behavior in the male northern elephant seal. *Behavior* **57**, 136-158.

Saunders, A. M. and Hubbard, R. C. (1966). Obstructive emphysema in an elephant seal, *Mirounga angustirostris*. *Laboratory Animal Care* **16**, 217-223.

Scammon, C. M. (1874). "The Marine Mammals of the Northwestern Coast of North America", Carmany, San Francisco.

Scheffer, V. B. (1964). Deep diving of elephant seals. *The Murrelet* **45**, 9.

Scheffer, V. B. (1958). "Seals, Sea Lions and Walruses", Stanford University Press, Stanford, CA.

Schusterman, R. J. (1968). Experimental laboratory studies of pinniped behavior. *In* "The Behavior and Physiology of Pinnipeds", (Eds. R. J. Harrison, R. C. Hubbard, R. S. Peterson, C. E. Rice and R. J. Schusterman), Appleton-Century-Crofts, New York.

Simpson, J. G., Gilmartin, W. G. and Ridgway, S. H. (1970). Blood volume and other hematologic values in young elephant seals *(Mirounga angustirostris) Am. J. Vet. Res.* **31**, 1449-1452.

Van Citters, R. L., Franklin, D. L., Smith, O. A. Jr., Watson, N. W. and Elsner, R. (1965). Cardivascular adaptations to diving in the northern elephant seal, *Mirounga angustirostris*. *Comp. Biochem. Physiol.* **16**, 267-276.

Willett, G. (1943). Elephant seal in South-eastern Alaska. *J. Mammal.* **24**, 500.

Index

Abundance; *see under* species
Aggression, 316, 336, 341
Air sac, tracheal, in *P. fasciata*, 104-105
Anaesthesia, 76
Anatomy; *see under* organ, Skull, system and species

Baculum, 41, 100, 310
Baikal seal, *Phoca sibirica*, 29-53
Bearded seal, 145-169; *see also Erignathus barbatus*
Behaviour, reproductive behaviour, social behaviour; *see under* species
Birth; *see under* species
Blood, 11, 45, 80-81, 105, 138, 165, 191, 227-229, 249, 283-285, 314, 340
Blubber, 152, 180
Bradycardia, 282, 340
Brain, 69, 281, 312
Brown adipose tissue, 79

Captivity, 12, 16-17, 48, 75-76, 138, 190, 199, 218, 271, 319, 346
Cardiovascular system, 11, 69-70, 227, 281, 311
Caribbean monk seal, 201-203; *see also Monachus tropicalis*
Caspian seal, *Phoca caspica*, 29-53
Caval sphincter, 71, 227, 249, 279, 311
Ciguatoxin, 213
Coloration; *see under* species
Common names of seals; *see under* each species
Common seal, 1-27; *see also Phoca vitulina*
Copulation; *see under* species

Crabeater seal, 221-235; *see also Lobodon carcinophagus*
Cystophora cristata, 171-194
  abundance, 177-178
  anatomy, 181
  behaviour, 185-187
  birth, 177, 187
  blood, 191
  captivity, 190
  coloration, 172-173
  conservation, 178-183
  copulation, 190
  delayed implantation, 189
  dental formula, 184
  dimensions, 173, 187-188
  diseases, 181
  distribution, 173-177
  diving, 185
  exploitation, 178-183
  external characteristics, 172-173
  feeding and food, 180
  genetics, 185
  genus, 171-172
  growth, 188
  lactation, 188
  life history, 177
  longevity, 181
  mating, 190
  maturity, 188-189
  migration, 173-176
  milk, 188
  morphology, 172-173
  mortality, 180-181
  moult, 176, 180
  parasites, 181
  population dynamics, 176-177
  pups, 187
  reproduction, 187-190
  reproductive behaviour, 190
  skeleton, 184

Cystophora cristata—cont.
  skull, 181-184
  sound production, 186
  species, 171-172
  swimming, 185-186
  teeth, 184
  thermoregulation, 186-187
  weights, 173

Delayed implantation, 17, 41, 76, 130, 155, 189, 251, 319, 322
Dental formula; *see under* species
Dimensions; *see under* species
Diseases, 18, 48, 79-80, 125-126, 166-167, 181, 213-214, 233, 319, 346
Distribution
  Cystophora, 173-177
  Erignathus, 148, 152-154
  Halichoerus, 115-118
  Hydrurga, 264-266
  Leptonychotes, 277-278
  Lobodon, 222-224
  Mirounga angustirostris, 333
  M. leonina, 298, 301
  Monachus, 200-204, 208
  Ommatophoca, 238, 249-250
  Phoca caspica, 31-32
  P. fasciata, 95-98
  P. groenlandica, 60-64
  P. hispida, 31-33
  P. largha, 5-8
  P. sibirica, 31-32
  P. vitulina, 5-8
Diving, 14, 46, 70-71, 138, 160, 165, 185, 199, 216, 249, 281-284, 286-288, 314, 339-340

Ear, 16, 74
EEG, of Weddell seal, 281
Elephant seals
  northern, 329-347
  southern, 297-327; *see also Mirounga angustirostris* and *M. leonina*
Endocrine organs, 11, 127, 281

Erignathus barbatus, 145-169
  abundance, 154-157
  anatomy, 162
  behaviour, 158-162
  birth, 154, 158
  blood, 165
  coloration, 147
  delayed implantation, 155
  dental formula, 162
  dimensions, 149-152
  diseases, 166-167
  distribution, 148, 152-154
  diving, 160, 165
  external characteristics, 146
  feeding and food, 161-162
  genus, 145
  growth, 147-152, 155
  lactation, 152, 154
  life history, 154-157
  locomotion, 158
  longevity, 155
  mating, 154
  migration, 152-154
  morphology, 146, 162-165
  morphometrics, 151-152
  moult, 156
  parasites, 166
  population dynamics, 157
  pups, 148-149
  reproduction, 154-155
  skeleton, 163
  skull, 162-164
  social behaviour, 159
  sound production, 160
  species, 145
  swimming, 158
  teeth, 149, 162
  weights, 149
External characteristics; *see under* species
Eye, 16, 74-75, 102, 242, 281, 340

Feeding and food; *see under* species

Grey seal, 111-144; *see also Halichoerus grypus*

INDEX

Halichoerus grypus, 111-144
  abundance, 119
  anatomy, 127-130
  behaviour, 132-135
  birth, 130-134
  blood, 138
  captivity, 138
  coloration, 113
  conservation, 135-136
  copulation, 130, 134-135
  delayed implantation, 130
  dental formula, 127
  dimensions, 112, 129
  diseases, 125, 126
  distribution, 115-118
  diving, 138
  external characteristics, 112-115
  feeding and food, 121-124
  genus, 111-112
  growth, 134
  hearing, 138
  karyotype, 130
  lactation, 130, 134
  life history, 119-125
  locomotion, 127
  longevity, 120-121
  mating, 130-132, 135
  maturity, 121
  migration, 115-118
  milk, 134
  morphology, 112-115
  morphometrics, 112
  moult, 113, 130
  parasites, 125-126
  polygyny, 113, 132
  population structure, 119-121
  pups, 113, 131-135
  reproduction, 121, 130-132
  reproductive behaviour, 132-135
  skull, 127
  sound production, 138
  species, 111-112
  swimming, 118
  teeth, 127
  territorial behaviour, 132-135
  weights, 112, 129

Harbour seal, 1-27; *see also Phoca vitulina*
Harems, 317, 323, 336, 343-344
Harp seal, 55-87; *see Phoca groenlandica*
Hawaiian monk seal, 203-220; *see also Monachus schauinslandi*
Heart, 11, 69-70, 227, 268, 279, 282
Hooded seal, 171-194; *see also Cystophora cristata*
Hydrurga leptonyx, 261-274
  abundance, 266
  anatomy, 266-268
  behaviour, 269-271
  birth, 271
  captivity, 271
  coloration, 262
  copulation, 271
  dimensions, 262, 264
  distribution, 264-266
  external characteristics, 262-264
  feeding and food, 269-270
  genus, 261-262
  growth, 271
  lactation, 272
  life history, 262
  mating, 271
  maturity, 271
  morphology, 262-264
  parasites, 272
  penguin catching, 269-270
  pups, 271-272
  reproduction, 271-272
  skull, 266
  sound production, 270-271
  species, 261-262
  teeth, 267
  weights, 262, 264, 271

Integument, 4-11, 205, 284, 312-314, 330, 338-339
Intestines, 11, 45, 102, 230, 249, 310

Kidney, 69, 247

Lactation; *see under* species

Largha seal, 1-27; see also *Phoca largha* under *P. vitulina*
Larynx, 279
Leopard seal, 261-274; see *Hydrurga leptonyx*
Leptonychotes weddelli, 275-296
  abundance, 278-279
  anatomy, 279-281
  behaviour, 285-292
  birth, 292
  blood, 283-285
  coloration, 277, 292
  copulation, 292
  dimensions, 277
  distribution, 277-278
  diving, 281-284, 286-288
  external characteristics, 277
  feeding and food, 286-288
  genus, 275
  growth, 292
  integument, 284
  lactation, 293
  mating, 292
  maturity, 292-293
  migration, 290-292
  milk, 285
  morphology, 277
  moult, 292
  pups, 284, 290
  reproduction, 292-293
  skeleton, 279
  skull, 279
  social behaviour, 290
  sound production, 288
  species, 275
  swimming, 288-290
  teeth, 279
  thermoregulation, 284-285
  weights, 292
Life history; *see under* species
Lobodon carcinophagus, 221-235
  abundance, 224-227
  anatomy, 227-230
  behaviour, 230-232
  birth, 232
  blood, 227-229

Lobodon carcinophagus—cont.
  coloration, 222
  copulation, 232
  dimensions, 222
  diseases, 233
  distribution, 222-224
  external characteristics, 222
  feeding and food, 231
  genus, 221
  lactation, 233
  locomotion, 224, 230-231
  maturity, 233
  morphology, 222
  moult, 222
  population census, 224-226
  pups, 232-233
  reproduction, 232-233
  skeleton, 229
  skull, 227
  social behaviour, 231-232
  sound production, 232
  species, 221
  teeth, 227-229
  weights, 222
Locomotion, 14-15, 71, 106, 127, 158, 224, 230-231, 314-316
Longevity, 9, 41, 77, 100, 120-121, 155, 181, 207, 251, 303-305
Lung, 44, 69, 129, 229, 282-284

Mating; *see under* species
Maturity; *see under* species
Mediterranean monk seal, 195-201; *see also Monachus monachus*
Mercury
  in *P. groenlandica*, 81
  in *Halichoerus*, 137
Milk, 78, 134, 188, 285, 324
Mirounga angustirostris, 329-349
  abundance, 334-337
  anatomy, 337-339
  behaviour, 339-342
  birth, 334, 343-344
  blood, 340
  captivity, 346

INDEX

Mirounga angustirostris—cont.
  coloration, 330
  copulation, 334, 345
  dimensions, 329-330
  diseases, 346
  distribution, 333
  diving, 339-340
  external characteristics, 329-332
  feeding and food, 340
  genus, 329
  integument, 330, 338-339
  lactation, 334-337
  life history, 334-337
  mating, 343, 345
  migration, 333
  moult, 330-332
  population dynamics, 334-336
  pups, 341, 344
  reproduction, 342
  reproductive behaviour, 342-346
  skull, 337
  social behaviour, 342-343
  sound production, 341
  species, 329
  swimming, 344
  teeth, 337
  thermoregulation, 339, 343
  vocalization, 341-342, 345
  weights, 329-330
Mirounga leonina, 297-327
  abundance, 303
  anatomy, 305-314
  behaviour, 314-319
  birth, 301-302, 323-324
  blood, 314
  captivity, 319
  coloration, 299, 312
  copulation, 317
  delayed implantation, 319, 322
  dental formula, 307-308
  dimensions, 299-300
  diseases, 319
  distribution, 298, 301
  diving, 314
  external characteristics, 297-300
  feeding and food, 317-318

Mirounga leonina—cont.
  genus, 297
  growth, 300
  integument, 312-314
  lactation, 302, 304
  life history, 301-305
  locomotion, 314, 316
  longevity, 303-305
  mating, 302, 319, 323
  migration, 301
  milk, 324
  morphology, 297-300
  morphometrics, 300
  moult, 299, 302, 312-317
  population dynamics, 303-307
  pups, 299, 301-302, 315, 324
  reproduction, 319-325
  reproductive behaviour, 316-317, 322
  sexual maturity, 302, 320
  skeleton, 305, 306
  skull, 307-310
  social behaviour, 314-317
  sound production, 318-319
  species, 297
  swimming, 314-315
  teeth, 307-310
  thermoregulation, 315-316
  weights, 300, 311-312
Monachus, 195-220: (*M. monachus, M. schauinslandi, M. tropicalis*)
  abundance, 201, 209
  behaviour, 214-218
  birth, 198-199, 211
  captivity, 199, 218
  coloration, 198, 204
  copulation, 210
  dental formula, 206
  dimensions, 198
  diseases, 213-214
  distribution, 200-204, 208
  diving, 199, 216
  external characteristics, 198, 204
  feeding and food, 199, 217
  genus, 195
  integument, 205

Monachus—cont.
  lactation, 211
  longevity, 207
  mating, 198, 210
  morphology, 198, 204
  mortality,
  causes of, 211-213
  moult, 198, 204
  pelage, 204
  polygyny, 198, 210
  population dynamics, 210-211
  pups, 204, 210-211, 217
  reproduction, 198-199, 210-211
  skull, 207-208
  social behaviour, 216
  species, 196-197
  swimming, 216
  teeth, 206
  thermoregulation, 217-218
  vocalization, 214
  weights, 198, 204
Monk seals, 195-220; *see also Monachus*
Morphology; *see under* species
Moult; *see under* species
Musculature, 11, 306

Nasal sac, in *Cystophora*, 185

Ommatophoca rossi, 237-259
  abundance, 238, 249-250
  anatomy, 243-249
  behaviour, 252-256
  birth, 251
  blood, 249
  coloration, 240-242
  copulation, 251
  delayed implantation, 251
  dental formula, 246
  dimensions, 240
  distribution, 238, 249-250
  diving, 249
  external characteristics, 239-243
  feeding and food, 251-252
  genus, 237-239

Ommatophoca rossi—cont.
  lactation, 251
  life history, 251-252
  locomotion, 256
  longevity, 251
  mating, 251
  maturity, 251
  morphology, 239-243
  moult, 240-242
  parasites, 251
  pelage, 242, 249
  pups, 251
  reproduction, 251
  skeleton, 243-246
  skull, 244-246
  social behaviour, 252-253
  sound production, 253-255
  species, 237
  swimming, 243
  teeth, 246
  thermoregulation, 253
  weights, 240
Organochlorine compounds
  Halichoerus, 136-137
  P. hispida, 48-49
Oesophagus, in *Ommatophoca*, 249
Ovary, 189, 321

Palate, in *Ommatophoca*, 247
Parasites, 10, 81-82, 125-126, 166, 181, 272
Pelage; *see under* species
Penguins, taken by leopard seals, 269-270
Phoca caspica, 29-53; *see also under P. hispida*
Phoca fasciata, 89-109
  abundance, 98-100
  anatomy, 102
  behaviour, 105-106
  birth, 99
  blood, 105
  coloration, 90-92, 100
  dental formula, 102
  dimensions, 92-95, 102

INDEX

Phoca fasciata—cont.
  distribution, 95-98
  external characteristics, 91-95
  feeding and food, 98-100
  genus, 89-91
  growth, 93-95
  lactation, 99, 101
  life history, 98-100
  locomotion, 106
  longevity, 100
  mating, 99
  maturity, 99
  morphology, 91-95
  moult, 91-100
  pups, 91-92, 99, 105
  reproduction, 99-100
  skeleton, 103
  skull, 102
  species, 89-91
  swimming, 99
  teeth, 102
  weights, 92-95
Phoca groenlandica, 55-87
  abundance, 64-65
  anatomy, 68-70
  behaviour, 71-73
  birth, 64, 77
  blood, 80-81
  captivity, 75-76
  coloration, 56-58
  copulation, 73
  delayed implantation, 76
  dental formula, 67
  dimensions, 60, 68-69, 78
  diseases, 79-80
  distribution, 60-64
  diving, 70-71
  external characteristics, 56-60
  feeding and food, 72-73, 75
  genus, 55
  hearing, 69
  karyotype, 81
  lactation, 78
  life history, 64-65
  locomotion, 71
  longevity, 77

Phoca groenlandica—cont.
  mating, 73, 77
  maturity, 77
  migration, 60-65
  milk, 78
  morphology, 56-60
  moult, 56-58, 65, 75
  parasites, 81-82
  pelage, 56-59
  population dynamics, 61-63
  pups, 58-60, 65, 75, 78-79
  reproduction, 76-79
  skeleton, 66
  skull, 65-66
  social behaviour, 73
  sound production, 73
  species, 55-56
  swimming, 70-71
  teeth, 67
  weights, 60, 69-70
Phoca hispida, P. caspica and *P. sibirica*, 29-53
  abundance, 33-35
  anatomy, 42-45
  behaviour, 46-48
  birth, 39-41
  blood, 43
  captivity, 48
  coloration, 35
  delayed implantation, 41
  dental formula, 42
  dimensions, 37-39
  diseases, 48
  distribution, 31-33
  diving, 46
  external characteristics, 35-38
  feeding and food, 46
  genus, 29-30
  growth, 37-39
  lactation, 40
  life history, 39-42
  longevity, 41
  mating, 39, 47
  maturity, 38-41
  morphology, 33-38
  morphometrics, 37-39

Phoca hispida—cont.
  moult, 35
  population dynamics, 33
  pups, 35, 44
  reproduction, 39-42
  reproductive behaviour, 47
  skeleton, 44
  skull, 42-43
  sound production, 47
  species and subspecies, 30-31, 49
  swimming, 47
  teeth, 42
  weights, 38-39, 45
Phoca largha, 1-27; see also *Phoca vitulina* for topics
Phoca sibirica, 29-53; see under *P. hispida*
Phoca vitulina, 1-27; (also *P. largha*)
  abundance, 8
  anatomy, 10-12
  behaviour, 12-14
  birth, 17
  blood, 11
  captivity, 12, 16-17
  coloration, 4
  delayed implantation, 17
  dimensions, 3
  diseases, 18
  distribution, 5-8
  diving, 14
  external characteristics, 3-4
  feeding and food, 9
  genus, 1-2
  growth, 4
  hearing, 16
  integument, 4, 11
  lactation, 12
  life history, 8
  locomotion, 14-15
  longevity, 9
  mating, 12
  morphology, 3-4
  morphometrics, 3
  moult, 17
  organs, 11-12
  parasites, 10

Phoca vitulina—cont.
  pelage, 4
  population dynamics, 8-9
  pups, 12, 14
  reproduction, 11, 17
  reproductive behaviour, 12
  skeleton, 11
  skull, 10-11
  social behaviour, 12
  sound production, 15
  species and subspecies, 1-2
  swimming, 15
  teeth, 11
  thermoregulation, 15
  vision, 16
  weights, 3
Phoca v. concolor, 2-27
Phoca v. mellonae, 2-27
Phoca v. richardsi, 2-27
Phoca v. stejnegeri, 2-27
Phoca v. vitulina, 2-27
Pollution, 136-137
Populations and population dynamics; see under species
Predators of seals, 42, 48, 125, 156, 180, 211, 242, 304-305
Pups: see under species

Reproduction and reproductive behaviour
  Cystophora, 187-190
  Erignathus, 154-155
  Halichoerus, 121, 130-135
  Hydrurga, 271-272
  Leptonychotes, 292-293
  Lobodon, 232-233
  Mirounga angustirostris, 342-346
  M. leonina, 316-325
  Monachus, 198-199, 210-211
  Ommatophoca, 251
  P. fasciata, 99-100
  P. groenlandica, 76-79
  P. hispida, 39-42, 47
  P. sibirica, 40
  P. vitulina, 11-12, 17
Respiratory system; see *Lung, Trachea*

INDEX 359

Retina, 16
Ribbon seal, 89-109; *see also Phoca fasciata*
Ringed sea, 29-53; *see also Phoca hispida*
Ross seal, 237-259; *see also Ommatophoca rossi*

Seals
  Baikal seal, 29-53
  bearded seal, 145-169
  Caspian seal, 29-53
  common seal, 1-27
  crabeater seal, 221-235
  elephant seal
   northern, 329-347
   southern, 297-327
  grey seal, 111-144
  harbour seal, 1-27
  harp seal, 55-87
  hooded seal, 171-194
  larga seal, 1-27
  leopard seal, 261-274
  monk seals, 195-220
  ribbon seal, 89-109
  ringed seal, 29-53
  Ross seal, 237-259
  Weddell seal, 275-296
Serology, 130, 177
Skeleton; *see under Skull*
Skull, teeth, skeleton
  Cystophora, 181-184
  Erignathus, 149, 162-164
  Halichoerus, 127
  Hydrurga, 266-267

Skull, teeth, skeleton—*cont.*
  Leptonychotes, 279
  Lobodon, 227-229
  Mirounga angustirostris, 337
  M. leonina, 307-310
  Monachus, 206-208
  Ommatophoca, 243-246
  P. fasciata, 102-103
  P. groenlandica, 65-67
  P. hispida, 42-44
  P. sibirica, 42
  P. vitulina, 10-11
Sleep, *Halichoerus*, 138
Social behaviour; *see under* species
Sound production; *see under* species
Swimming, 15, 47, 70-71, 99, 118, 144, 158, 185-186, 216, 243, 288-290, 314-315

Teeth; *see under* species
Testis, 11, 40-41, 321
Thermoregulation, 15, 186-187, 217-218, 253, 284-285, 315-316, 339-343
Trachea, 44, 69, 103, 165, 184, 229, 247, 268, 279, 310
Twins, 320

Urine, 285

Vibrissae, 242, 249
Vision, 16, 74-75

Weddell seal, 275-296; *see also Leptonychotes weddelli*
Weights; *see under* species

$48.50